Asymptotic Methods in Queuing Theory

Probability and Mathematical Statistics (Continued)

TJUR • Probability Based on Radon Measures
WILLIAMS • Diffusions, Markov Processes, and Martingales, Volume I: Foundations
ZACKS • Theory of Statistical Inference

Applied Probability and Statistics

ABRAHAM and LEDOLTER • Statistical Methods for Forecasting
AICKIN • Linear Statistical Analysis of Discrete Data
ANDERSON, AUQUIER, HAUCK, OAKES, VANDAELE, and WEISBERG • Statistical Methods for Comparative Studies
ARTHANARI and DODGE • Mathematical Programming in Statistics
BAILEY • The Elements of Stochastic Processes with Applications to the Natural Sciences
BAILEY • Mathematics, Statistics and Systems for Health
BARNETT • Interpreting Multivariate Data
BARNETT and LEWIS • Outliers in Statistical Data
BARTHOLOMEW • Stochastic Models for Social Processes, *Third Edition*
BARTHOLOMEW and FORBES • Statistical Techniques for Manpower Planning
BECK and ARNOLD • Parameter Estimation in Engineering and Science
BELSLEY, KUH, and WELSCH • Regression Diagnostics: Identifying Influential Data and Sources of Collinearity
BHAT • Elements of Applied Stochastic Processes
BLOOMFIELD • Fourier Analysis of Time Series: An Introduction
BOX • R. A. Fisher, The Life of a Scientist
BOX and DRAPER • Evolutionary Operation: A Statistical Method for Process Improvement
BOX, HUNTER, and HUNTER • Statistics for Experimenters: An Introduction to Design, Data Analysis, and Model Building
BROWN and HOLLANDER • Statistics: A Biomedical Introduction
BROWNLEE • Statistical Theory and Methodology in Science and Engineering, *Second Edition*
BURY • Statistical Models in Applied Science
CHAMBERS • Computational Methods for Data Analysis
CHATTERJEE and PRICE • Regression Analysis by Example
CHOW • Analysis and Control of Dynamic Economic Systems
CHOW • Econometric Analysis by Control Methods
COCHRAN • Sampling Techniques, *Third Edition*
COCHRAN and COX • Experimental Designs, *Second Edition*
CONOVER • Practical Nonparametric Statistics, *Second Edition*
CONOVER and IMAN • Introduction to Modern Business Statistics
CORNELL • Experiments with Mixtures: Designs, Models and The Analysis of Mixture Data
COX • Planning of Experiments
DANIEL • Biostatistics: A Foundation for Analysis in the Health Sciences, *Third Edition*
DANIEL • Applications of Statistics to Industrial Experimentation
DANIEL and WOOD • Fitting Equations to Data: Computer Analysis of Multifactor Data, *Second Edition*
DAVID • Order Statistics, *Second Edition*
DAVISON • Multidimensional Scaling
DEMING • Sample Design in Business Research
DODGE and ROMIG • Sampling Inspection Tables, *Second Edition*
DOWDY and WEARDEN • Statistics for Research
DRAPER and SMITH • Applied Regression Analysis, *Second Edition*
DUNN • Basic Statistics: A Primer for the Biomedical Sciences, *Second Edition*
DUNN and CLARK • Applied Statistics: Analysis of Variance and Regression
ELANDT-JOHNSON and JOHNSON • Survival Models and Data Analysis

continued on back

Asymptotic Methods in Queuing Theory

A. A. BOROVKOV
Institute of Mathematics
Novosibirsk, USSR

Translated by

D. NEWTON
University of Sussex, UK

JOHN WILEY & SONS

Chichester · New York · Brisbane · Toronto · Singapore

Translated from Asimptoticheskie Metody V
Teorii Massovogo Obsluzhivaniya by A. A. Borovkov
Nauka Publishing House, Moscow 1980.

Library of Congress Cataloging in Publication Data:

Borovkov, Aleksandr Alekseevich.
Asymptotic methods in queuing theory.
Translation of: Asimptoticheskie metody v teorii massovogo obsluzhivaniya.
Includes index.
1. Queuing theory. 2. Asymptotic expansions.
I. Title.
T57.9.B66713 1984 519.8′2 83–12557
ISBN 0 471 90286 1

British Library Cataloguing in Publication Data:

Borovkov, A. A.
Asymptotic methods in queuing theory.—(Wiley series in probability and mathematical statistics)
1. Queuing theory
I. Title II. Newton, D.
III. Asimptoticheskie metodi v teorii massovogo obsluzhivaniya. *English*
519.8′2 T57.9

ISBN 0 471 90286 1

Filmset by Mid-County Press, London SW15.
Printed by the Pitman Press Ltd., Bath, Avon.

Contents

* See Foreword, p. x.

Foreword

This book is devoted to the presentation of general asymptotic methods in the theory of queues. It is in some sense a sequel to Borovkov (1976a) but may be read independently. The fundamental objects here are the methods of asymptotic analysis of queuing processes. Our aim is to make these methods as general as possible and to give effective means of investigation of fairly complex systems. We have not attempted here to consider the full spectrum of different types of systems since our main aim is the discussion of general methods which the reader may use for himself in various concrete situations.

What do we mean by asymptotic methods in queuing theory? This question needs to be made precise since, with a broad enough interpretation of the notion 'asymptotic methods', almost the entire theory may be included; at least such fundamental areas as ergodic theorems, the search for stationary distributions and the study of convergence to them. The connection between these areas and asymptoticity is entirely natural when we take into account that in them we study the limiting behaviour of processes for use in 'in the limit' conditions.

In this book asymptotic methods will not have such a broad meaning. We have in mind the following three directions of research (the choice is very conventional) whose basic aim is the study of queuing processes (including stationary) by means of suitable approximations.

1. Asymptotic analysis of explicit formulae or equations which describe the distribution (usually stationary) of some characteristic or other of the system. For the implementation of such an analysis it is naturally assumed that there are such explicit formulae or equations and, in addition, that the system comes arbitrarily close to one of its critical states. It was in just this direction that Kingman in 1961 obtained the first results on the behaviour of a single-server system in heavy traffic (see Kingman, 1961, 1962; and also Gnedenko and Kovalenko, 1966; Borovkov, 1976a; Köllerström, 1974; Newell, 1968). It is possible to quote several other papers in this direction; for example, papers which study the behaviour of systems with refusals and a large number of service channels (Viskov and Prokhorov, 1964; Borovkov, 1976a), etc. The regularity noted here in relatively simple examples then turned out to be valid under much more general conditions in which the explicit formulae were absent. Since today in queuing theory it is

evident that almost all cases of 'explicit solvability' have been studied, then the search for other means of asymptotic analysis becomes topical.

2. The second direction is much broader and is connected with the study of the limit behaviour of the random processes which define the system (as well as the approach of the system to one of its critical states). Prokhorov (1963) revealed that at the basis of the phenomena which arise in the above-mentioned busy systems there lies the well-known invariance principle due to Donsker–Prokhorov. This principle appears to be very effective and, with the help of a Wiener process, allows one to approximately describe quite complicated processes generated by sums of random variables.

The main aim of the second direction is the establishment of the so-called 'collective' limit theorems and the clarification of the broadest and most general conditions under which these collective asymptotic laws hold (for example, the law on convergence of normalized queue lengths to a diffusion process). It is obvious that the methods in this direction differ substantially from the methods of part 1 and are based, as a rule, on the use of general convergence theorems for processes.

3. The third direction is somewhat specialized and is connected with the so-called stability (or continuity) theorems. Here again we are discussing essentially limit theorems for processes; however, they are no longer 'collective' but are 'individual'. Namely, conditions are given under which service systems will be close (distributionally via some characteristic process of the system) to a given concrete system. The term 'stability' is connected with the fact that such theorems allow conclusions to be drawn about small changes in stationary characteristics under small deviations of the parameters of the system.

In this book we consider mainly the latter two directions. It is clear that the first direction is well represented in existing journals and monographs.

The book consists of four chapters. The first chapter contains the preliminaries. Here we collect the ideas and theorems to be used later. In particular, in sections 1.4–1.10 we prove theorems on convergence to degenerate processes and to diffusion processes (with and without boundaries). In a separate section of chapter 1 we isolate examples which illustrate the results of the chapter. These examples embrace, in particular, the basic types of input streams and show that the convergence conditions to be placed on them later (chapters 2 and 3), will be satisfied under very broad assumptions.

The material in chapter 1 may sometimes be more than is immediately required for an understanding of chapters 2–4 and the footnotes. This is done for the benefit of the reader who is not fully prepared, and also to give a connected exposition of the material.

Chapter 2 is devoted to the investigation of the class of multichannel systems where the number of channels is large and the input stream is intensive. Depending on the relation between the intensity of the input stream and the number of channels, three regimes are possible: 'supercritical' (when the number of channels is asymptotically infinite), 'subcritical' (almost all channels are constantly busy), and 'critical' (intermediate between these two). In all three regimes we study the asymptotic behaviour of the number of busy lines. As one might expect, the laws obtained here are qualitatively different. As limit processes we obtain some fairly complex processes. Convergence to a Markov diffusion process (in the critical and supercritical regimes) is observed only in the simplest cases when the service time has an exponential distribution. In the second chapter we also consider the connection with the problems which arise in branching processes with intensive immigration. Some results of this chapter have been published in Borovkov (1977a, 1967c) and Borisov and Borovkov (1980).

As the limit processes in queuing theory, apparently, the Markov diffusion processes appear more often than other processes. In the fifties Kolmogorov made a conjecture about the approximate description of the number of busy channels in systems with refusals using diffusion processes with reflections from an upper boundary. In Prokhorov (1963) and also in Barrer (1957), Blomqvist (1973), Borovkov (1964), Gaver (1968), Geza (1974), Gnedenko and Kovalenko (1966), Iglehart (1965), Iglehart and Whitt (1970), Kyprianou (1971), Loulou (1973), Samandarov (1963), Szcotha (1976), and many other papers, the question of approximating busy systems with waiting by diffusion processes is discussed (for a more detailed bibliography, see the surveys by Iglehart, 1973, Whitt, 1974, and Cohen, 1973). It was established, in particular, that for the simplest systems the normalized queue length (or waiting time), as a process, tends to a diffusion process with reflection from a lower boundary at zero.

In chapter 3 we give the broadest (within natural limits) conditions under which the queue length (or 'occupation') of the system can be described using diffusion processes.

These results are applied to various concrete situations: the so-called loaded systems, systems with stochastic refusals, multichannel systems, etc.

In both the second and third chapters we use a more general than usual definition of the service process, as a three-dimensional random process

$$\mathbf{S}(t) = (e(t), r(t), s(t); t \geqslant 0),$$

where all the components are nonnegative and monotone, and such that $q(t) = e(t) - r(t) - s(t) \geqslant 0$.

$e(t)$ is the number of requests received by the system up to time t, $r(t)$ is the number of requests which are refused, and $s(t)$ is the number of requests which are served. As characteristics of the system which are usually studied we mention the 'queuing' process $q(t)$ and the 'refusal' process $\pi(t) = r(t)/e(t)$.

In real systems the joint distribution of the components e, r, and s is usually given by 'local' properties (the distribution of service times, etc.), and by algorithms which determine the nature of the service processes.

However, from the asymptotic viewpoint, it is often enough to characterize $\mathbf{S}(t)$ by means of the properties of the increments of e, r, and s over relatively large intervals of time (this also is an assignment of 'local' properties, but no longer as a time process). The conditions imposed on these increments are usually in a form which is very simple and easy to verify. The generality of the results obtained is the undoubted advantage of the asymptotic approach.

The fourth chapter deals with stability theorems. Currently the investigation of stability (we have in mind first and foremost the stability of stationary characteristics) is carried out by several essentially different approaches. The bases of these approaches are explained in the papers by Franken (1970) (who applies the theory of point processes), Zolotarev (1975b) (general metric methods), and Kalashnikov (1977, 1978) and Kalashnikov and Tsitsiashvili (1972) (the method of trial functions). In this book we explain another method, which seems to us to be very natural and which we call the 'renewal method'. This method was used earlier by the author (Borovkov, 1972e, 1976a). Its advantage is that it allows a unified approach to stability and ergodic theorems (they are closely connected) for all the basic types of service systems. For the simplest systems the method gives the stability theorem under minimal assumptions. In addition, the renewal method allows an estimation of the speed of convergence, again for all basic types of service systems. This is illustrated in section 4.4.

A more detailed survey of the contents of chapters 2–4 can be found in the introductions to the chapters.

Naturally we do not claim to cover all aspects of the subject 'asymptotic methods in queuing theory'. For example, outside our point of view lies the important area of collective limit theorems for the input process (or input stream of requests). This area may be included in the second direction of research mentioned above. We have omitted it because our basic discussion is devoted to asymptotic methods and approximations connected with '*solutions*' of the service systems, that is, with processes which describe interesting characteristics of the system when the input process and other processes which determine the operation of the system are already *given*. To some extent we touch upon limit theorems for input processes only in chapter 1, in sections 1.5, 1.11, and 1.12.

For almost all the results here this is the first publication in book form.

We recommend that sections 2.4–2.6, marked with an asterisk, should be omitted on the first reading. Because of the complexity of the objects under discussion the account in sections 2.4 and 2.5 is not everywhere rigorous. The results in section 2.6 are completely natural and intelligible, but mastering their proofs requires significant effort.

In selecting the references the author, where possible, has preferred to quote a book rather than the original article, since a reader working with this book will find the former more convenient. In this connection it must be noted that the list of references is not claimed to be complete. However, the union of the bibliographies of all the quoted works will, probably, be fairly complete.

We assume that the reader is acquainted with basic probability theory and the theory of random processes. For this the coverage of these subjects in the books

Probability Theory, by A. A. Borovkov, and *Introduction to the Theory of Random Processes*, by I. I. Gikhman and A. V. Skorokhod (see Borovkov, 1976c; Gikhman and Skorokhod, 1965), for example, is ample.

A. Borovkov

CHAPTER 1

On the Convergence of Random Processes

In this chapter we present the basic ideas and general results relating to convergence of random processes, which will be needed in the following chapters. We abide by the following rules. Results which can be found in well-known monographs on the theory of random processes, such as Doob (1953), Gikhman and Skorokhod (1965; 1971, 1973, 1976), and Billingsley (1968), are not proved and are simply accompanied by appropriate references. The remaining results are given with proofs or with explanations which allow the proofs to be easily reconstructed. Because of this the exposition cannot be called homogeneous but, clearly, within the limits of this book it is quite adequate.

1.1. General definitions and theorems

Let $(\Omega, \mathfrak{A}, \mu)$ be a probability space. A *random process* (or *stochastic process*) is a family of random variables $\{\xi(u) = \xi(u, \omega);\ u \in U\}$ depending on a parameter $u \in U$. In other words, $\xi(u, \omega)$ is a real-valued function on $U \times \Omega$ which is $\mathfrak{A}$-measurable with respect to $\omega \in \Omega$. As parameter set U we mostly consider either sequences of the form $(0, 1, 2, \ldots)$, $(\ldots, -1, 0, 1, \ldots)$, or subsets, or the whole, of the real line $(-\infty, \infty)$. In the first case a random process is called a *random sequence* and in the second case a *process with continuous time*.

Let $R(U)$ be a space of functions $r(u)$ on U, and let $\mathfrak{A}_{R(U)}$ be a σ-algebra of subsets of $R(U)$ which contains the sets of the form

$$\{r(u)\colon r(u_1) \in B, \ldots, r(u_k) \in B_k\}, \tag{1.1}$$

called *cylinders*, where $u_j \in U$ and B_j is a Borel set on the line, $j = 1, \ldots, k$. Then, obviously, any measurable mapping of $(\Omega, \mathfrak{A})$ into $(R(U), \mathfrak{A}_{R(U)})$ defines a random process (corresponding to ω is the function $\xi(u, \omega) \in R(U)$, where $\{\omega\colon \xi(u, \omega) \in B\} \in \mathfrak{A}$). The measure $\mathbf{P}$ on $(R(U), \mathfrak{A}_{R(U)})$, to which μ passes is also called the *distribution of the process* $\xi(u)$, and the probability space $(R(U), \mathfrak{A}_{R(U)}, \mathbf{P})$ is called the *sample space* for $\xi(u)$.

It is often convenient, in the study of random processes, to regard as initial probability space the sample space

$$(R(U), \mathfrak{A}_{R(U)}, \mathbf{P}).$$

As U we will mostly consider segments $[0, U]$, $U > 0$. Evidently we are also using U to denote a number. This will not lead to any misunderstandings in what follows.

For sets U of the form $[0, U]$ the following function spaces $R(U)$ are often considered in probability theory.

1. $R(U) = R^U = \prod_{u \in U} R^u$, where $R^u = (-\infty, \infty)$ is the set of values of $\xi(u, \omega)$. R^U is the space of all real functions on U. It is convenient to consider this space coupled with the σ-algebra $\mathfrak{C}_U$, generated by the cylinder sets (that is, sets of the form (1.1)). By Kolmogorov's theorem, the assignment of consistent finite-dimensional distributions (probabilities of sets of the form (1.1)) uniquely defines a random process on $(R^U, \mathfrak{C}_U)$ (a distribution $\mathbf{P}$ on $(R^U, \mathfrak{C}_U)$) (see, for example, Gikhman and Skorokhod, 1965).

2. $R(U) = C(0, U)$; the space of all continuous functions on $[0, U]$. In this space distributions are sometimes given also on the *Borel* σ-algebra $\mathfrak{B}_{C(0, U)}$, generated by the open sets in the uniform metric

$$\rho(r_1, r_2) = \sup_u |r_1(u) - r_2(u)|. \tag{1.2}$$

In this connection it must be noted that a measure on $\mathfrak{B}_{C(0, U)}$ defines a process since $\mathfrak{C}_{C(0, U)} \subset \mathfrak{B}_{C(0, U)}$, where $\mathfrak{C}_{C(0, U)}$ is the σ-algebra generated by the sets (1.1) from $C(0, U)$. In fact, it turns out that $\mathfrak{C}_{C(0, U)} = \mathfrak{B}_{C(0, U)}$ (Gikhman and Skorokhod, 1965; 1971, 1973, 1976).

3. $R(U) = D(0, U)$ is the space of functions on $[0, U]$, with no discontinuities of the second kind. For definiteness, when speaking of D, we will usually have in mind the set of functions such that for each u the limits $r(u-0)$ and $r(u+0)$ exist and, in addition, $r(u)$ coincides with either $r(u-0)$ or with $r(u+0)$; $r(0) = r(+0)$, $r(U) = r(U-0)$. The space D with the property $r(u) = r(u+0)$, $0 \leqslant u < U$, will be denoted $D^+ = D^+(0, U)$.

As with C, the distributions of the process are sometimes given on the Borel σ-algebra $\mathfrak{B}_{D(0, U)}$, generated by the open sets in the Skorokhod–Prokhorov metric

$$\rho(r_1, r_2) = \inf\left[\sup_{u \in [0, U]} |r_1(u) - r_2(\lambda(u))| + \sup_{u \in [0, U]} |u - \lambda(u)|\right] \tag{1.3}$$

where inf is taken over all continuous monotone functions $\lambda(u)$, for which $\lambda(0) = 0$, $\lambda(U) = U$ (functions r_1 and r_2 are close in the metric (1.3) if after suitable small 'displacements' along the u-axis they are close in the uniform metric).

For D also it turns out that $\mathfrak{B}_{D(0, U)}$ coincides with the σ-algebra $\mathfrak{C}_{D(0, U)}$ of cylinder sets* of D (see Billingsley, 1968).

Thus, if $\xi(u)$ is a process in the spaces C or D, then, for example, the functions

$$f_1(\xi) = \sup_{u \in U} \xi(u) \quad \text{or} \quad f_2(\xi) = \int_0^U \phi(\xi(u))\, du \tag{1.4}$$

* Instead of 'σ-algebra generated by the cylinder sets' we will sometimes, for brevity, write simply 'σ-algebra of cylinder sets' or even 'cylindrical σ-algebra'.

where ϕ is continuous, are random variables, since the sets

$$f_1(r) < a, \quad f_2(r) < a$$

are open and, consequently, belong to the cylindrical σ-algebra.

However, for processes given in an arbitrary $R(U)$, functionals such as (1.4) and others, which are important in applications, will not, in general, be measurable relative to a σ-algebra $\mathfrak{A}_{R(U)}$ containing the cylinder sets or generated by them.

To avoid this problem in the general case, and to enlarge the class of random variables (or events), we will always assume that the *measure* **P** *of the process* $\xi(u)$ *is complete, and the process* $\xi(u)$ *itself is separable.*

The first means that, together with a set $A \in \mathfrak{A}_{R(U)}$, all sets A' which differ from A by a subset of a set of measure 0 will also belong to $\mathfrak{A}_{R(U)}$. It is natural here to suppose that $\mathbf{P}(A') = \mathbf{P}(A)$.

Separability of $\xi(u)$ means that there is a countable dense set S in U such that for any interval $I \subset U$

$$\mathbf{P}\left[\sup_{\substack{u\in I\\ u\in S}} \xi(u) = \sup_{u\in I} \xi(u);\ \inf_{\substack{u\in I\\ u\in S}} \xi(u) = \inf_{u\in I} \xi(u)\right] = 1.$$

This definition of separability may be written equivalently in the form

$$\mathbf{P}\left[\limsup_{\substack{t\to u\\ t\in S}} \xi(t) \geqslant \xi(u) \geqslant \liminf_{\substack{t\to u\\ t\in S}} \xi(t) \text{ for all } u \in U\right] = 1.$$

Two processes $\xi(u, \omega)$ and $\xi^*(u, \omega)$ are called *stochastically equivalent* if, for all $u \in U$,

$$\mathbf{P}(\xi(u, \omega) = \xi^*(u, \omega)) = 1.$$

Obviously, if $\xi(u)$ and $\xi^*(u)$ are stochastically equivalent, then their distributions agree on the σ-algebra of cylinder sets of $R(U)$.

By Doob's theorem (Doob, 1953), for any process $\xi(u, \omega)$ there is a process $\xi^*(u, \omega)$ which is separable and stochastically equivalent to $\xi(u, \omega)$. Such processes $\xi^*(u)$ are called *separable modifications of* $\xi(u)$.

For example, the process $\xi(u) \equiv \xi$, where ξ is a random variable, not depending on u, is continuous and, hence, separable. Alongside the space $(\Omega, \mathfrak{A})$ on which ξ is defined, we consider the space $(\Omega \times \Omega_1, \mathfrak{A} \times \mathfrak{A}_1)$, where $\Omega_1 = [0, U]$ and $\mathfrak{A}_1$ is the σ-algebra of Borel sets in $[0, U]$. In this new space we may define a random variable η, not depending on ξ and uniformly distributed on $[0, U]$. We put

$$\xi_1(u) = \begin{cases} 0 & u \neq \eta \\ 1 & u = \eta \end{cases}.$$

Then it is easy to see that the random process $\zeta(u) = \xi(u) + \xi_1(u)$ is not separable (this would not hold for nonrandom η). A separable modification of $\zeta(u)$, obviously, coincides with $\xi(u)$.

Separabilization of $\xi(u, \omega)$ (that is, the transition from $\xi(u, \omega)$ to a separable process $\xi^*(u, \omega)$) also signifies the extension of the σ-algebra $\mathfrak{A}_{R(U)}$ by the addition of uncountable intersections of the form

$$A = \bigcap_{u \in [u_1, u_2] \subset U} \{\xi(u) \in [a, b]\}$$

$$= \left\{ \sup_{u \in [u_1, u_2]} \xi(u) \leqslant b, \quad \inf_{u \in [u_1, u_2]} \xi(u) \geqslant a \right\}$$

and the extension of $\mathbf{P}$ by the equality

$$\mathbf{P}(A) = \mathbf{P}\left[\bigcap_{\substack{u \in [u_1, u_2] \\ u \in S}} \{\xi(u) \in [a, b]\} \right]$$

where under the probability symbol on the right there is an element of the initial σ-algebra $\mathfrak{A}_{R(U)}$.

It may be noted that, for separable processes, such sets as the set of all increasing functions, the set of all continuous functions, the set of all functions from D, and so on, are *events*, that is, are elements of $\mathfrak{A}_{R(U)}$. We recall also that processes given in C or D are automatically separable and, consequently, separabilization is not required.

Further, if the trajectories of a separable process ξ, defined say in $(R^U, \mathfrak{C}_U)$, belong to C or D with probability 1, then we can construct a process ξ^* with, roughly speaking, 'the same distributions', but now in $(C(0, U), \mathfrak{B}_{C(0, U)})$ or in $(D(0, U), \mathfrak{B}_{D(0, U)})$ respectively. For example, let $\mathbf{P}(C) = 1$ and let $A \in \mathfrak{C}_U$ be an arbitrary cylinder. Then the set A_C of functions from C defined by the same relations on the projections of $\xi(u)$ is $A_C = A \cap C$. We obtain the required distribution $\mathbf{P}^*$ in $(C(0, U), \mathfrak{B}_{C(0, U)})$, if for any cylinder set A_C from $\mathfrak{B}_{C(0, U)}$ or $\mathfrak{C}_{C(0, U)}$ we put

$$\mathbf{P}^*(A_C) = \mathbf{P}(A \cap C) = \mathbf{P}(A).$$

It then remains to make use of the extension theorem for measures.

There are simple sufficient conditions for the trajectory of a process to belong to C or D respectively with probability 1.

THEOREM 1. (*Kolmogorov*) *The sample trajectories of a separable process* $\xi(u, \omega)$ *on* $[0, U]$ *are continuous with probability* 1, *if for some* $\alpha > 0$, $\varepsilon > 0$, $0 < c < \infty$ *and all* $u, u + h$ *from* $[0, U]$

$$\mathbf{E}|\xi(u + h) - \xi(u)|^\alpha < c|h|^{1 + \varepsilon}. \tag{1.5}$$

THEOREM 2. (*Kolmogorov–Chentsov*) *The sample trajectories of a separable process* $\xi(u, \omega)$ *on* $[0, U]$ *belong to* $D(0, U)$ *with probability* 1, *if for some* $\alpha \geqslant 0$, $\beta \geqslant 0$, $\varepsilon > 0$, $0 < c < \infty$ *and all* $u_3 \geqslant u_2 \geqslant u_1$ *from* $[0, U]$

$$\mathbf{E}|\xi(u_3) - \xi(u_2)|^\alpha|\xi(u_2) - \xi(u_1)|^\beta < c|u_3 - u_1|^{1+\varepsilon}. \tag{1.6}$$

The following generalizations of conditions (1.5) and (1.6) are given in Billingsley (1968), Cramer and Leadbetter (1967), and Loeve (1962).

THEOREM 1A. *Suppose that for all* $u, u + h$ *from* $[0, U]$

$$\mathbf{P}(|\xi(u + h) - \xi(u)| \geqslant g(h)) \leqslant q(h) \tag{1.7}$$

where g and q *are even functions of* h, *nonincreasing as* $h \downarrow 0$ *and such that*

$$\sum_{n=1}^{\infty} g(2^{-n}) < \infty, \quad \sum_{n=1}^{\infty} 2^n q(2^{-n}) < \infty.$$

Then there exists a random process $\xi^*(u)$, *stochastically equivalent to* $\xi(u)$, *whose trajectories are continuous on* $[0, U]$ *with probability* 1.

THEOREM 1B. *If instead of* (1.7) *it is required that there exists a nondecreasing function* $F(t)$ *on* $[0, U]$ *such that*

$$\mathbf{P}(\xi(u + h) - \xi(u) > \lambda) \leqslant \lambda^{-\alpha}|F(u + h) - F(u)|^{1+\beta}$$

for some $\alpha > 0$, $\beta > 0$, *then the result of theorem* 1A *also holds.*

THEOREM 2A. *Assume that for all* $0 \leqslant u_1 < u_2 < u_3 \leqslant U$, $u_3 - u_1 = h$ *we have*

$$\mathbf{P}(|\xi(u_3) - \xi(u_2)|\,|\xi(u_2) - \xi(u_1)| \geqslant g^2(u)) \leqslant q(h) \tag{1.8}$$

where g *and* q *are as defined in theorem* 1A. *Then there is a random process* $\xi^*(t)$, *stochastically equivalent to* $\xi(u)$, *whose sample trajectories belong to* $D(0, U)$ *with probability* 1.

THEOREM 2B. *The result of theorem* 2A *remains valid if* (1.8) *is replaced by the condition*

$$\mathbf{P}(|\xi(u_3) - \xi(u_2)| \geqslant \lambda;\ |\xi(u_2) - \xi(u_1)| \geqslant \lambda) \leqslant \lambda^{-2\alpha}|F(u_3) - F(u_1)|^{1+\beta}$$

for some $\alpha > 0$, $\beta > 0$, *where* $F(t)$, *as before, is a continuous nondecreasing function.*

1.2. Fundamental aspects of convergence of processes

The basic problems considered in this book are connected with the approximation of the individual random processes which arise in queuing theory, with the help of others which are simpler and better known. Here we must clarify first which processes are to be regarded as nearby and which are not. Naturally, in different problems the requirements for proximity will differ.

Thus, we look at various notions of convergence of random processes. Suppose we have a sequence of random processes $\xi_n(u, \omega)$ and a process $\xi(u, \omega)$ (all given on $(R(U), \mathfrak{A}_{R(U)})$).

One of the broadest notions of convergence is *weak convergence of the finite-dimensional distributions* of $\xi_n(u, \omega)$ to the corresponding distributions of $\xi(u, \omega)$:

$$\mathbf{P}(\xi_n(u_j, \omega) < x_j; j = 1, \ldots, k) \to \mathbf{P}(\xi(u_j, \omega) < x_j; j = 1, \ldots, k)$$

for all k, $u_1, \ldots, u_k$ and $x_1, \ldots, x_k$ such that

$$\mathbf{P}(\xi(u_j, \omega) = x_j) = 0, \quad j = 1, \ldots, k.$$

This convergence will be denoted

$$\xi_n \Rightarrow \xi$$

or

$$\mathbf{P}_n \Rightarrow \mathbf{P}$$

when referring to distributions $\mathbf{P}_n$ and $\mathbf{P}$ of ξ_n and ξ. The symbol $\Rightarrow$ will also be retained for the usual weak convergence of distributions of random variables (convergence at points of continuity).

If, for example, $\xi_n(u)$ describes the behaviour of the queue length at time u for some queuing system, and we are interested only in the distribution of ξ_n at time u_0, then, obviously, for large n we can approximate this distribution by the distribution of $\xi(u_0)$ provided that ξ_n converges to ξ in this broad sense.

If, however, we are interested in the behaviour of some functional $f(\xi_n)$ of ξ_n, say the average queue length

$$\frac{1}{U}\int_0^U \xi_n(u)\,du \tag{1.9}$$

or the maximal size of the queue

$$\sup_{u \leqslant U} \xi_n(u) \tag{1.10}$$

then convergence of the finite-dimensional distributions is no longer adequate. It is not difficult to give examples which show this. Let, for example

$$\xi_n(u) = \xi(u) + \alpha_n(u)$$

where $\alpha_n(u)$ does not depend on $\xi(u)$ and is a random process constructed as follows: $\alpha_n(u) = 0$ outside the interval $\left[\eta - \frac{1}{n}, \eta + \frac{1}{n}\right]$, where η is uniformly distributed on $[0, U]$. On $\left[\eta - \frac{1}{n}, \eta + \frac{1}{n}\right]$ we put $\alpha_n(u) = n$. Then it is obvious that the distributions of the functionals (1.9) and (1.10) do not converge whereas their finite-dimensional distributions do converge ($\xi_n \Rightarrow \xi$), since the probability that the interval $\left[\eta - \frac{1}{n}, \eta + \frac{1}{n}\right]$ contains at least one point from a fixed set $(u_1, \ldots, u_k)$ tends to zero as $n \to \infty$.

On the other hand, it is clear, for example, that a requirement of the form

$$\sup_n |\xi_n(u) - \xi(u)| \to 0$$

for almost all ω (if ξ_n and ξ are given on the same probability space) is too strong, although it will ensure the proximity of $f(\xi_n)$ and $f(\xi)$ for a broad class of f.

The definition of convergence of processes generally depends on *the distributions of which functionals $f(\xi_n)$ we wish to approximate by the distributions of $f(\xi)$.*

For this book we restrict ourselves to two (apart from $\Rightarrow$) modes of convergence which are, clearly, quite sufficient for the needs of concrete problems in queuing theory.

I. *C-convergence*—used where the limit process ξ is continuous with probability 1 ($\mathbf{P}(\xi \in C) = 1$). Consider separable processes $\xi(u)$ and $\xi_n(u)$, $n = 1, 2, \ldots$ in $(R(U), \mathfrak{A}_{R(U)})$, $R(U) \supset C(0, U)$. We say that the sequence ξ_n C-converges to ξ:

$$\xi_n \underset{C}{\Rightarrow} \xi$$

(or, the sequence of distributions $\mathbf{P}_n$ of the ξ_n C-converges

$$\mathbf{P}_n \underset{C}{\Rightarrow} \mathbf{P}$$

to the distribution $\mathbf{P}$ of ξ), if for any $\mathfrak{A}_{R(U)}$-measurable functional f, continuous at the 'points' of C relative to the uniform metric (1.2), we have weak convergence of distributions

$$\mathbf{P}(f(\xi_n) < x) \Rightarrow \mathbf{P}(f(\xi) < x).$$

It is easy to see that C-convergence $\xi_n \Rightarrow_C \xi$ implies convergence of finite-dimensional distributions ($\Rightarrow$), convergence of the distributions of the functionals

$$\sup_{u \in [u_1, u_2]} \xi_n(u), \quad \int_{u_1}^{u_2} \phi(\xi_n(u))\, du \tag{1.11}$$

where ϕ is continuous, and so on.

II. *D-convergence* is defined in a similar way to C-convergence and is used where the limit process belongs to $D(0, U)$ with probability 1. Again let $\xi_n(u)$ and $\xi(u)$ be separable processes in $(R(U), \mathfrak{A}_{R(U)})$, $R(U) \supset D(0, U)$. We will say that the sequence ξ_n (or $\mathbf{P}_n$) D-converges to ξ (or $\mathbf{P}$):

$$\xi_n \underset{D}{\Rightarrow} \xi \quad (\text{or } \mathbf{P}_n \underset{D}{\Rightarrow} \mathbf{P})$$

if for any $\mathfrak{A}_{R(U)}$-measurable functional f, continuous at the 'points' of D in the metric (1.3), we have convergence

$$\mathbf{P}(f(\xi_n) < x) \Rightarrow \mathbf{P}(f(\xi) < x).$$

It is clear that the class of functionals continuous relative to the metric (1.3) is smaller than the class of functionals continuous relative to the uniform metric. Therefore C-convergence implies D-convergence but not conversely. D-convergence does not imply the convergence $\Rightarrow$ or convergence of the first functional in (1.11). However, convergence for the functional $\sup_{u\in[0,U]} \xi_n(u)$ and convergence of the finite-dimensional distributions for some everywhere dense set $S \subset U$ of values u do hold (see theorem 2).

Conditions for C- and D-convergence are given in the following two theorems (see Billingsley, 1968; Borovkov, 1972b, 1976a; Gikhman and Skorokhod, 1965).

THEOREM 1. *In order that* $\mathbf{P}_n \underset{C}{\Rightarrow} \mathbf{P}$, $\mathbf{P}(C) = 1$ *it is necessary and sufficient that:*

(1) *there is an everywhere dense set S in U such that the finite-dimensional distributions of* $\{\xi_n(u)\colon u \in S\}$ *converge weakly to the distributions of* $\{\xi(u); u \in S\}$;

(2) *for any* $\varepsilon > 0$

$$\lim_{\Delta\to 0} \lim_{n} \sup \mathbf{P}(\omega_\Delta^C(\xi_n) > \varepsilon) = 0$$

where

$$\omega_\Delta^C(r) = \sup_{|u'-u''|\leqslant\Delta} |r(u') - r(u'')|$$

is the 'modulus of continuity' in C.

By the separability of the processes the functions $\omega_\Delta^C(\xi_n)$ will be random variables. The set S in the first condition can be replaced by $[0, U]$.

In exactly the same way we have a theorem on D-convergence.

THEOREM 2. *In order that* $\mathbf{P}_n \underset{D}{\Rightarrow} \mathbf{P}$, $\mathbf{P}(D) = 1$ *it is necessary and sufficient that:*

(1) *there is an everywhere dense set S in U such that the finite-dimensional distributions of* $\{\xi_n(u); u \in S\}$ *weakly converge to the distributions of* $\{\xi(u); u \in S\}$;

(2) *for any* $\varepsilon > 0$

$$\lim_{\Delta\to 0} \lim_{n} \sup \mathbf{P}(\omega_\Delta^D(\xi_n) > \varepsilon) = 0$$

where

$$\omega_\Delta^D(r) = \sup_{u\in U} \min\,[\omega^+(u,\Delta), \omega^-(u,\Delta)] + \omega^+(0,\Delta) + \omega^-(U,\Delta)$$

$$\omega^{\pm}(u, \Delta) = \sup_{\substack{0<t<\Delta \\ u\pm t\in[0,U]}} |r(u \pm t) - r(u)|.$$

If $R(U) = C(0, U)$ in theorem 1 and $R(U) = D(0, U)$ in theorem 2, then the proofs of the above results are in Billingsley (1968) and Gikhman and Skorokhod (1965).

It is possible to give various conditions which are sufficient for the second conditions in theorems 1 and 2. We mention here conditions connected with theorems 1B and 2B of section 1.1 on whether trajectories belong to C and D respectively.

Suppose there exists a continuous nondecreasing function $F(u)$ and positive numbers α and β such that

$$\sup_n \mathbf{P}(|\xi_n(u+\Delta) - \xi_n(u)| \geqslant \lambda) \leqslant \lambda^{-\alpha}|F(u+\Delta) - F(u)|^{1+\beta} \tag{1.12}$$

or, for $\delta > 0$, $\Delta > 0$,

$$\sup_n \mathbf{P}(|\xi_n(u-\delta) - \xi_n(u)| \geqslant \lambda, |\xi_n(u+\Delta) - \xi_n(u)| \geqslant \lambda)$$

$$\leqslant \lambda^{-2\alpha}|F(u+\Delta) - F(u-\delta)|^{1+\beta}. \tag{1.13}$$

Then (1.12) implies (2) of theorem 1 and (1.13) implies (2) of theorem 2 (see Billingsley, 1968; Gikhman and Skorokhod, 1971, 1973, 1976).

The well-known Donsker–Prokhorov *invariance principle* is an example of C-convergence. We give it in simplified form for identically distributed random variables. Let $\xi_{1,n}, \ldots, \xi_{n,n}$, $n = 1, 2, \ldots$ be a double sequence (series array) of independent random variables, identically distributed in each series, which satisfy the Lindberg condition:

$$\mathbf{E}(|\xi_{1,n}|^2; |\xi_{1,n}| > \varepsilon\sqrt{n}) \to 0$$

for any $\varepsilon > 0$, where, without loss of generality, we assume that $\mathbf{E}\xi_{1,n} = 0$, $\mathbf{E}\xi_{1,n}^2 = 1$. We form a continuous polygon $\xi_n(t)$ on the segment $[0, 1]$, with vertices at the points $(k/n, X_k/\sqrt{n})$, $k = 0, 1, \ldots, n$, where $X_k = \sum_{j=1}^{k} \xi_{j,n}$. Then the random variables $\xi_n(t)$ on $[0, 1]$ C-converge as $n \to \infty$ to a Wiener process $w(t)$. This is a corollary of the fact that the $\xi_n(t)$, constructed as polygons, satisfy the conditions of theorem 1 (see Billingsley, 1968; Prokhorov, 1956).

The conditions of the theorem will be satisfied even by discontinuous step (piecewise-constant) random functions constructed relative to $(k/n, X_k/\sqrt{n})$, $k = 0, 1, \ldots, n$.

1.3. The fundamental types of limit processes

We consider here three classes of processes which appear particularly frequently as limits in very varied problems.

1.3.1. *Homogeneous processes with independent increments.*
1.3.2. *Diffusion processes.*
1.3.3. *Gaussian processes.*

1.3.1. A process $\xi(u)$, $u \in [0, U]$, is called a *homogeneous process with independent increments*, if for any set $0 \leqslant u_0 \leqslant \cdots \leqslant u_k \leqslant U$, the random variables $\xi(u_{j+1}) - \xi(u_j)$, $j = 0, 1, \ldots, k-1$, are independent and the distribution of $\xi(u'') - \xi(u')$ depends only on the difference $u'' - u'$. If $\xi(0) = 0$, then for such processes

$$\mathbf{E} \exp\{\mathrm{i}\lambda\xi(u)\} = \mathrm{e}^{u\psi(\lambda)} \tag{1.14}$$

where

$$\psi(\lambda) = \mathrm{i}\lambda\gamma - \frac{\sigma^2\lambda^2}{2} + \int\left(\mathrm{e}^{\mathrm{i}\lambda x} - 1 - \frac{\mathrm{i}\lambda x}{1 + x^2}\right)\Phi(\mathrm{d}x) \tag{1.15}$$

(Levy's formula (Gikhman and Skorokhod, 1965)). Here Φ is the so-called *spectral measure*, with the properties

$$\Phi(\{0\}) = 0, \quad \int \frac{x^2}{1 + x^2}\,\Phi(\mathrm{d}x) < \infty.$$

These formulae obviously determine any finite-dimensional distribution of $\xi(u)$. On the right-hand side of (1.15) we may pick out three components.

The term $\mathrm{i}\lambda\gamma$ defines the *drift* of the process (in magnitude, γt after time t). The term $-\sigma^2\lambda^2/2$ defines the *diffusion* or *Wiener component*. Finally, the last term in (1.15) defines, roughly speaking, the *jump component*. If Φ is such that $\int_{-1}^{1} |x|\Phi(\mathrm{d}x) < \infty$, then the last term may be written

$$\lambda\gamma_1 + \int(\mathrm{e}^{\mathrm{i}\lambda x} - 1)\Phi(\mathrm{d}x). \tag{1.16}$$

If Φ is a finite measure, then, obviously, such a description is possible and the second term in (1.16) defines a so-called *compound Poisson process* having, with probability 1, a finite number of jumps on $[0, 1]$ each with distribution function $\Phi(-\infty, x)/\Phi(-\infty, \infty)$.

The trajectories of a separable process with independent increments have no discontinuities of the second kind, with probability 1 (Gikhman and Skorokhod, 1971, 1973, 1976). Therefore, if we introduce a stochastically equivalent process, continuous on the right (left at U), then its trajectories will belong to D with probability 1. This means that a process with independent increments can be regarded as a process on $(D(0, U), \mathfrak{C}_{D(0, U)})$.

If we consider a separable process for which $\Phi(-\infty, \infty) = 0$, then we obtain a Wiener process whose trajectories are continuous with probability 1. Such a process can be considered as a process in $(C(0, U), \mathfrak{C}_{C(0, U)})$. If $\Phi(-\infty, \infty) > 0$, then the trajectories of $\xi(u)$, with positive probability, do not belong to $C(0, U)$

(for more detail see, for example, Gikhman and Skorokhod, 1965; 1971, 1973, 1976).

1.3.2. Homogeneous diffusion processes, like any *Markov* processes, can be given by transition functions. A function $\mathbf{P}(z, t, A) \leqslant 1$ ($z \in R$, $t \leqslant U$, $A \in \mathfrak{B}$ a Borel set on the line) is called a *transition function* if:

(1) for fixed z and t, $\mathbf{P}(z, t, A)$ is a probability measure on $(R, \mathfrak{B})$; $\mathbf{P}(z, 0, A) = I(A)$ is the indicator of A;
(2) for fixed t and A, $\mathbf{P}(z, t, A)$ is a $\mathfrak{B}$-measurable function;
(3) for any z, $u \leqslant t$ and A we have the Kolmogorov–Chapman equation

$$\mathbf{P}(z, t, A) = \int \mathbf{P}(z, u, \mathrm{d}y)\mathbf{P}(y, t - u, A).$$

Suppose there is given an initial distribution $\mathbf{P}_0(A)$. The *homogeneous Markov process* corresponding to the transition function $\mathbf{P}(\cdot, \cdot, \cdot)$ and initial distribution $\mathbf{P}_0$ is the process $\xi(u, \omega)$ whose consistent finite-dimensional measures (probabilities of cylinder sets) are given by

$$\mathbf{P}(\xi(0) \in A_0, \xi(u_1) \in A_1, \ldots, \xi(u_k) \in A_k)$$
$$= \int_{A_0} \mathbf{P}_0(\mathrm{d}z_0) \int_{A_1} \mathbf{P}(z_0, u_1, \mathrm{d}z_1) \cdots \int_{A_k} \mathbf{P}(z_{k-1}, u_k - u_{k-1}, \mathrm{d}z_k).$$

The process $\xi(u)$ for which the measure $\mathbf{P}_0$, corresponding to $\xi(0)$, is concentrated at a point a (that is, $\xi(0) = a$ with probability 1), is denoted $\xi^{(a)}(u)$.

The most important property of these processes is the *Markov* property. Let $(R(U), \mathfrak{A}_{R(U)}, \mathbf{P})$ be the sample space and $\mathfrak{M}(u) \subset \mathfrak{A}_{R(U)}$ the least σ-algebra containing all ω-sets of the form $\{\xi(t, \omega) \in B\}$ for $t \leqslant u$, $B \in \mathfrak{B}$.

Denote by $\sigma(\xi)$ the σ-algebra generated by the random variable ξ, and by $\mathbf{P}_{\mathfrak{F}}$ the conditional probability relative to a σ-algebra $\mathfrak{F}$. The Markov property is that the conditional probability relative to $\mathfrak{M}(u)$

$$\mathbf{P}_{\mathfrak{M}(u)}(\xi(u + t) \in B) = \mathbf{P}_{\sigma(\xi(u))}(\xi(u + t) \in B) = \mathbf{P}(\xi(u), t, B)$$

that is, if the state of the system at time u is fixed, $\xi(u) = a$, then the subsequent evolution is the same as for the process $\xi^{(a)}(t)$ and does not depend on other events from $\mathfrak{M}(u)$.

What are the properties of the trajectories of a Markov process? (In what follows Markov process will always mean *homogeneous* Markov process, as defined above.)

Suppose that the following natural conditions on the transition function are satisfied.

(I) *For any* $\varepsilon > 0$ *as* $t \to 0$

$$\mathbf{P}(z, t, R - V_\varepsilon(z)) \to 0$$

where $V_\varepsilon(z)$ *is the* ε*-neighbourhood of* z.

(II) *For each compact set* K

$$\lim_{y \to \infty} \sup_{t \leqslant u} \mathbf{P}(y, t, K) = 0.$$

Then there is a Markov process $\xi(u)$ *with transition function* $\mathbf{P}(z, t, B)$ *whose trajectories, with probability* 1, *belong to* D^+ (bounded, no discontinuities of the second kind, and continuous to the right (Dynkin, 1963; Gikhman and Skorokhod, 1971, 1973, 1976)). *Here for each* u

$$\mathbf{P}(\xi(u - 0) = \xi(u + 0)) = 1.$$

If, instead of condition (I), *we require*

(III) $\dfrac{1}{t}\mathbf{P}(z, t, R - V_\varepsilon(z)) \to 0$

as $t \to 0$, *then there will exist a Markov process* $\xi(u)$ *with transition function* $\mathbf{P}(z, t, B)$ *and continuous trajectories* (Dynkin, 1963; Feller, 1950, 1966; Gikhman and Skorokhod, 1971, 1973, 1976).

We return to the Markov property, which asserts that for any fixed t

$$\mathbf{P}_{\mathfrak{M}(t)}(\xi(t + u) \in B) = \mathbf{P}(\xi(t), u, B). \tag{1.17}$$

Will this property hold if t is random? For arbitrary random times the answer to the question is obviously negative. However, we can give an important class of random variables, $\tau = \tau(\omega)$, for which (1.17) is preserved for a broad class of processes.

A random variable $\tau = \tau(\omega) \geqslant 0$ is called a *stopping time* relative to the family of* σ*-algebras* $\{\mathfrak{M}(t)\}$ (or, in our case, simple a stopping time) if for each $t \geqslant 0$

$$\{\tau(\omega) \leqslant t\} \in \mathfrak{M}(t).$$

This relation means that whether τ is bigger than t or not depends only on the trajectories of the process in $[0, t]$.

We consider those sets $A \in R(U)$ such that for any $t \geqslant 0$

$$A \cap \{\tau \leqslant t\} \in \mathfrak{M}(t) \tag{1.18}$$

and we denote by $\mathfrak{M}(\tau)$ the σ-algebra generated by these sets.

A Markov process $\xi(u)$ is called *strictly Markov* if for any stopping time τ and any $u \geqslant 0$

$$\mathbf{P}_{\mathfrak{M}(\tau)}(\xi(\tau + u) \in B) = \mathbf{P}_{\sigma(\xi(\tau))}(\xi(\tau + u) \in B) = \mathbf{P}(\xi(\tau), u, B). \tag{1.19}$$

* In the Russian literature the term 'Markov Epoch' is sometimes used instead of 'stopping time'.

We note that if the stopping time τ takes no more than countably many values, then (1.19) always holds. In fact, let $t_1, t_2, \ldots$ be the set of values of τ. Then

$$\mathbf{P}_{\mathfrak{M}(\tau)}(\xi(\tau+u)\in B) = \sum_k I(\tau = t_k)\mathbf{P}_{\mathfrak{M}(\tau)}(\xi(t_k+u)\in B)$$

where $I(A)$ denotes the indicator of A. But the intersections of $\mathfrak{M}(\tau)$ and $\mathfrak{M}(t_k)$ with the event $\{\tau = t_k\}$ coincide. Therefore

$$\mathbf{P}_{\mathfrak{M}(\tau)}(\xi(\tau+u)\in B) = \sum_k I(\tau = t_k)\mathbf{P}_{\mathfrak{M}(t_k)}(\xi(t_k+u)\in B)$$

$$= \sum_k I(\tau = t_k)\mathbf{P}(\xi(t_k), u, B) = \mathbf{P}(\xi(\tau), u, B).$$

In general extra assumptions are needed. Let $\xi(u)$ satisfy one of the conditions

(IVa) *For any continuous and bounded function ϕ and any $t > 0$ the function*

$$\mathbf{E}\phi(\xi^{(a)}(t)) = \int \phi(y)\mathbf{P}(a, t, \mathrm{d}y)$$

is continuous relative to a.

This is often called the *Feller* property.

(IVb) *The set of values of* $\xi(u)$ (the phase space) *is discrete, that is, is either finite or countable.*

Conditions (IVa) and (IVb) may not be distinguishable (denoting their union as condition (IV), since in (IVb) $\mathbf{E}\phi(\xi^{(a)}(t))$ is continuous relative to the discrete topology.

The following result holds. *Every Markov process which is continuous from the right* ($\mathbf{P}(\xi(u+0) = \xi(u)) = 1$) *and satisfies* (IV) *is strictly Markov* (Dynkin, 1963).

We may add also that for this class of processes (Dynkin, 1963)

$$\mathfrak{M}(u) = \mathfrak{M}(u+0) = \bigcap_{n=1}^{\infty} \mathfrak{M}\left(u + \frac{1}{n}\right).$$

A Markov process $\xi(u)$ is called a *process with unbounded diffusion* if its transition function satisfies the conditions:

(1°) *For every $z \in R$ and $\varepsilon > 0$ as $t \to 0$*

$$\int_{|z-y|>\varepsilon} \mathbf{P}(z, t, \mathrm{d}y) = o(t)$$

(2°) *There exist functions $a(z)$ and $b(z)$ such that for every z and $\varepsilon > 0$*

$$\int_{|z-y|\leqslant\varepsilon} (z-y)\mathbf{P}(z, t, \mathrm{d}y) = ta(z) + o(t)$$

$$\int_{|z-y|\leqslant \varepsilon} (z-y)^2 \mathbf{P}(z, t, dy) = tb(z) + o(t).$$

The coefficient $a(z)$ is called the *drift coefficient* and $b(z)$ is called the *diffusion coefficient*. Their physical meanings are clear from the definitions.

If a and b satisfy a Lipshits condition, then the diffusion process $\xi(u)$ *is strictly Markov and its trajectories are continuous with probability* 1 (Dynkin, 1963; Gikhman and Skorokhod, 1971, 1973, 1976).

Thus a Markov process satisfying (I) and (II) may be regarded as a process in D^+, and the diffusion process (for, obviously, (I) and (II) are again satisfied) may be regarded as a process in C.

It is important to note that the coefficients a and b, in a broad class of cases, are uniquely determined by the transition function $\mathbf{P}(\cdot, \cdot, \cdot)$ and, consequently, by the distribution of the Markov process. The following theorem holds.

THEOREM 1. (*Kolmogorov* (*Gikhman and Skorokhod, 1965; Kolmogorov, 1958; Skorokhod, 1961*)) *Let the bounded function* $\phi \in C(-\infty, \infty)$ *be such that*

$$v(t, x) = \int \phi(y)\mathbf{P}(x, t, dy)$$

has continuous and bounded derivatives $\dfrac{\partial v}{\partial x}, \dfrac{\partial^2 v}{\partial x^2}$. *Then* $\dfrac{\partial v}{\partial t}$ *exists and* v *satisfies the so-called backward Kolmogorov equation*

$$\frac{\partial v}{\partial t} = a\frac{\partial v}{\partial x} + \frac{b}{2}\frac{\partial^2 v}{\partial x^2} \tag{1.20}$$

and the condition

$$\lim_{t\to 0} v(t, x) = \phi(x). \tag{1.21}$$

Thus, if the solution of (1.20) is unique, then we can define the function

$$\int \phi(y)\mathbf{P}(x, t, dy)$$

which, in its turn, determines the transition function $\mathbf{P}(\cdot, \cdot, \cdot)$, as ϕ runs through some everywhere dense class in $C(-\infty, \infty)$, in the sense of the uniform metric on finite intervals.

We will denote by C_k the class of k-times differentiable functions whose kth derivatives satisfy a uniform Hölder condition. It is obvious, for example, that C_2 is dense in C.

If for any y the transition function $\mathbf{P}(x, t, dy)$ takes the form $p(x, t, y)\, dy$, then we will say that a *transition density* exists.

THEOREM 2. (*Eidel'man, 1964; Il'in et al., 1962*) *Let $a(x)$ and $b(x)$ be differentiable functions such that $b(x) > b_0 > 0$ and which satisfy one of the two conditions*

(A) *a and b belong to* C_0 (this means, in particular, that a and b cannot grow more quickly than $|x|^\alpha$, $\alpha \leqslant 1$) *or*

(B) *The functions a', b', a'', b'' belong to C_0 and are bounded* (in this case a and b can grow like $c|x|$).

Then there exists a unique transition density $p(x, t, y)$ (and, consequently, a unique process) *for which the functions a and b are the drift and diffusion coefficients respectively. For each $t > 0$, as a function of x, the density is of class C_2, and as a function of t is of class C_1. Here*

$$\frac{\partial^k p}{\partial x^k} \leqslant c_k t^{-(1+k)/2} \exp\left\{\frac{-c(x-y)^2}{t}\right\}.$$

In addition, for any bounded function $\phi \in C$ the function

$$v(t, x) = \int \phi(y) p(x, t, y)\, dy$$

is the unique solution to the Cauchy problem (1.20), (1.21) ($v(t, \cdot) \in C_1, v(\cdot, x) \in C_2$ *for $t > 0$*). *If ϕ belongs to C_2, then $v(t, x)$ satisfies*

$$\lim_{t \to 0} \frac{\partial v}{\partial x} = \phi', \quad \lim_{t \to 0} \frac{\partial^2 v}{\partial x^2} = \phi''.$$

It follows from this theorem that a and b have the property

$$a(x) = \lim_{t \to 0} \frac{1}{t} \mathbf{E}(\xi^{(x)}(t) - x)$$

$$b(x) = \lim_{t \to 0} \frac{1}{t} \mathbf{E}(\xi^{(x)}(t) - x)^2$$

where $\xi^{(x)}(t)$ is a process with drift $a(x)$ and diffusion $b(x)$.

This theorem also implies that under condition (A) or (B) the functions $\partial v/\partial t$, $\partial v/\partial x$, $\partial^2 v/\partial x^2$ are uniformly continuous in $0 \leqslant t \leqslant U$, $x \in R$.

Diffusion processes may also be characterized by an *infinitesimal operator* $\mathscr{A}$ generated by the transition function. $\mathscr{A}$ is defined by

$$\mathscr{A}\phi(x) = \lim_{t \to 0} \frac{1}{t} \int (\phi(y) - \phi(x)) \mathbf{P}(x, t, dy) \tag{1.22}$$

on some suitable set of functions ϕ. If ϕ is twice continuously differentiable and the functions a and b in condition (2°) of the definition of a diffusion process are continuous, then

$$\mathscr{A}\phi(x) = \left(a(x)\frac{\mathrm{d}}{\mathrm{d}x} + \frac{b(x)}{2}\frac{\mathrm{d}^2}{\mathrm{d}x^2}\right)\phi(x). \tag{1.23}$$

This is just another description of the backward equation (1.20). In fact, apply $\mathscr{A}$ to the function $v(t, x) = \mathscr{A}_t\phi(x)$, where $\mathscr{A}_t$ are the operators

$$\mathscr{A}_t\phi(x) = \int \phi(y)\mathbf{P}(x, t, \mathrm{d}y)$$

generating a subgroup $\mathscr{A}_t\mathscr{A}_u\phi(x) = \mathscr{A}_{t+u}\phi(x)$. We then obtain

$$\mathscr{A}v(t, x) = \lim_{\Delta\to 0}\frac{1}{\Delta}(v(t + \Delta, x) - v(t, x)) = \frac{\partial v(t, x)}{\partial t}.$$

To obtain (1.20) it remains to apply (1.23) to $v(t, x)$.

The simplest diffusion process is the *Brownian motion process* (*Wiener process*) which is at the same time a one-dimensional process with independent increments. Here $a(x) \equiv 0$, $b(x) \equiv 1$. The transition density is

$$p(x, t, y) = \frac{1}{\sqrt{2\pi t}}\left\{\exp\frac{-(x - y)^2}{2t}\right\} \tag{1.24}$$

and satisfies (together with the functions $v(t, x) = \mathscr{A}_t\phi(x) \to \phi(x)$ as $t \to 0$) the equation

$$\frac{\partial p}{\partial t} = \frac{1}{2}\frac{\partial^2 p}{\partial x^2}. \tag{1.25}$$

It is also not difficult to write down the transition density for the diffusion process with coefficients $a(x) = a = \text{const.}$, $b(x) = \sigma^2 = \text{const.} > 0$. This process may be represented as

$$\xi(t) = at + \sigma w(t)$$

where $w(t)$ is a Wiener process. For the increments of $\xi(t)$ we have

$$\mathrm{d}\xi(t) = a\,\mathrm{d}t + \sigma\,\mathrm{d}w(t)$$

which may be generalized to the case of arbitrary $a(x)$ and $b(x)$. Namely, if $\xi(t)$ is a diffusion process with coefficients a and $b = \sigma^2$, then (Gikhman and Skorokhod, 1965)

$$\mathrm{d}\xi(t) = a(\xi(t))\,\mathrm{d}t + \sigma(\xi(t))\,\mathrm{d}w(t).$$

This equality nicely illustrates the significance of the coefficients a and b. It is, by definition, another way of writing

$$\xi(t) = \int_0^t a(\xi(u))\,\mathrm{d}u + \int_0^t \sigma(\xi(u))\,\mathrm{d}w(u)$$

where $\int_0^t \sigma(\xi(u))\,dw(u)$ is the *Ito stochastic integral*. The construction of the Ito integral and the theory of stochastic differential equations may be found in Gikhman and Skorokhod (1965; 1971, 1973, 1976). Since we will sometimes use this integral we recall its definition and basic properties.

A random function $f(t, \omega)$ is said to *not depend on the future* if, for each t, $f(t, \omega)$ is measurable relative to $\mathfrak{M}(t)$. We consider the class L of functions $f(t, \omega)$ which are measurable relative to $\mathfrak{B} \times \mathfrak{M}(U)$, do not depend on the future, and have the property

$$\mathbf{P}\left(\int_0^U f^2(u, \omega)\,du < \infty\right) = 1.$$

A function $f(t, \omega) \in L$ is called *simple*, if there exists a finite partition of $[0, U)$ into semi-intervals $[t_k, t_{k+1})$, $k = 0, 1, \ldots, N$, such that $f(t, \omega) = f_k(\omega)$ for $t \in [t_k, t_{k+1})$, $0 \leqslant k \leqslant N$.

The stochastic integral of a simple function is defined by

$$\int_0^t f(u, \omega)\,dw = \sum_{k=0}^{m} f_k(\omega)[w(t_{k+1}) - w(t_k)] + f_{m+1}(\omega)[w(t) - w(t_{m+1})]$$

where $m = \max\{k : t_{k+1} < t\}$.

For an arbitrary function from L the integral is defined as the limit over a sequence of simple functions, which exists and does not depend on the choice of sequence.

The fundamental properties of the stochastic integral are:

1. It is a linear functional of f.
2. As a function of the upper limit the integral is additive and continuous with probability 1.
3. $\mathbf{E}_{\mathfrak{M}(u)} \int_0^t f(v, \omega)\,dw(v) = \int_0^u f(v, \omega)\,dw(v)$.
4. $\mathbf{E}\left(\int_0^t f(u, \omega)\,dw(u)\right)\left(\int_0^t g(u, \omega)\,dw(u)\right) = \int_0^t \mathbf{E} fg\,du$.
5. If $f(t, \omega) = f(t)$ is a nonrandom differentiable function, then

$$\int_0^t f(u)\,dw(u) = f(t)w(t) - f(0)w(0) - \int_0^t w(u)f'(u)\,du$$

(usual integration by parts; on the right we have the usual integral).
6. If $f(t, \omega) = f(t)$, then $\int_0^t f(u)\,dw(u)$ is a process with independent increments, whose distribution coincides with the distribution of $w(\int_0^t f^2(u)\,du)$.

We mention now, alongside $w(t)$, certain other diffusion processes for which the transition function may be written down explicitly. Knowing the fundamental solution (1.24) of equation (1.25), it is not difficult to give the form of the solution of the equation

$$\frac{\partial v}{\partial t} = \alpha^2(x)\frac{\partial^2 v}{\partial x^2} + \alpha'(x)\alpha(x)\frac{\partial v}{\partial x}.$$

For this we make the substitution

$$y = \int_0^x \frac{du}{\alpha(u)}$$

(this integral is assumed to be finite), which leads to (1.25).

In applications there frequently occur processes for which $a(x) = ax$, $b(x) = \sigma^2 = \text{const.}$ (for $a < 0$, these are the so-called *Ornstein–Uhlenbeck processes* (Feller, 1950, 1966)). The equation

$$\frac{\partial v}{\partial t} = \frac{\sigma^2}{2}\frac{\partial^2 v}{\partial x^2} + ax\frac{\partial v}{\partial x}$$

corresponding to these coefficients may also be solved explicitly (see Feller, 1950, 1966), and has the fundamental solution (transition density)

$$p(x, t, y) = \frac{1}{\sqrt{2\pi}\sigma(t)}\exp\left\{-\frac{(y - x\,\mathrm{e}^{at})^2}{2\sigma^2(t)}\right\}$$

$$\sigma^2(t) = \frac{\sigma^2}{2a}(\mathrm{e}^{2at} - 1).$$

The process $\xi^{(x)}(t)$ itself (with initial value x at $t = 0$) may be represented in the form

$$\begin{aligned}\xi^{(x)}(t) &= x\,\mathrm{e}^{at} + \sigma\int_0^t \mathrm{e}^{a(t-v)}\,\mathrm{d}w(v)\\ &= \mathrm{e}^{at}\left(x + \sigma\int_0^t \mathrm{e}^{-av}\,\mathrm{d}w(v)\right)\end{aligned} \tag{1.26}$$

or, what is the same thing,

$$\xi^{(x)}(t) = x\,\mathrm{e}^{at} + \sigma\,\mathrm{e}^{at}w\left(\frac{1 - \mathrm{e}^{-2at}}{2a}\right).$$

A direct calculation establishes that there are homogeneous Markov diffusion processes with coefficients $a(x) = ax$, $b(x) = \sigma^2$. Hence, as a corollary, we may obtain that the transition density $p(x, t, y)$ of this process has the above form.

For $a < 0$ there is a *stationary* process $\xi(t)$ with coefficients ax and σ^2, given in the form

$$\xi(t) = \sigma \int_{-\infty}^{t} e^{a(t-v)}\, dw(v)$$

or

$$\xi(t) = \sigma\, e^{at} w\left(-\frac{e^{-2at}}{2a}\right).$$

For $a \neq 0$, alongside (1.26), we now consider the process

$$\bar{\xi}^{(x)}(t) = \xi^{(x+A/a)}(t) - \frac{A}{a} \tag{1.26'}$$

which, obviously, will also be a diffusion process with constant diffusion coefficient σ^2. The drift coefficient of this process is

$$\lim_{t\to 0} \mathbf{E}\,\frac{1}{t}\,(\bar{\xi}^{(x)}(t) - x) = \lim_{t\to 0} \mathbf{E}\,\frac{1}{t}\left(\xi^{(x+A/a)}(t) - x - \frac{A}{a}\right)$$
$$= \left(x + \frac{A}{a}\right)a.$$

Therefore (1.26′) together with (1.26) gives us a diffusion process with coefficients $a(x) = ax + A$, $b(x) = \sigma^2$.

We now consider *diffusion processes with reflections.*

A Markov process $\xi(u)$ is called a *homogeneous diffusion process with reflection at the barrier* $x = 0$, *if*

(1R) $\xi(u) \geqslant 0$ *and for every* $x > 0$ *conditions* (1°), (2°) *are satisfied, where a and b are defined on* $[0, \infty]$.

(2R) $\displaystyle\int_{0\leqslant y\leqslant \varepsilon} y^2 P(0, t, dy) = tb(0) + o(t), \quad 0 < b(0) < \infty.$

If, for example, $\xi^{(x)}(u)$ is a Wiener process:

$$\mathbf{P}(\xi^{(x)}(t) < y) = \mathbf{P}(x, t, (-\infty, y)) = \frac{1}{\sqrt{2\pi t}} \int_{-\infty}^{x} e^{-[(u-x)^2/2t]}\, du$$

then the process $\eta(u) = |\xi(u)|$ is, obviously, a diffusion process with reflection. Its transition function $\mathbf{Q}(x, t, B)$, $B \in [0, \infty)$, has the form

$$\mathbf{Q}(x, t, B) = \mathbf{P}(x, t, B) + \mathbf{P}(x, t, -B).$$

The study of diffusion processes with reflection, to a large extent, reduces to the study of ordinary diffusion processes. Here, as for a usual diffusion, the coefficients a and b uniquely determine the transition function in a broad class of cases.

THEOREM 3. *Let* $\mathbf{P}(\cdot,\cdot,\cdot)$ *be the transition function of a process with reflection and let the bounded function* $\phi \in C(0, \infty)$ *be such that*

$$v(t, x) = \int \phi(y)\mathbf{P}(x, t, dy)$$

has derivatives $\partial v/\partial x, \partial^2 v/\partial x^2$ *which are continuous and bounded in* $[0, \infty)$. *Then in the domain* $x > 0, t \geqslant 0$, $\partial v/\partial t$ *exists and* v *satisfies*

$$\frac{\partial v}{\partial t} = a\frac{\partial v}{\partial x} + \frac{b}{2}\frac{\partial^2 v}{\partial x^2} \tag{1.27}$$

and

$$\lim_{t\to 0} v(t, x) = \phi(x), \quad x \geqslant 0. \tag{1.28}$$

In addition, if $b(0) > 0$ *and for* $x = 0$ $\partial v/\partial t$ *exists and is finite, then*

$$\frac{\partial v(t, 0)}{\partial x} = 0. \tag{1.29}$$

This is the so-called second boundary value problem for equation (1.27).

Proof. The equality $v(+0, x) = \phi(x)$ follows from the relations

$$v(t, x) = \phi(x) + \int [\phi(y) - \phi(x)]\mathbf{P}(x, t, dy)$$

$$= \phi(x) + \int_{|y-x|\leqslant\varepsilon} [\phi(y) - \phi(x)]\mathbf{P}(x, t, dy) + o(t)$$

and the continuity of $\phi(x)$. The proof of the theorem in $x > 0$ reiterates the proof of Kolmogorov's theorem (Gikhman and Skorokhod, 1965; Kolmogorov, 1958): by the properties of the transition function we have

$$v(t, x) = \int \mathbf{P}(x, \Delta, dy)v(t - \Delta, y), \quad t > \Delta > 0.$$

Therefore

$$v(t, x) - v(t - \Delta, x)$$

$$= \int [v(t - \Delta, y) - v(t - \Delta, x)]\mathbf{P}(x, \Delta, dy)$$

$$= \int_{|x-y|\leqslant\varepsilon} + o(\Delta)$$

$$= \int_{|x-y|\leqslant\varepsilon} \left[(y - x)\frac{\partial v(t - \Delta, x)}{\partial x} + (y - x)^2 \frac{\partial^2 v(t - \Delta, x)}{2\,\partial x^2}\right.$$

$$\left. + \frac{\beta}{2}(y-x)^2\right]\mathbf{P}(x, \Delta, dy) + o(\Delta),$$

where

$$|\beta| \leqslant \sup_{\substack{u, z \\ |z-x|\leqslant \varepsilon \\ 0\leqslant t-u\leqslant 1}} \left| \frac{\partial^2 v(u, z)}{\partial x^2} - \frac{\partial^2 v(u, x)}{\partial x^2} \right| \equiv \beta_\varepsilon.$$

Hence, by the continuity of v'_x, v''_x, we get

$$\frac{v(t, x) - v(t-\Delta, x)}{\Delta} = o(1) + \frac{\partial v(t, x)}{\partial x} ax + \frac{\partial^2 v(t, x)}{\partial x^2} \frac{b(x)}{2} + \frac{b(x)}{2} \beta'_\varepsilon$$

where $|\beta'_\varepsilon| \leqslant \beta_\varepsilon$. Passing to the limit as $\Delta \to 0$, $\varepsilon \to 0$ we obtain (1.27).

Further

$$v(t, 0) - v(t-\Delta, 0)$$
$$= \int_{y\leqslant\varepsilon} \left[y \frac{\partial v(t-\Delta, 0)}{\partial x} + \frac{y^2}{2} \left(\frac{\partial^2 v(t-\Delta, 0)}{\partial x^2} + \beta \right) \right] \mathbf{P}(0, \Delta, \mathrm{d}y) + o(\Delta).$$

If we suppose that $\partial v(t, 0)/\partial x \neq 0$, then this equality and the finiteness of $\partial v(t, 0)/\partial t$ will mean that for any $\varepsilon > 0$ there is a finite limit

$$\lim_{\Delta\to 0} \frac{1}{\Delta} \int_{y\leqslant\varepsilon} y\mathbf{P}(0, \Delta, \mathrm{d}y) \frac{\partial v(t, 0)}{\partial x}$$

not depending on ε.

But

$$\int_{y\leqslant\varepsilon} y^2 \mathbf{P}(0, \Delta, \mathrm{d}y) \geqslant \frac{1}{\varepsilon} \int_{y\leqslant\varepsilon} y^2 \mathbf{P}(0, \Delta, \mathrm{d}y) = \frac{1}{\varepsilon} (\Delta b(0) + o(\Delta)).$$

Consequently, this limit is greater than or equal to

$$\frac{v(t, 0)}{x} \cdot \frac{b(0)}{\varepsilon}.$$

This contradiction completes the proof of the theorem. □

Thus, if the coefficients a and b are such that they guarantee uniqueness for the problem (1.27), (1.28), (1.29), and the necessary smoothness of the derivatives of the function v, which is obtained via a fundamental solution of (1.27), then relative to these coefficients we can reconstruct the transition function of a diffusion process with reflection.

THEOREM 4. *Let the functions a and b on* $[0, \infty]$ *satisfy conditions* (A) *or* (B) *of theorem* 2. *Then theorem* 2 *remains true if throughout its statement the transition density* $p(x, t, y)$, $x \geqslant 0$, $y \geqslant 0$, *corresponds to a diffusion process with reflection.*

Proof. We consider initially the case when a and b satisfy condition (A) and

$a(0) = 0$. Then we extend the functions a, b, and $\phi \in C_2(0, \infty)$ to the semiaxis $[-\infty, 0]$ as follows:

$$\begin{aligned} a(-x) &= -a(x) \\ b(-x) &= b(x) \\ \phi(-x) &= \phi(x), \quad x \geqslant 0. \end{aligned} \tag{1.30}$$

These new functions will satisfy the conditions of theorem 2.

This means that there will exist a transition density $p(x, t, y)$ of an unbounded diffusion process which, amongst others, will have the property $p(-x, t, -y) = p(x, t, y)$. The function $v(t, x) = \mathscr{A}_t\phi(x)$, constructed relative to this density and the function ϕ of (1.30), will be a symmetric function of x. This means (smoothness of v being ensured by theorem 2), that $\partial v(t, 0)/\partial x = 0$. This, together with theorem 2, means that v will be the unique solution of the second mixed problem (1.27)–(1.29).

We note also that, if we consider a process of unbounded diffusion $\xi(u)$ with coefficients satisfying (1.30), then the process $\eta(u)$ of diffusion with reflection at the barrier $x = 0$ can be written

$$\eta(t) = |\xi(t)|.$$

This means that, if $\mathbf{P}_\xi(x, t, B)$ is the transition function of ξ, then the transition function $\mathbf{P}_\eta(x, t, B)$ of η is

$$\mathbf{P}_\eta(x, t, B) = \mathbf{P}_\xi(x, t, B) + \mathbf{P}_\xi(\mathrm{x}, \mathrm{t}, -\mathrm{B}),$$

where $B \subset (0, \infty)$.

If a and b satisfy condition (B), then in order to retain the smoothness properties of (B) under passage to the extension (1.30) of a and b, we must assume that

$$a(0) = 0, \quad a''(0) = 0, \quad b'(0) = 0. \tag{1.31}$$

Under these conditions the construction of a process with reflection proceeds quite analogously.

Now let a and b satisfy condition (A) and $a(0) \neq 0$. We note initially that, if $\xi(u)$ is a diffusion process with coefficients a_ξ and b_ξ, then, as is easily verified, the process

$$\eta(t) = \psi(\xi(t)) \tag{1.32}$$

will also be a diffusion, provided ψ is a sufficiently smooth monotone function. Its coefficients will be the functions

$$\begin{aligned} a_\eta &= a\psi' + \frac{b}{2}\psi'' \\ b_\eta &= b(\psi')^2 \end{aligned} \tag{1.33}$$

(this is easy to establish by making the substitution $y = \psi(x)$ in (1.27)).

We consider an analytic transformation ψ, with the properties

$$\psi(0) = 0, \quad \psi'(x) \geqslant \psi_0 > 0, \quad \psi''(0) = -\frac{2a(0)\psi'(0)}{b(0)}. \tag{1.34}$$

In this case the process with reflection ξ will generate a process with reflection η and ψ may always be chosen so that (A) and (B) also hold for a_η and b_η.

Thus, if we make the transformation (1.32) to a process with reflection $\xi(u)$, then we obtain a process with reflection $\eta(u)$, for which

$$a_\eta(0) = 0, \quad b_\eta = b(\psi')^2 > b_0\psi_0^2 > 0.$$

We now consider the functions a_η and b_η. From the above there is a unique diffusion process with reflection $\eta^*(u)$, having them as coefficients. Since the mapping $y = \psi(x)$ is one-to-one and analytic, then we can now construct a process $\xi^*(u) = \psi^{-1}(\eta^*(u))$, which, by the above remark, will be the unique process with reflection having the coefficients a, b. The remaining results of theorem 2 are obvious.

Finally let a and b satisfy condition (B) but let at least one of the equalities in (1.31) not hold. Again we make the transformation (1.32). For $\psi(\xi)$ to satisfy (1.31) it is necessary and sufficient that

$$\left[a\psi' + \frac{b}{2}\psi''\right]_{x=0} = 0$$

$$[b(\psi')^2]'_{x=0} = 0$$

$$\left[a\psi' + \frac{b}{2}\psi''\right]''_{x=0} = 0.$$

The last equality can always be guaranteed at the expense of the choice of the third and fourth derivatives of ψ. However, the first two may be satisfied simultaneously only if

$$b'(0) = 4a(0).$$

If this condition is not satisfied, then the transformation $\eta = \psi(\xi)$ cannot reduce our problem to the problem of continuous diffusion with coefficients having property (B). In this case for the required reduction we must introduce a *nonhomogeneous* process

$$\eta(u) = \psi(\xi(u), u),$$

where

$$\psi(0, u) = 0, \quad \psi'_x(x, u) \geqslant \psi_0 > 0.$$

In this case the coefficients of drift and diffusion of η will depend on u (nonhomogeneity of the process). Here

$$a_\eta(x, u) = a\psi'_x + \psi'_u + \psi''_x \frac{b}{2}$$

$$b_\eta(x, u) = b(\psi'_x)^2$$

and it is always possible to choose an analytic function ψ such that

$$a_\eta(0, t) = 0, \quad \frac{\partial^2 a_\eta(0, t)}{\partial x^2} = 0, \quad \frac{\partial b_\eta(0, t)}{\partial x} = 0.$$

However, we must omit the remainder of the proof since, for simplicity, we have stated and proved the theorems that we would need here only in the homogeneous case. In essence the proof is quite similar to the preceding proof, since the generalization of these theorems to the nonhomogeneous case presents no qualitatively new problem and requires no further restrictions on a and b as functions of x. It is necessary only to require their smoothness with respect to t (Eidel'man, 1964; Il'in *et al.*, 1962). □

A diffusion process with two reflecting barriers (say at $x = 0$ and $x = R$) is defined as a process $0 \leqslant \xi(u) \leqslant R$, whose transition function satisfies condition (1R) in the domain $x \in (0, R)$, and condition (2R) at $x = 0$ and $x = R$. The function

$$v(t, x) = \mathscr{A}_t \phi(x)$$

for such a process satisfies (under the conditions of theorem 3), equations (1.27) and (1.28) for $x \in [0, R]$ and

$$\frac{\partial v(t, 0)}{\partial x} = \frac{\partial v(t, R)}{\partial x} = 0.$$

1.3.3. *Gaussian processes* are defined as processes whose finite-dimensional distributions are normal. A normal multidimensional distribution is completely determined by the vector of means and the matrix of second moments. Therefore the distribution of a Gaussian process $\xi(u)$ is completely determined by two functions

$$m(u) = \mathbf{E}\xi(u)$$

$$R(u, t) = \mathbf{E}(\xi(u) - m(u))(\xi(t) - m(t)) = R(t, u).$$

$R(u, t)$ is called the *correlation* function. If no linear combination of 1, $\xi(u_1), \ldots, \xi(u_k)$ is equal to 0, with probability 1, then $R(u, t)$ is necessarily positive definite (the quadratic form $\sum R(u_i, u_j)x_i x_j$ is positive definite for any $u_1, \ldots, u_k$). If $A = M^{-1}$ is the inverse of the matrix $M = \|R(u_i, u_j)\|$, then the joint distribution of $\xi(u_1), \ldots, \xi(u_k)$ is defined by a k-dimensional density

$$p_{u_1,\ldots,u_k}(x_1, \ldots, x_k) = \frac{(2\pi)^{-n/2}}{\sqrt{|A|}} \exp\{-\tfrac{1}{2}(x - m)A(x - m)^{\mathrm{T}}\},$$

where $(x - m)$ is the row vector $(x_1 - m(u_1), \ldots, x_k - m(u_k))$ and $(x - m)^{\mathrm{T}}$ is a column vector.

If

$$R(u, t) = R(u - t) \tag{1.35}$$

then the process is called *stationary* or *homogeneous.* (An arbitrary process $\xi(t, \omega)$ is called *strictly stationary* if the joint distribution of $\xi(u_1 + t, \omega)$, $\xi(u_2 + t, \omega), \ldots, \xi(u_k + t, \omega)$ does not depend on t. Stationarity in the *wide sense* means (1.35). For Gaussian processes these notions coincide.)

A function $R(t)$ is positive definite if and only if it can be represented in the form

$$\int_0^\infty \cos(\lambda t) F(\mathrm{d}\lambda)$$

(or $\int e^{i\lambda t} F(\mathrm{d}\lambda)$ for complex-valued processes), where $F(B)$ is a finite measure on the line. $F(B)$ is called the *spectral function.*

If $\xi(u)$ is stochastically continuous in the mean square, that is,

$$\mathbf{E}[\xi(u) - \xi(t)]^2 \to 0 \quad \text{for } u \to t$$

then $R(u, t)$ is continuous. In fact, supposing, without loss of generality, that $m(u) \equiv 0$, we obtain

$$\begin{aligned} R(u, t) - R(u, t + \Delta) &= \mathbf{E}\xi(u)[\xi(t) - \xi(t + \Delta)] \\ &\leqslant \sqrt{R(u, u)\mathbf{E}(\xi(t) - \xi(t + \Delta))^2}. \end{aligned}$$

And, conversely, if $R(u, t)$ is continuous then

$$\mathbf{E}(\xi(u) - \xi(t))^2 = R(u, u) + R(t, t) - 2R(u, t) \equiv D(u, t) \to 0$$

as $u \to t$. If $R(u, t)$ uniformly in each of its arguments satisfies a Hölder condition, then $D(t, t + \Delta) < H|\Delta|^\alpha$ for some $H > 0, 1 \geqslant \alpha > 0$. In this case the trajectories of the separable Gaussian process $\xi(u)$ will be continuous with probability 1. By Kolmogorov's theorem for this it is sufficient to show that for some $\alpha > 0, \varepsilon > 0$, $c < \infty$

$$\mathbf{E}|\xi(u) - \xi(u + \Delta)|^\alpha \leqslant c|\Delta|^{1+\varepsilon}.$$

But

$$\mathbf{E}|\xi(u) - \xi(u + \Delta)|^{2k} \leqslant c_k D^k(u, u + \Delta) \leqslant c_k H^k |\Delta|^{\alpha k}.$$

Consequently, it is sufficient to choose $k > 1/\alpha$.

Using theorem 1A of section 1.1 it is easy to give a weaker condition sufficient for the continuity of $\xi(u)$, namely

$$D(t, t + \Delta) \leqslant H\left(\log\frac{1}{\Delta}\right)^a, \quad a > 3$$

which, obviously, will be guaranteed by the same type of inequality for $R(t, t + \Delta) - R(t, t)$ (Cramer and Leadbetter, 1967).

The subsequent increase of smoothness of $R(u, t)$ implies an increase of smoothness of $\xi(u)$. For example, if R is twice continuously differentiable, then (in the mean square) the derivative $\xi'(u)$ exists (Cramer and Leadbetter, 1967).

If $\xi(u)$ is stationary and $R(t) = R(u, u + t)$ has the property

$$\left| R(\Delta) - R(0) - \frac{R''(0)}{2}\Delta^2 \right| \leqslant H\Delta^2\left(\log\frac{1}{\Delta}\right)^a, \quad a > 3$$

then the trajectories of $\xi(u)$, with probability 1, will have a derivative in the usual sense and this derivative will be continuous (Cramer and Leadbetter, 1967).

1.4. Conditions for convergence to degenerate processes

In accordance with the accepted notation let $\{\xi(u); 0 \leqslant u \leqslant U\}$ be an arbitrary separable process given in some space $(R(U), \mathfrak{A}_{R(U)})$ of real functions on $[0, U]$, with the σ-algebra $\mathfrak{A}_{R(U)}$ containing the cylinder sets. We will consider a sequence of such processes $\xi_T(u)$, depending on a parameter $T \to \infty$, and given on lengthening intervals of time $U = vT$, where $v > 0$ is fixed. (Without loss of generality we could consider the length of U itself as the parameter.) We will be interested in the condition for C-convergence of the sequence

$$y_T(t) = \frac{\xi_T(t)}{B(T)}, \quad 0 \leqslant t \leqslant v$$

where $B(T)$ is a normalizing factor, to some deterministic process $d(t) \in C(0, v)$. In other words, we will be interested in conditions under which

$$\mathbf{P}\left(\sup_{t\to v} |y_T(t) - d(t)| > \gamma\right) \to 0$$

as $T \to \infty$ and for any $\gamma > 0$. This corresponds to the convergence of the distribution of measurable functionals $f(y_T)$, continuous in the uniform metric at d, to the distribution of $f(d)$, concentrated at one point.

Such problems arise, for example, in the study of ergodic theorems for busy multichannel queuing systems.

We note that, if $\xi(t)$ is given on the whole axis and is such that $\xi(t)/t \to d$ as $t \to \infty$, with probability 1, then the process

$$y_T(t) = \frac{\xi(tT)}{T}, \quad 0 \leqslant t \leqslant v$$

will C-converge to the function $d(t) = td$. In fact, we must estimate

$$\mathbf{P}\left(\sup_{0\leqslant t\leqslant v}\left|\frac{\xi(tT)}{T}-td\right|>\gamma\right)$$
$$\leqslant \mathbf{P}\left(\sup_{0\leqslant u\leqslant N}\left|\frac{\xi(u)-ud}{T}\right|>\gamma\right)+\mathbf{P}\left(\sup_{N\leqslant u\leqslant vT}\left|\frac{\xi(u)-ud}{T}\right|>\gamma\right)$$
$$\leqslant \mathbf{P}\left(\sup_{0\leqslant u\leqslant N}|\xi(u)-ud|>\gamma T\right)+\mathbf{P}\left(\sup_{u\geqslant N}\left|\frac{\xi(u)}{u}-d\right|>\frac{\gamma}{d}\right).$$

Here the second term, by the ergodicity of $\xi(u)$ and the choice of N, can be made arbitrarily small; the first term, for each N, converges to 0 as $T\to\infty$. Consequently

$$\mathbf{P}\left(\sup_{0\leqslant t\leqslant v}\left|\frac{\xi(tT)}{T}-td\right|>\gamma\right)\to 0$$

as $T\to\infty$, which means C-convergence of $y_T(t)$ to the function td.

We return to the general case. For simplicity we suppose that $d(t)\equiv 0$ and $B(T)=T$. We also assume that there is a right continuous modification of the processes $\xi_T(u)$; by the *jump of the process at* u we mean the quantity

$$\mathrm{Jp}\,(u)=\lim_{\delta\to+0}\ \sup|\xi_T(u)-\xi_T(u-\delta)|$$

and we will denote by H_δ the event

$$H_\delta=\{\mathrm{Jp}\,(u)<\delta T\ \textit{for all}\ 0\leqslant u\leqslant U\}.$$

The assumptions made are not, in principle, required and are largely of a technical nature.

Further we put

$$z_{u,t}=\xi_T(u+t)-\xi_T(u)$$

and consider a collection $\mathfrak{M}(u)$, $0\leqslant u\leqslant U$, of nested σ-algebras, $\mathfrak{M}(u_1)\subset\mathfrak{M}(u_2)\subset\mathfrak{A}_{R(U)}$ for $u_1\leqslant u_2$, such that $\xi_T(t)$ is measurable with respect to $\mathfrak{M}(u)$ for any $t\leqslant u$.

By θ we will denote a function of T such that $\theta\to\infty$, $\theta=o(T)$ as $T\to\infty$. In this sense θ (as other functions of T) will often be called a *sequence*, since we will regard the increasing values of T (and hence of θ) as forming a sequence.

The symbols ε and c (with or without indices) will denote respectively arbitrarily small numbers and constants. Finally, Γ_u will denote the event

$$\Gamma_u=\{\varepsilon_1 T<|\xi_T(u)|<c_1 T\}.$$

THEOREM 1. *Suppose there is a sequence θ and a system of sets $\Omega_{u,\theta}\in\mathfrak{M}(u)$ such that*

$$\Omega(u)=\bigcap_{s\leqslant u}\Omega_{s,\theta}\in\mathfrak{M}(u),\quad \mathbf{P}(\Omega(u))\to 1$$

and on $\Omega_{u,\theta} \cap \Gamma_u \in \mathfrak{M}(u)$ *for* $t \leqslant \theta$ *and fixed* ε, ε_1, c_1 *we have*

(I) $\mathbf{E}_{\mathfrak{M}(u)} z_{u,t} \leqslant \varepsilon\theta(u), \quad \text{if } \xi_T(u) > \varepsilon_1 T$

$\mathbf{E}_{\mathfrak{M}(u)} z_{u,t} \geqslant -\varepsilon\theta(u), \quad \text{if } \xi_T(u) < -\varepsilon_1 T$

(II) *for any fixed* $\delta > 0$ *and* $T \to \infty$

$$\mathbf{P}(H_\delta) \to 1, \quad \mathbf{E}_{\mathfrak{M}(u)}(|z_{u,t}|; |z_{u,t}| > \delta\sqrt{\theta T}) < \varepsilon\theta(u)$$

where $\theta(u)$ *is such that*

$$\sum_{0 \leqslant k \leqslant T/\theta} \theta(k\theta) \leqslant T.$$

If in addition $y_T(t) = \xi_T(tT)/T$ *has the property:* $\mathbf{P}(|y_T(0)| > \delta) \to 0$ *as* $T \to \infty$ *and for any* $\delta > 0$, *then*

$$y_T(t) \underset{C}{\Rightarrow} 0.$$

These conditions have a more homogeneous nature if we put $\theta(u) = \theta$.

We preface the proof of the theorem by two lemmas.

Let $\tau = \tau(\omega)$ be a stopping time relative to the family of σ-algebras $\{\mathfrak{M}(u)\}$, that is, a random variable such that $\{\tau \leqslant u\} \in \mathfrak{M}(u)$, and let $\mathfrak{M}(\tau)$ be the σ-algebra defined by (see (1.18))

$$A \in \mathfrak{M}(\tau) \quad \text{if } A \cap \{\tau \leqslant u\} \in \mathfrak{M}(u) \tag{1.36}$$

for any $u \geqslant 0$. It is easy to verify that if $A \in \mathfrak{M}(\tau)$, then $A \cap \{\tau < s\} \in \mathfrak{M}(u)$, $A \cap \{\tau = s\} \in \mathfrak{M}(u)$ for any $s \leqslant u$ (Lipster and Shiryaev, 1974). Consider the function

$$\phi_n(t) = \begin{cases} \dfrac{k}{n}, & \text{if } \dfrac{k-1}{n} < t \leqslant \dfrac{k}{n} \\ 0, & \text{if } t = 0 \end{cases}$$

and denote by τ_n the random variable $\tau_n = \phi_n(\tau)$. Obviously, $\tau_n \downarrow \tau$ as $n \to \infty$ and τ_n is a stopping time, since

$$\{\tau_n \leqslant t\} = \{\phi_n(\tau) \leqslant t\} = \begin{cases} \{\tau \leqslant t\} & \text{if } t = k/n \\ \{\tau \leqslant [tn]/n\} & \text{if } t \neq k/n \end{cases}.$$

Both events on the right obviously belong to $\mathfrak{M}(t)$.

Further, let $\eta(u) = \eta(u, \omega)$, $u \geqslant 0$, be an arbitrary real-valued random process given on the original probability space and having trajectories which are right continuous with probability 1.

LEMMA 1. *Suppose that* $\mathbf{E}_{\mathfrak{M}(t)}\eta(t) \leqslant c$ *for all t and* $\eta(\tau_n) < \eta$, $\mathbf{E}|\eta| < \infty$. *Then* $\mathbf{E}_{\mathfrak{M}(\tau)}\eta(\tau) \leqslant c$.

The dominated integrability condition for $\eta(\tau_n)$ in this theorem will obviously be satisfied if $\eta(t)$ is bounded by a constant.

Proof. First we note that $\eta(\tau) = \eta(\tau(\omega), \omega)$ is a random variable. In fact, since the trajectory of $\eta(t)$ is right continuous almost surely, then for almost all ω

$$\eta(\tau) = \lim_{n\to\infty} \eta_n, \quad \eta_n = \eta(\tau_n) = \sum_k I\left(\tau_n = \frac{k}{n}\right)\xi\left(\frac{k}{n}\right),$$

where $I(A)$ is the indicator function of the event A. Thus $\eta(t)$ is the limit (almost surely) of the sequence of random variables η_n and, consequently, is itself a random variable.

We must show that for any $B \in \mathfrak{M}(\tau)$

$$\mathbf{E}(\eta(\tau); B) \leqslant c\mathbf{P}(B). \tag{1.37}$$

We note first of all that for any $B \in \mathfrak{M}(\tau_n)$

$$\mathbf{E}(\eta_n; B) \leqslant c\mathbf{P}(B).$$

In fact, by the conditions of the lemma, $\mathbf{E}(\eta(u); A) \leqslant c\mathbf{P}(A)$ for any $A \in \mathfrak{M}(u)$. Since $\{\tau_n = k/n\}B \in \mathfrak{M}(k/n)$ if $B \in \mathfrak{M}(\tau_n)$, then for any $B \in \mathfrak{M}(\tau_n)$ we have

$$\begin{aligned}\mathbf{E}(\eta_n; B) &= \sum_k \mathbf{E}\eta\left(\frac{k}{n}\right)I\left(\tau = \frac{k}{n}\right)I(B)\\ &\leqslant \sum_k c\mathbf{P}\left(\tau_n = \frac{k}{n}; B\right) = c\mathbf{P}(B).\end{aligned}$$

To obtain (1.37) for any $B \in \mathfrak{M}(\tau)$ we need to use $\mathfrak{M}(\tau) \subset \mathfrak{M}(\tau_n)$ (Gikhman and Skorokhod, 1965), the convergence $\eta_n \to \eta$ a.s., and Fatou's lemma. The latter is possible by the uniform integrability of η_n stipulated in the lemma. □

LEMMA 2. *Let*

$$z_{u,t} = \xi(u + t) - \xi(u)$$

be the increments of a random process $\xi(u)$, continuous on the right and $\xi(u)$ measurable relative to $\mathfrak{M}(u)$. Let $\Omega(u)$ be a nested family of sets

$$\Omega(u) = \bigcap_{s \leqslant u} \Omega(s) \in \mathfrak{M}(u)$$

such that for $0 \leqslant u \leqslant U - t$ and $t \leqslant \theta$

$$I(\Omega(u))\mathbf{P}_{\mathfrak{M}(u)}(z_{u,t} \leqslant -x) < \tfrac{1}{2}.$$

Then for all s, $0 \leqslant s \leqslant U - \theta$,

$$\mathbf{P}(\sup_{t \leqslant \theta} z_{s,t} > 2x; \Omega(s + \theta)) \leqslant 2\mathbf{P}(z_{s,\theta} > x; \Omega(s)).$$

Proof. Let τ be the first passage time of the process $\{z_{s,t}, t \geqslant 0\}$ through the level $2x$:

$$\tau = \inf\{t > 0: z_{s,t} \geqslant 2x\}.$$

Since

$$\{\tau \leqslant t\} = \left\{\sup_{v \leqslant t} z_{s,v} \geqslant 2x\right\} \in \mathfrak{M}(s+t)$$

then this is a stopping time relative to the family of σ-algebras $\mathfrak{N}(t) = \mathfrak{M}(s+t)$, $t \geqslant 0$. The family of random variables

$$\eta_1(u) = I(\Omega(u)), \quad \eta_2(u) = I(z_{s+u,\theta-u} \leqslant -x)$$

as u takes values from 0 to θ form a random process continuous from the right. Therefore $\eta(u) = \eta_1(u)\eta_2(u)$ satisfies the conditions of lemma 1, for $c = \frac{1}{2}$, if we take $\mathfrak{N}(u)$ as the family of σ-algebras. Using this lemma we obtain

$$\begin{aligned}
&\mathbf{P}(z_{s,0} > x; \Omega(s)) \\
&\quad \geqslant \mathbf{E}\mathbf{E}_{\mathfrak{N}(\tau)} I(\tau \leqslant \theta) I(\Omega(s+\tau)) I(z_{s,\theta} > x) \\
&\quad \geqslant \mathbf{E} I(\tau \leqslant \theta) I(\Omega(s+\tau))[1 - \mathbf{E}_{\mathfrak{N}(\tau)} I(z_{s+\tau,\theta\tau} \leqslant -x)] \\
&\quad \geqslant \mathbf{E} I(\tau \leqslant \theta)[I(\Omega(s+\tau)) - \tfrac{1}{2} I(\Omega(s+\tau))] \\
&\quad \geqslant \tfrac{1}{2}\mathbf{P}(\tau \leqslant \theta; \Omega(s+\tau)) \geqslant \tfrac{1}{2}\mathbf{P}\left(\sup_{t \leqslant \theta} z_{s,t} > 2x; \Omega(s+\theta)\right).
\end{aligned}$$

The lemma is proved. □

We now proceed to the proof of the theorem. We note first that by conditions (I), (II) there exist sequences $\varepsilon = \varepsilon(T)$, $\varepsilon_1 = \varepsilon_1(T)$, and $\delta = \delta(T)$, tending to 0 as $T \to \infty$, such that (I) and (II) hold. Since ε_1 can be chosen arbitrarily slowly converging to 0, then we can suppose that $\delta \leqslant \varepsilon_1$. In addition, without loss of generality, we can suppose that $\xi_T(0) = 0$, $\xi_T(u) \geqslant 0$, and consider only the first of the conditions (I).

We subdivide $[0, U]$ into intervals of length θ and suppose for simplicity that θ divides U without remainder. Alongside the increments $z_{k\theta,\theta}$, $k = 0, 1, \ldots, n-1$, where $n = U/\theta$, we consider random variables z_k, z_k^*, and z_k^0, $k = 0, 1, \ldots, n-1$, constructed as follows. Let

$$\tau_k = \begin{cases} \theta & \text{if } \sup_{0 \leqslant u \leqslant \theta} \xi_T(k\theta + u) \leqslant \varepsilon_1 \\ \inf\{u \geqslant 0: \xi_T(k\theta + u) > \varepsilon_1\} & \text{otherwise.} \end{cases}$$

Put

$$\mathfrak{M}_k = \mathfrak{M}(k\theta)$$

$$z_k = \min[cT, \xi_T((k+1)\theta) - \xi_T(k\theta + \tau_k)]$$

$$z_k^* = I(\Omega(k\theta))z_k$$
$$z_k^0 = z_k^* - \mathbf{E}_{\mathfrak{M}_k} z_k^*, \quad k = 0, 1, \ldots, n-1.$$

The random variables z_k^0 satisfy

$$\begin{gathered}\mathbf{E}_{\mathfrak{M}_k} z_k^0 = 0 \\ \mathbf{E}_{\mathfrak{M}_k}(|z_k^0|; |z_k^0| > 3\delta\sqrt{\theta T}) \leqslant 4(\varepsilon\theta(k\theta) + cT\kappa_k)\end{gathered} \tag{1.38}$$

where $\kappa_k = \mathbf{P}_{\mathfrak{M}_k}(\Omega(k\theta)\bar{\Omega}((k+1)\theta))$. To prove this inequality we note that by lemma 1 (the random variables to which lemma 1 is applied are bounded by cT)

$$\begin{aligned}&\mathbf{E}_{\mathfrak{M}_k}(|z_k^*|; |z_k^*| > \delta\sqrt{\theta T}) \\ &\quad\leqslant \mathbf{E}_{\mathfrak{M}_k}\mathbf{E}_{\mathfrak{M}(k\theta+\tau_k)}[I(\Omega(k\theta+\tau_k))|z_k|; |z_k| > \delta\sqrt{\theta T}] \\ &\quad\quad + \mathbf{E}_{\mathfrak{M}_k}\mathbf{E}_{\mathfrak{M}(k\theta+\tau_k)}[I(\Omega(k\theta)\bar{\Omega}(k\theta+\tau_k))|z_k|; |z_k| > \delta\sqrt{\theta T}] \\ &\quad\leqslant \varepsilon\theta(k\theta) + cT\kappa_k.\end{aligned}$$

Therefore

$$\mathbf{E}_{\mathfrak{M}_k}|z_k^*| \leqslant \delta\sqrt{\theta T} + \varepsilon\theta(k\theta) + cT\kappa_k$$

and for $\varepsilon\theta(k\theta) + cT\kappa_k < \delta\sqrt{\theta T}$

$$\begin{aligned}&\mathbf{E}_{\mathfrak{M}_k}(|z_k^0|; |z_k^0| > 3\delta\sqrt{\theta T}) \\ &\quad\leqslant \mathbf{E}_{\mathfrak{M}_k}(|z_k^*|; |z_k^*| > \delta\sqrt{\theta T}) + \mathbf{E}_{\mathfrak{M}_k}|z_k^*| \,.\, \mathbf{P}_{\mathfrak{M}_k}(|z_k^0| > 3\delta\sqrt{\theta T}) \\ &\quad\leqslant \varepsilon\theta(k\theta) + cT\kappa_k + 2\delta\sqrt{\theta T}\,\mathbf{P}_{\mathfrak{M}_k}(|z_k^0| > 3\delta\sqrt{\theta T}) \equiv A_k.\end{aligned} \tag{1.39}$$

Since, by Chebyshev's inequality

$$\mathbf{P}_{\mathfrak{M}_k}(|z_k^0| > 3\delta\sqrt{\theta T}) \leqslant \frac{A_k}{3\delta\sqrt{\theta T}}$$

then

$$\frac{1}{3}\mathbf{P}_{\mathfrak{M}_k}(|z_k^0| > 3\delta\sqrt{\theta T}) \leqslant \frac{\varepsilon\theta(k\theta) + cT\kappa_k}{3\delta\sqrt{\theta T}}.$$

Substituting this into (1.39) we obtain (1.38). If $\varepsilon\theta(k\theta) + cT\kappa_k \geqslant \delta\sqrt{\theta T}$, then

$$\mathbf{E}_{\mathfrak{M}_k}|z_k^0| \leqslant 2\mathbf{E}_{\mathfrak{M}_k}|z_k^*| \leqslant 4(\varepsilon\theta(k\theta) + cT\kappa_k)$$

which also implies (1.38).

With the help of (1.38) we may show that, uniformly with respect to k and l, $0 \leqslant k < n$, $0 \leqslant l \leqslant n-k-1$

$$\mathbf{P}_{\mathfrak{M}_k}\left(\left|\sum_{j=k}^{k-l} z_j^0\right| > \alpha T\right) \xrightarrow{P} 0 \tag{1.40}$$

for any $a > 0$ as $T \to \infty$. For this consider the characteristic function

$$M_{k,l} = \mathbf{E}_{\mathfrak{M}_k} \exp\left\{i\lambda \sum_{j=k}^{k+l} z_j^0\right\}$$

$$= \mathbf{E}_{\mathfrak{M}_k} \exp\left\{i\lambda \sum_{j=k}^{k+l-1} z_j^0\right\} . \mathbf{E}_{\mathfrak{M}_{k+l}} \exp\{i\lambda z_{k+l}^0\}.$$

The last factor can be written

$$1 + \mathbf{E}_{\mathfrak{M}_{k+l}}(e^{i\lambda z_{k+l}^0} - 1 - i\lambda z_{k+l}^0)$$

where the second term, in absolute value, does not exceed

$$\mathbf{E}_{\mathfrak{M}_{k+l}}\left[\frac{(\lambda z_{k+l}^0)^2}{2}; |z_{k+l}^0| < 3\delta\sqrt{\theta T}\right] + \mathbf{E}_{\mathfrak{M}_{k+l}}[2|\lambda z_{k+l}^0|; |z_{k+l}^0| > 3\delta\sqrt{\theta T}]$$

$$\leqslant \frac{\lambda^2}{2} 9\delta^2\theta T + 8\lambda(\varepsilon\theta((k+l)\theta) + cT\kappa_{k+l}).$$

Putting $\lambda = \mu/T$, we obtain

$$\log M_{k,l} \leqslant \frac{9l\mu^2}{2} \cdot \frac{\delta^2\theta}{T} + \frac{8\mu}{T} \sum_{j=0}^{l} (\varepsilon\theta((k+j)\theta) + cT\kappa_{k+j})$$

$$\leqslant \varepsilon_2 + c\mu \mathbf{E}_{\mathfrak{M}_k} \sum_{j=0}^{l} \kappa_{k+j}$$

which implies (1.40).

The theorem will be proved if we show for any $\gamma > 0$ and $T \to \infty$, the convergence to zero of

$$\mathbf{P}\left(\sup_{0 \leqslant k \leqslant U} \xi_T(u) \geqslant 2\gamma T\right)$$

$$\leqslant \mathbf{P}(\bar{H}_\delta) + \mathbf{P}(\bar{\Omega}(u)) + \mathbf{P}(\bar{C}) + \mathbf{P}\left(\max_{k \leqslant n} \xi_T(k\theta) > \gamma T; H_\delta, \Omega(U), C\right)$$

$$+ \mathbf{P}\left(\sup_{0 \leqslant u \leqslant U} \xi_T(u) \geqslant 2\gamma T; \max_{k \leqslant n} \xi_T(k\theta) < \gamma T, H_\delta, \Omega(u), C\right) \tag{1.41}$$

where $C = \{\max_{0 \leqslant k \leqslant n} |z_{k\theta,\theta}| < cT\}$. There is an obvious estimate of the first three terms on the right since, by condition (II),

$$\mathbf{P}(\bar{C}) \leqslant \sum_k \frac{\varepsilon\theta(k\theta)}{cT} \leqslant \frac{\varepsilon}{c}.$$

To estimate the fourth term we construct the trajectory

$$\tilde{\xi}_k = \delta_1 T + \sum_{j=0}^{k-1} \tilde{z}_j$$

where $\delta_1 > 0$ will be chosen later

$$\tilde{z}_k = z_k^* - \mathbf{E}_{\mathfrak{M}_k} z_k^* + \varepsilon\theta(k\theta) + cT\kappa_k.$$

Here $\tilde{z}_k \geqslant z_k^*$ since, by lemma 1,

$$\begin{aligned}\mathbf{E}_{\mathfrak{M}_k} z_k^* &= \mathbf{E}_{\mathfrak{M}_k} \mathbf{E}_{\mathfrak{M}(k\theta+\tau_k)}[I(\Omega(k\theta + \tau_k))z_k + I(\Omega(k\theta)\bar{\Omega}(k\theta + \tau_k))z_k] \\ &\leqslant \varepsilon\theta(k\theta) + cT\kappa_k.\end{aligned}$$

Denoting $A = \{\min \bar{\xi}_k > (\varepsilon_1 + \delta)T\}$, we obtain

$$\mathbf{P}\left(\max_{0\leqslant k\leqslant n} \xi_T(k\theta) > \gamma T; H_\delta, \Omega(U), C\right)$$

$$\leqslant \mathbf{P}(\bar{A}) + \mathbf{P}\left(\max_{0\leqslant k\leqslant n} \xi_T(k\theta) > \gamma T; H_\delta, \Omega(U), C, A\right). \quad (1.42)$$

But on $H_\delta\Omega(U)CA$ we have $\xi_T(k\theta) \leqslant \bar{\xi}_k$ for all k. If $\xi_T(k\theta) \leqslant \varepsilon_1$, then this is obvious. If $\xi_T(k\theta) > \varepsilon_1$, then we consider an epoch $j \leqslant k$ close to k for which $\xi_T(j\theta) > \varepsilon_1$, $\xi_T((j-1)\theta) \leqslant \varepsilon_1$. Then

$$\xi_T(j\theta) \leqslant \varepsilon_1 + \delta + z_{j-1} = \varepsilon_1 + \delta + z_{j-1}^* \leqslant \bar{\xi}_{j-1} + \tilde{z}_{j-1} = \bar{\xi}_j.$$

But on the above set the increments $z_j, \ldots, z_{k-1}$ coincide with the increments $\xi_T((j+1)\theta) - \xi_T(j\theta), \ldots, \xi_T(k\theta) - \xi_T((k-1)\theta)$. Consequently

$$\xi_T(k\theta) = \xi_T(j\theta) + z_j + \cdots + z_{k-1} \leqslant \bar{\xi}_j + \tilde{z}_j + \cdots + \tilde{z}_{k-1} = \bar{\xi}_k.$$

Therefore

$$\mathbf{P}\left(\max_{0\leqslant k\leqslant n} \xi_T(k\theta) > \gamma T; H_\delta, \Omega(U), C, A\right)$$

$$\leqslant \mathbf{P}\left(\max_{0\leqslant k\leqslant n} \bar{\xi}_k > \gamma T\right)$$

$$\leqslant \mathbf{P}\left(\max_{0\leqslant k\leqslant n} \left(\delta_1 T + \sum_{j=0}^k z_j^0 + \varepsilon \sum_{j=0}^k \theta(j\theta) + cT \sum_{j=0}^k \kappa_j\right) > \gamma T\right)$$

$$\leqslant \mathbf{P}\left(\max_{0\leqslant k\leqslant n} \sum_{j=0}^k z_j^0 > (\gamma - \delta_1)T - \varepsilon \sum_{j=0}^k \theta(j\theta) - cT \sum_{j=0}^k \kappa_j\right).$$

Since $\mathbf{E} \sum_{j=0}^k \kappa_j = \mathbf{P}(\bar{\Omega}(U)) \to 0$ as $T \to \infty$, then for $\delta_1 = \gamma/2$ this expression does not exceed

$$\mathbf{P}\left(\max_{0\leqslant k\leqslant n} \sum_{j=0}^k z_j^0 > \frac{\gamma}{3} T\right) + o(1).$$

On the other hand, in a similar way, we find

$$\mathbf{P}(\Lambda) = \mathbf{P}\left(\min_{0\leqslant k\leqslant n} \tilde{\xi}_k \leqslant (\varepsilon_1 + \delta)T\right)$$
$$= \mathbf{P}\left(\min_{0\leqslant k\leqslant n}\left(\delta_1 T + \sum_{j=0}^{k} z_j^0 + \varepsilon \sum_{j=0}^{k} \theta(j\theta) + cT\sum_{j=0}^{k}\kappa_j\right) \leqslant (\varepsilon_1 + \delta)T\right)$$
$$\leqslant \mathbf{P}\left(\min_{0\leqslant k\leqslant n}\sum_{j=0}^{k} z_j^0 \leqslant \frac{-\gamma}{3}T\right) + o(1).$$

Therefore we will obtain the required estimate for (1.42) if, as $T \to \infty$, we have

$$\mathbf{P}\left(\max_{0\leqslant k\leqslant n}\left|\sum_{j=0}^{k} z_j^0\right| > \frac{\gamma}{3}T\right) \to 0. \tag{1.43}$$

But, as we have seen, for any k, l, and $\alpha > 0$

$$\mathbf{P}_{\mathfrak{M}(k\theta)}\left(\left|\sum_{j=k}^{k+l} z_j^0\right| > \alpha T\right) \xrightarrow{P} 0$$

as $T \to \infty$, uniformly in k and l. Therefore, putting $\tau = \min\{k\colon \sum_{j=0}^{k} z_j^0 > 2\alpha T\}$, we obtain, for sufficiently large T

$$o(1) = \mathbf{P}\left(\sum_{j=0}^{n-1} z_j^0 > \alpha T\right) \geqslant \sum_{k=0}^{n-1}\mathbf{P}(\tau = k)\mathbf{P}\left(\sum_{j=k+1}^{n-1} z_j^0 > -\alpha T \mid \tau = k\right)$$
$$\geqslant \tfrac{1}{2}\sum \mathbf{P}(\tau = k) = \tfrac{1}{2}\mathbf{P}\left(\max_{k}\sum_{j=0}^{k} z_j^0 > 2\alpha T\right).$$

Hence as $T \to \infty$

$$\mathbf{P}\left(\max_{k}\sum_{j=0}^{k} z_j^0 > \frac{\gamma}{3}T\right) \to 0.$$

A similar estimate for $\min_k \sum z_j^0$ completes the proof of (1.43).

We turn, finally, to the last term in (1.41). By condition (II) and lemma 2 (for $x = \gamma T/2$) it does not exceed

$$\mathbf{P}\left(\sup_{u\leqslant U}\xi_T(u) > 2\gamma T; \max_{k\leqslant n}\xi_T(k\theta) < \gamma T, \Omega(u)\right)$$
$$\leqslant \sum_{k=0}^{n}\mathbf{P}\left(\sup_{t\leqslant\theta} z_{k\theta,t} > \gamma T, \Omega((k+1)\theta)\right)$$
$$\leqslant 2\sum_{k=0}^{n}\mathbf{P}\left(z_{k\theta,\theta} > \frac{\gamma T}{2}; \Omega(k\theta)\right) \leqslant 2\sum_{k=0}^{n}\frac{2\varepsilon\theta(k\theta)}{\gamma T} \leqslant \frac{4\varepsilon}{\gamma}.$$

The convergence to 0 in probability of the left-hand side of (1.41), and with it the theorem, is proved. □

1.5. Conditions for convergence to a process of unbounded diffusion

In this and the following sections we consider conditions for convergence of a sequence of arbitrary* processes to a Markov diffusion process. Since the most convenient and effective method of assignment of diffusion processes is the 'infinitesimal' method, then naturally conditions of convergence must be connected with the increments of the process over relatively small intervals of time; more precisely with the asymptotic behaviour (just as a diffusion process) of the first two moments of these increments. These are, from the point of view of the moments, conditions relative to the σ-algebras generated by the pasts. Consequently the question is of the asymptotic behaviour of a sequence of random variables. But there are several methods of characterizing the asymptotic properties of random sequences which then generate different convergence conditions. In order to make these conditions as general as possible, we will impose bounds on the moments of the increments, not in the whole space of elementary events—this is too strict—but only on certain ω-sets of sufficiently high probability. We point out here two methods of construction of convergence conditions.†

1. Conditions 'in probability', in which the above ω-sets of high probability are given explicitly.
2. Conditions 'in the mean', in which the required asymptotic regularity is in 'averaged' form.

These conditions give us the most natural approach to the study of convergence to a diffusion process (regarding this, see remark 1 to section 1.8). Here we discussed conditions 'in probability'; conditions 'in the mean' are discussed in section 1.8.

Conditions 'in probability'. As before, let $(R(U), \mathfrak{U}_{R(U)}, \mathbf{P})$ be the sample space. For simplicity we will supplose that $C(0, U) \subset R(U)$ and that contraction of the argument by α transforms $R(U)$ into $R(\alpha U)$.

We consider a series array, that is, a sequence of processes $\{\xi_T(u); 0 \leqslant u \leqslant U\}$, depending on a parameter $T \to \infty$, and given on lengthening intervals of time $U = vT$, where $v > 0$ is fixed. We will study conditions for C-convergence to a diffusion process of the processes

$$y(u) = y_T(u) = \frac{\xi_T(uT)}{B(T)}$$

($B(T)$ is a normalizing factor), each given on $(R(U), \mathfrak{A}_{R(U)})$.

* On the convergence of Markov processes to a diffusion process, see Khinchin (1936), Skorokhod (1964), Gikhman and Skorokhod (1965; 1971, 1973, 1976), and others.

† Naturally other methods are possible. In Borovkov (1967a, 1970), for example, there are used sets of states, hits on which weaken the 'non-Markov' dependence of the moments of the increments on the past. A very general method is explained in Gikhman (1973).

We will assume that in $(R(U), \mathfrak{A}_{R(U)})$ there is a family of nested σ-algebras $\mathfrak{M}(u)$: $\mathfrak{M}(u_1) \subset \mathfrak{M}(u_2)$ for $u_1 \leqslant u_2$, so that any event connected with the trajectories of $\xi(s)$ on $[0, u]$ belongs to $\mathfrak{M}(u)$. In a number of cases it is convenient to consider $\mathfrak{M}(u)$ simply as the σ-algebras generated by the flow of $\xi(s)$ on $[0, u]$, that is, generated by the events

$$\{\xi(s) \in B_s\}, \quad s \leqslant u$$

where B_s is a Borel set on the line.

Thus, let there be given a sequence $\xi_T(u)$. Put

$$z_{u,t} = \xi_T(u+t) - \xi_T(u), \quad \zeta(u) = \frac{\xi_T(u)}{\sqrt{T}}.$$

In what follows the letter t will denote a sequence $t = t(T) \to \infty$, $t = o(T)$ as $T \to \infty$.

Finally, let $\Gamma_u^N \in \mathfrak{M}(u)$ denote the event

$$\Gamma_u^N = \{|\zeta(u)| < N\}.$$

Conditions (PI)–(PIII). *Let there exist a sequence $t = t(T)$ and a system of sets $\Omega_{u,t} \in \mathfrak{M}(u)$, $0 \leqslant u \leqslant U$ such that, as $T \to \infty$*

$$\mathbf{P}\left(\bigcap_{k=0}^{[U/t]} \Omega_{kt,t}\right) \to 1.$$

For each fixed $N > 0$, $\varepsilon > 0$, and $u \leqslant U - t$ on $\Omega_{u,t} \cap \Gamma_u^N$ we have

(PI) $\mathbf{E}_{\mathfrak{M}(u)} z_{u,t} = \dfrac{t}{\sqrt{T}}[a(\zeta(u)) + r_1(\omega)]$

(PII) $\mathbf{E}_{\mathfrak{M}(u)} z_{u,t}^2 = t[b(\zeta(u)) + r_2(\omega)]$,

where $|r_1(\omega)| < \varepsilon$, $|r_2(\omega)| < \varepsilon$ for sufficiently large T; a and b are functions whose properties will be described later.

(PIII) *For any fixed $\varepsilon > 0$, $\delta > 0$, and all sufficiently large T*

$$\mathbf{E}_{\mathfrak{M}(u)}(z_{u,t}^2; |z_{u,t}| > \delta\sqrt{T}) < \varepsilon t.$$

Regarding (PI)–(PIII) *we will also assume that being once satisfied they remain valid for sequences t_1 of the form $t_1 = ct$, $\frac{1}{2} \leqslant c \leqslant 2$.*

Sometimes instead of the phrase '$\psi(T) < \varepsilon$ for all sufficiently large T' we will use the shorter: $\psi(T) < \varepsilon$ as $T \to \infty$.

We see that in the above conditions the asymptotic regularity of moments (the nature of the growth depending on t and the nature of the dependence on ω) is required not everywhere but only on sets $\Omega_{u,t} \cap \Gamma_u^N$, where $\bar{\Omega}_{kt,t} = R(U) - \Omega_{kt,t}$ is any system of sets in which we rarely find ourselves.

If conditions (PI)–(PIII) are satisfied for any N and ε, then it can be assumed that there are sequences $N = N(T) \to \infty$ and $\varepsilon = \varepsilon(T) \to 0$ as $T \to \infty$, such that

(PI)–(PIII), as before, remain valid. In what follows, whenever necessary, we will understand N and ε to mean such sequences (not always one and the same) keeping these numbers fixed in the conditions of the theorems.

We note that for (PIII) it is sufficient that on $\Omega_{u,t} \cap \Gamma_u^N$.

$$\mathbf{E}_{\mathfrak{M}(u)} |z_{u,t}|^{2+2\gamma} < ct^{1+\gamma}$$

for some $\gamma > 0$, $c > 0$.

We now introduce condition (PIV), which relates to the functions a and b. As before, let C_k denote the space of k-times differentiable functions whose kth derivatives satisfy a uniform Hölder condition.

(PIV) *There is a transition function* $\mathbf{P}(x, u, B)$ *of some Markov process, such that the function*

$$V(x, u) = \int \phi(y)\mathbf{P}(x, u, \mathrm{d}y), \quad \phi \in C_2,$$

has derivatives $\dfrac{\partial V}{\partial u}, \dfrac{\partial V}{\partial x}, \dfrac{\partial^2 V}{\partial x^2}$, *which are continuous and bounded in* $0 \leqslant u \leqslant v$, $-\infty < x < \infty$, *and is a solution of the Cauchy problem*

$$V(x, 0) = \phi(x)$$

for the equation

$$\frac{\partial V}{\partial u} = a\frac{\partial V}{\partial x} + \frac{b}{2}\frac{\partial^2 V}{\partial x^2}.$$

In addition, for some $c > 0$

$$|a(x)| < c(1 + |x|), \quad b(x) < c.$$

Under these conditions there will be a diffusion process w, for which a and b will be respectively the coefficients of drift and diffusion.

The following conditions are sufficient for (PIV) (see section 2.3):

A. *The functions* a *and* $b > b_0 > 0$ *belong to class* C_0

or

B. *The derivatives* a', b', a'', b'' *are bounded and belong to* C_0.

Under these assumptions the following result holds. Let

$$y_T(u) = \frac{\xi_T(uT)}{\sqrt{T}} = \zeta(uT), \quad 0 \leqslant u \leqslant v$$

and let there be a limit distribution of $y_T(0)$, which we denote by p_0. In addition, let $\{w(u), 0 \leqslant u \leqslant v\}$ be a diffusion process on $C(0, v)$ with the σ-algebra of Borel sets, which is constructed relative to the transition function $\mathbf{P}(x, u, B)$ and with initial distribution p_0.

THEOREM 1. *Under conditions* (PI)–(PIV) *the finite-dimensional distributions of* $y_T(u)$ *converge to the finite-dimensional distributions of* $w(u)$.

We note that, without loss of generality, in (PI)–(PIII) we may assume that t divides U. Then for theorem 1 we need these conditions only for

$$u = 0, t, 2t, \ldots, (n-1)t; \quad a = U/t$$

and as $\mathfrak{M}(kt)$ we mean the σ-algebras generated by $\xi(0), \ldots, \xi(kt)$.

It may also be noted that (PI)–(PIV), which represent conditions for homogeneous convergence to a homogeneous diffusion, can be given a more general form by demanding, for example, instead of (PI), (PII), that there exist a partition $0 = t_0 < t_1 < \cdots < t_n = U$ of the segment U, such that $\max_{k<n}(t_{k+1} - t_k) = o(U)$ and for which on $\Omega_{t_k,t} \cap \Gamma_{t_k}^N$

$$\mathbf{E}_{\mathfrak{M}(t_k)} \frac{z_{t_k,t}}{T} = \frac{t}{T} a(t_k, \zeta(t_k)) + r_{1,k}$$

$$\mathbf{E}_{\mathfrak{M}(t_k)} \frac{z_{t_k,t}^2}{T} = \frac{t}{T} b(t_k, \zeta(t_k)) + r_{2,k}$$

where $t = t_{k+1} - t_k$, $\sum_{k<n} |r_{i,k}| < \varepsilon$ and $a(u, x)$ and $b(u, x)$ are functions having properties of the form (PIV). The introduction of such a family of conditions complicates somewhat the statements and proofs without introducing any essential difficulties. Therefore, in this and in later theorems, for simplicity we will restrict ourselves to the homogeneous case.

The result of theorem 1 can be strengthened up to C-convergence by introducing additional conditions. First of all, it is convenient to assume that $\xi(u)$ is continuous from the right, that is, for all u with probability 1

$$\lim_{\delta \to 0+} \xi(u + \delta) = \xi(u).$$

This is needed only for technical reasons. In accordance with this the jump of $\xi_T(u)$ at u is the quantity

$$\mathrm{Jp}_\xi(u) = \limsup_{\delta \to 0+} |\xi_T(u) - \xi_T(u - \delta)|.$$

We denote by H_u^δ the event

$$H_u^\delta = \{\mathrm{Jp}_\xi(u) < \delta\sqrt{T}\}$$

and introduce the condition

(PV) *The system of sets* $\Omega_{u,t}$ *appearing in conditions* (PI)–(PIII) *and the sets* H_u^δ *are such that**

$$\mathbf{P}\left(\bigcap_{u\leqslant U}\Omega_{u,t}\right)\to 1,\quad \mathbf{P}\left(\bigcap_{u\leqslant U}H_u^\delta\right)\to 1$$

as $T\to\infty$ *for any* $\delta>0$.

THEOREM 2. *Under conditions* (PI)–(PV) *the distribution of* y_T *C-converges to the distribution of* w.

Regarding the second condition of (PV), we must remark that it is necessary for theorem 2 and does not follow from (PI)–(PIII). This is easily proved with the help of the following example. Let $\xi_T(u)=w(u)+\pi(u)-\pi(u-\mathrm{e}^{-T})$, where $w(u)$ is a standard Wiener process and $\pi(u)$ is a Poisson process with parameter T^{-1} and jumps of size T, which does not depend on $w(u)$. If we put

$$\Omega_{u,t}=\left\{\sup_{s\leqslant u}|w(s)|<T^{\frac{3}{4}}\right\}$$

then (PI)–(PIII) are satisfied for $N<T^{\frac{3}{4}}$, but (PV) and theorem 2 are not satisfied.

We note also the obvious fact that theorems 1, 2 remain true if their conditions are satisfied in the probability space $(R(U),\mathfrak{A}_{R(U)},\mathbf{P}^*)$, where $\mathbf{P}^*(\mathrm{d}\omega)=\mathbf{P}(\mathrm{d}\omega/\Omega^*)$, for any set Ω^* with the property $\mathbf{P}(\Omega^*)\to 1$ as $T\to\infty$.

At the end of section 1.9 of this chapter there is a remark showing that theorems 1, 2 still hold if we allow (PI)–(PIII) to be violated in a neighbourhood $|\zeta(u)-D|<\delta$ of some point D, where the fixed $\delta>0$ can be chosen arbitrarily small. These violations consist of replacing (PI)–(PIII) in $|\zeta(u)-D|<\delta$ by conditions for sufficiently rapid exit of $\zeta(u)$ from the neighbourhood of D, which are close to the reflection conditions (PRIV) of section 1.9.

1.6. The proof of theorem 1 of section 1.5

1.6.1. Without loss of generality, we may suppose that t divides U without remainder and that the distribution of $y_T(0)=\xi_T(0)/T$ is entirely concentrated on some finite interval $[-r,r]$. Put

$$n=U/t;\quad \xi_k=\xi_T(kT),\quad k=\overline{0,n}$$

$$z_k=\xi_k-\xi_{k-1}=z_{(k-1)t,t},\quad k=\overline{1,n}.$$

* Here, and in what follows, for simplicity we assume that the products

$$\omega(r,s)=\bigcap_{r<u\leqslant s}\Omega_{u,t},\quad h(r,s)=\bigcap_{r<u\leqslant s}H_u^\delta \tag{1.44}$$

are events. The precise formulation of condition (PV) requires the existence of measurable sets from $\mathfrak{M}(s)$, contained in these products, and having probability converging to 1. The changes which we would have to introduce in the text if we used the more precise formulation, are trivial.

LEMMA 1. *As* $T \to \infty$, $x \to \infty$

$$\mathbf{P}\left(\max_{k \leqslant n} |\xi_k| > x\sqrt{T}\right) \to 0.$$

Proof. For the proof we introduce a number of auxiliary sequences. Since the behaviour of z_k in $|\xi_{k-1}| > N\sqrt{T}$ is unknown, we introduce the sequence

$$z_k^* = \begin{cases} (z_k)_{2N\sqrt{T}} & \text{if } \max_{i<k} |\xi_i| < N\sqrt{T} \\ 0 & \text{if } \max_{i<k} |\xi_i| \geqslant N\sqrt{T} \end{cases}$$

where $(\xi)_N$ denotes the truncation of ξ at the level N:

$$(\xi)_N = \begin{cases} \xi & \text{if } |\xi| \leqslant N \\ 0 & \text{if } |\xi| \geqslant N. \end{cases}$$

We denote

$$\xi_k^* = \xi_0 + \sum_{i=1}^{k} z_i^*, \quad \mathfrak{M}_k = \mathfrak{M}(kt).$$

Since by (PIII) on $\Omega_{kt,t} \cap \Gamma_{kt}^N$

$$|\mathbf{E}_{\mathfrak{M}_k}(z_{k+1}; |z_{k+1}| > N\sqrt{T})| = \frac{o(t)}{N\sqrt{T}} = o\left(\frac{t}{\sqrt{T}}\right),$$

then everywhere on $\Omega_{kt,t}$ ($I(A_k)$ is the indicator of the event $A_k = \{\max_{i<k} |\xi_i| < N\sqrt{T}\}$)

$$\begin{aligned} |\mathbf{E}_{\mathfrak{M}_k} z_{k+1}^*| &= I(A_{k+1})\mathbf{E}_{\mathfrak{M}_k}(z_{k+1})_{2N\sqrt{T}} \\ &\leqslant I(A_{k+1})\frac{ct}{\sqrt{T}}\left(1 + \frac{|\xi_k|}{\sqrt{T}}\right) \leqslant \frac{ct}{\sqrt{T}}\left(1 + \frac{|\xi_k^*|}{\sqrt{T}}\right) \end{aligned} \tag{1.45}$$

$$\mathbf{E}_{\mathfrak{M}_k}(z_{k+1})^2 < ct.$$

In addition, with probability 1 all $|\xi_k^*| < 3N\sqrt{T}$.

We divide the proof of lemma 1 into several stages.

We denote $\omega_k = \bigcap_{j=0}^{k} \Omega_{jt,t}$.

LEMMA 2. *For any* $k \leqslant n = \dfrac{U}{t} = v\dfrac{T}{t}$ *we have*

$$D_k \equiv \mathbf{E}_{\mathfrak{M}_0}((\xi_k^*)^2; \omega_{k-1}) \leqslant TD$$

$$D = (2cv + r^2)\,e^{3cv}.$$

Proof. By (1.45) we have

$$\begin{aligned} D_{k+1} &= \mathbf{E}_{\mathfrak{M}_0}\mathbf{E}_{\mathfrak{M}_k}[(\xi_k^* + z_{k+1}^*)^2; \omega_k] \\ &= D_k + 2\mathbf{E}_{\mathfrak{M}_0}\xi_k^* I(\omega_k)\mathbf{E}_{\mathfrak{M}_k} z_{k+1}^* + \mathbf{E}_{\mathfrak{M}_0} I(\omega_k)\mathbf{E}_{\mathfrak{M}_k}({z_{k+1}^*}^2) \\ &\leqslant D_k + \frac{2ct}{\sqrt{T}}\mathbf{E}_{\mathfrak{M}_0}|\xi_k^*|\left(1 + \frac{|\xi_k^*|}{\sqrt{T}}\right)I(\omega_k) + ct. \end{aligned}$$

Since $\mathbf{E}_{\mathfrak{M}_0}|\xi_k^*|I(\omega_k) \leqslant \sqrt{D_k}$, then

$$D_{k+1} \leqslant D_k\left(1 + \frac{2ct}{\sqrt{T}}\right) + 2ct\sqrt{\frac{D_k}{T}} + ct.$$

Estimating the second term here using

$$ab \leqslant \tfrac{1}{2}(a^2 + b^2) \tag{1.46}$$

we obtain

$$D_{k+1} \leqslant D_k(1 + \alpha) + \beta \tag{1.47}$$

where $\alpha = 3ct/T$, $\beta = 2ct$.

LEMMA 3. *If* $d_{k+1} \leqslant d_k(1 + \alpha) + \beta$, $k \geqslant 0$, $\alpha > 0$, $\beta > 0$, *then*

$$d_k \leqslant (d_0 + \beta k)(1 + \alpha)^k.$$

Proof. The equation $d_{k+1} = d_k(1 + \alpha) + \beta$ has the solution

$$\begin{aligned} d_k &= d_0(1 + \alpha)^k + \beta \sum_{j=0}^{k-1} (1 + \alpha)^j \\ &\leqslant d_0(1 + \alpha)^k + \beta k(1 + \alpha)^k. \end{aligned}$$

Lemma 3 is proved. □

Since for $\alpha = 3ct/T$

$$(1 + \alpha)^n = \left(1 + \frac{3ct}{T}\right)^{Tv/t} \leqslant e^{3cv}$$

then by (1.47) and lemma 3 for $k = n = U/t$ we obtain

$$\begin{aligned} D_n &\leqslant (\xi_0^2 + 2cTv)\, e^{3cv} \leqslant TD \\ D &= (r^2 + 2cv)\, e^{3cv}. \end{aligned}$$

Lemma 2 is proved. □

In what follows it will be convenient for us to 'merge' the random variables z_k^* so that they have positive expectation. For this we form a new sequence X_k,

$k = \overline{0, n}$, in the following way:

$$X_0 = \xi_0^* = \xi_0, \quad X_{k+1} = X_k + x_{k+1}$$
$$x_{k+1} = z_{k+1}^* + \Delta_{k+1}$$

where

$$\Delta_{k+1} = \frac{ct}{\sqrt{T}}\left(1 + \frac{|\xi_k^*|}{\sqrt{T}}\right) - \mathbf{E}_{\mathfrak{M}_k} z_{k+1}^* \tag{1.48}$$

so that on $\Omega_{kt,t}$

$$\frac{2ct}{\sqrt{T}}\left(1 + \frac{|\xi_k^*|}{\sqrt{T}}\right) > \Delta_{k+1} > 0. \tag{1.49}$$

This means that on $\omega_k = \bigcap_{j=0}^{k} \Omega_{jt,t}$

$$X_{k+1} \geqslant \xi_{k+1}^*, \quad \Delta_{k+1} < \frac{2ct}{\sqrt{T}}(1 + 3N) \tag{1.50}$$

and that everywhere

$$\mathbf{E}_{\mathfrak{M}_k} x_{k+1} = \frac{ct}{\sqrt{T}}\left(1 + \frac{|\xi_k^*|}{\sqrt{T}}\right) > 0 \tag{1.51}$$

$$\begin{aligned} I(\omega_k)\mathbf{E}_{\mathfrak{M}_k} x_{k+1}^2 &\leqslant I(\omega_k)\mathbf{E}_{\mathfrak{M}_k}[(z_{k+1}^*)^2 + \Delta_{k+1}^2 + 2\Delta_{k+1} z_{k+1}^*] \\ &\leqslant ct + \frac{\gamma c^2 t^2}{T}\left(1 + \frac{|\xi_k^*|}{\sqrt{T}}\right)^2. \end{aligned}$$

LEMMA 4. *There is a constant d, depending only on c, v, and r, such that for all* $k \leqslant n$

$$d_k \equiv \mathbf{E}_{\mathfrak{M}_0}(X_k^2; \omega_{k-1}) \leqslant Td.$$

Proof. Proceeding in exactly the same way as in the proof of lemma 2 we obtain by (1.45), (1.48)–(1.51)

$$\begin{aligned} d_{k+1} &\leqslant d_k + \frac{2ct}{\sqrt{T}} \mathbf{E}_{\mathfrak{M}_0} I(\omega_k)|X_k|\left(1 + \frac{|\xi_k^*|}{\sqrt{T}}\right) \\ &\quad + \mathbf{E}_{\mathfrak{M}_0} I(\omega_k)\left[ct + \frac{\gamma c^2 t^2}{T}\left(1 + \frac{|\xi_k^*|}{\sqrt{T}}\right)^2\right] \\ &\leqslant d_k + \frac{2ct}{\sqrt{T}}\left(\sqrt{d_k} + \sqrt{\frac{d_k D_k}{T}}\right) + ct\left(1 + \frac{16ct}{T}\left(1 + \frac{D_k}{T}\right)\right). \end{aligned}$$

Therefore by lemma 2 and for sufficiently large T

$$d_{k+1} \leqslant d_k + 2ct(1 + \sqrt{D})\sqrt{\frac{d_k}{T}} + 2ct$$

$$\leqslant d_k(1 + \alpha) + \beta$$

where $\alpha = c(1 + \sqrt{D})t/T$, $\beta = ct(3 + \sqrt{D})$. We again have the conditions of lemma 3 and we use it as in lemma 2. Lemma 4 is proved. □

LEMMA 5.

$$\mathbf{P}_{\mathfrak{M}_0}\left(\max_{k \leqslant n} X_k > x\sqrt{T}; \omega_{n-1}\right) \leqslant \frac{2d}{(x-h)^2} + 3\mathbf{P}_{\mathfrak{M}_0}(\bar{\omega}_{n-1})$$

$$h = \sqrt{\gamma c v}.$$

Proof. The inequality is proved roughly as in lemma 2 of section 1.4.

We denote

$$\tau_{x\sqrt{T}} = \min\{k: X_k \geqslant x\sqrt{T}\}$$

$$B_l = \{\tau_{x\sqrt{T}} = l\}, \quad A = \{X_n > (x-h)\sqrt{T}\}.$$

Then for $x > 0$, $h > 0$

$$\mathbf{P}_{\mathfrak{M}_0}(X_n > (x-h)\sqrt{T}) \geqslant \mathbf{P}_{\mathfrak{M}_0}\left(\bigcup_{l=0}^{n} AB_l\right) = \sum_{l=0}^{n} \mathbf{P}_{\mathfrak{M}_0}(AB_l)$$

$$= \sum_{l=0}^{n} \mathbf{P}_{\mathfrak{M}_0}(B_l - B_l\bar{A}).$$

But $B_l\bar{A} \subset B_l\{X_n - X_l \leqslant -h\sqrt{T}\}$, therefore the last sum is larger than

$$\sum_{l=0}^{n} \mathbf{E}_{\mathfrak{M}_0} I(B_l) \mathbf{E}_{\mathfrak{M}_l}(1 - I(X_n - X \leqslant -h\sqrt{T}))$$

$$\geqslant \sum_{l=0}^{n} \mathbf{E}_{\mathfrak{M}_0} I(B_l) \mathbf{E}_{\mathfrak{M}_l}\left(1 - I(X_n - X \leqslant -h\sqrt{T}; \omega_{n-1}) - \mathbf{P}_{\mathfrak{M}_0}\left(\bigcup_{0}^{n} B_l; \omega_{n-1}\right)\right). \tag{1.52}$$

Here we must estimate from above

$$\mathbf{P}_{\mathfrak{M}_l}(X_n - X_l \leqslant -h\sqrt{T}; \omega_{n-1}). \tag{1.53}$$

Doing this with the help of lemmas 2 and 4 does not succeed, since on B_l the initial value $X_l > x\sqrt{T}$. Therefore we proceed as follows. We only increase this probability if the terms $x_{l+1}, \ldots, x_n$ decrease, replacing them by (see (1.51))

$$x_j^* = x_j - \mathbf{E}_{\mathfrak{M}_{j-1}} X_j.$$

But the sum of these terms forms a martingale and has the necessary estimate for the variance. Put

$$d_{l,s} = \mathbf{E}_{\mathfrak{M}_l}[(x_{l+1}^* + \cdots + x_{l+s+1}^*)^2, \omega_{l+s+1}].$$

Then

$$d_{l,s+1} \leqslant d_{l,s} + 2\mathbf{E}_{\mathfrak{M}_l}(x_{l+1}^* + \cdots + x_{l+s}^*)I(\omega_{l+s})\mathbf{E}_{\mathfrak{M}_{l+s}} x_{l+s+1}^*$$
$$+ \mathbf{E}_{\mathfrak{M}_l} I(\omega_{l+s})\mathbf{E}_{\mathfrak{M}_{l+s}}(x_{l+s+1}^*)^2.$$

Here the second term is equal to zero. In the last term the difference between x_{l+s+1}^* and z_{l+s+1}^* is $\mathfrak{M}_{l+s}$-measurable and, by (1.45) and (1.51), on ω_{l+s} is no larger than

$$\frac{ct}{\sqrt{T}}(1 + 3N).$$

Consequently, if we choose N so that $1 + 3N \leqslant \sqrt{T/ct}$, then the square of the difference is less than ct. Thus, by (1.45)

$$d_{l,s+1} \leqslant d_{l,s} + 4ct, \quad d_{l,s} \leqslant 4sct.$$

Therefore for (1.53) we find

$$\mathbf{P}_{\mathfrak{M}_l}(X_n - X_l \leqslant -h\sqrt{T}; \omega_{n-1}) \leqslant \frac{d_{l,n-l}}{h^2 T} \leqslant \frac{4cU}{h^2 T} = \frac{4cv}{h^2} = \frac{1}{2}$$

where $h = \sqrt{8cv}$. Consequently, for this choice of h, we obtain from (1.52) that

$$\mathbf{P}_{\mathfrak{M}_0}(X_n > (x - h)\sqrt{T})$$
$$\geqslant \tfrac{1}{2}\mathbf{P}_{\mathfrak{M}_0}\left(\max_{k \leqslant n} X_k \geqslant x\sqrt{T}\right) - \mathbf{P}_{\mathfrak{M}_0}\left(\max_{k \leqslant n} X_k \geqslant x\sqrt{T}; \bar{\omega}_{n-1}\right).$$

And further, by lemma 4

$$\mathbf{P}_{\mathfrak{M}_0}\left(\max_{k \leqslant n} X_k \geqslant x\sqrt{T}; \omega_{n-1}\right) \leqslant 2\mathbf{P}_{\mathfrak{M}_0}(X_n > (x - h)\sqrt{T}; \omega_{n-1}) + 3\mathbf{P}_{\mathfrak{M}_0}(\bar{\omega}_{n-1})$$
$$\leqslant \frac{2d}{(x - h)^2} + 3\mathbf{P}_{\mathfrak{M}_0}(\bar{\omega}_{n-1}). \quad \square$$

We turn now to the proof of lemma 1. By (1.50)

$$\mathbf{P}_{\mathfrak{M}_0}\left(\max_{k \leqslant n} \xi_k^* > x\sqrt{T}; \omega_{n-1}\right) \leqslant \mathbf{P}_{\mathfrak{M}_0}\left(\max_{k \leqslant n} X_k > x\sqrt{T}; \omega_{n-1}\right).$$

Since, in addition, similar inequalities hold for $\min_{k \leqslant n} \xi_k^*$, then we obtain

$$\mathbf{P}_{\mathfrak{M}_0}\left(\max_{k \leqslant n} |\xi_k^*| > x\sqrt{T}; \omega_{n-1}\right) \leqslant \frac{a(c, r, v)}{x^2} + 6\mathbf{P}_{\mathfrak{M}_0}(\bar{\omega}_{n-1}),$$

where $a(c, r, v)$ is a known function of its arguments. We note further that the trajectories $\{\xi_k\}$ and $\{\xi_k^*\}$ coincide in $|z| < N\sqrt{T}$. Therefore for $x \leqslant N$ we obtain the same estimate for $\mathbf{P}_{\mathfrak{M}_0}(\max_{k \leqslant n} |\xi_k| > x\sqrt{T};\ \omega_{n-1})$.

Since $\mathbf{P}_{\mathfrak{M}_0}(\bar{\omega}_{n-1}) \to 0$ as $n \to \infty$ and since N is arbitrary, then lemma 1 is proved. □

1.6.2. We may now proceed to the proof of the convergence of the finite-dimensional distributions of $y(t) = y_T(t)$. It is enough for us to show that for any $t_1, \ldots, t_k$ from $[0, v]$ and bounded functions $\phi_1, \ldots, \phi_k$ from $C(-\infty, \infty)$

$$\mathbf{E}\phi_1(y(t_1)) \cdots \phi_k(y(t_k)) \to \mathbf{E}\phi_1(w(t_1)) \cdots \phi_k(w(t_k)) \tag{1.54}$$

as $T \to \infty$. This is because (1.54) implies the convergence of the characteristic functions of the vectors $\{y(t_j)\}_{j=1}^k$. Further, it is easy to show that for (1.54), in its turn, it is sufficient that for any u, s, and bounded $\phi \in C$

$$\mathbf{E}_{\mathfrak{M}(uT)}\phi(y(u+s)) - \mathbf{E}\phi(w^{(y)}(s))|_{y=y(u)} \underset{P}{\to} 0 \tag{1.55}$$

where

$$\mathbf{E}\phi(w^{(y)}(s)) = \int \phi(x)\mathbf{P}(y, s, \mathrm{d}z)$$

and $w^{(y)}(s)$ is a diffusion process with initial value $w^{(y)}(0) = y$.

In fact,

$$\mathbf{E}\prod_{j=1}^{k} \phi_j(y(t_j)) = \mathbf{E}\phi_1(y(t_1))\mathbf{E}_{\mathfrak{M}(t_1)}\phi_2(y(t_2)) \cdots \mathbf{E}_{\mathfrak{M}(t_{k-1})}\phi_k(y(t_k)).$$

But

$$\phi_{k-1}(y(t_{k-1}))\mathbf{E}_{\mathfrak{M}(t_{k-1})}\phi(y(t_k)) - \phi_{k-1}(y)\mathbf{E}\phi_k(w^{(y)}(t_k - t_{k-1})) \underset{P}{\to} 0$$

for $y = y(t_{k-1})$ and the function

$$\phi_{k-1}^*(x) = \phi_{k-1}\mathbf{E}\phi_k(w^{(x)}(t_k - t_{k-1}))$$

by virtue of the properties of a diffusion process, again belongs to $C(-\infty, \infty)$.

Therefore, in a similar way, we may find an expression asymptotically equivalent to

$$\phi_{k-2}(y(t_{k-2}))\mathbf{E}_{\mathfrak{M}(t_{k-2})}\phi_{k-1}^*(y(t_{k-1}))$$

etc. As a result we obtain (1.54).

Thus, it is sufficient for us to prove (1.55). For this it is clear that we can limit the discussion to functions ϕ from any dense subclass of C. For example, C_2.

To avoid the introduction of new notation it is convenient to regard uT as the initial epoch ($\xi_T^*(s) = \xi_T(uT + s)$ has all the properties of $\xi_T(s)$), to reduce (1.55) to the proof of

$$\mathbf{E}_{\mathfrak{M}(0)}\phi(y(v)) - \mathbf{E}\phi(w^{(y)}(v))|_{y=y(0)} \to 0, \quad \phi \in C_2 \tag{1.56}$$

for the initial process $y(u) = y_T(u)$.

1.6.3. We denote

$$s_k = z_k/\sqrt{T}, \quad \zeta_k = y(0) + s_1 + \cdots + s_k, \quad \Delta = t/T.$$

Then for the values of t, chosen in lemma 1,

$$n = U/t = v/\Delta; \quad y(v) = y(0) + s_1 + \cdots + s_n = \zeta_n.$$

Here, by (PI)–(PIII), on $\Omega_{kt,t} \cap \Gamma_{kt}^N \equiv \Omega_k^N$

$$\begin{aligned} \mathbf{E}_{\mathfrak{M}_k} s_{k+1} &= \Delta(a(\zeta_k) + r_1(\omega)) \\ \mathbf{E}_{\mathfrak{M}_k} s_{k+2}^2 &= \Delta(b(\zeta_k) + r_2(\omega)) \\ \mathbf{E}_{\mathfrak{M}_k}(s_{k+1}^2, |s_{k+1}| > \delta) &= \Delta r_3(\omega) \end{aligned} \tag{1.57}$$

where $|r_i(\omega)| < \varepsilon$.

We denote

$$V(x, u) = \mathbf{E}\phi(w^{(x)}(u))$$

and show for any $M > 0$ that on Ω_k^M

$$|\mathbf{E}_{\mathfrak{M}_k} V(\zeta_{k+1}, u) - V(\zeta_k, u + \Delta)| \leqslant \varepsilon^* \Delta \tag{1.58}$$

where $\varepsilon^* \to 0$ as $T \to \infty$, and does not depend on k, u, ω. In fact, by (PIV) the derivatives $\partial V/\partial u, \partial V/\partial x, \partial^2 V/\partial x^2$ are uniformly continuous and bounded in the rectangle $0 \leqslant u \leqslant v$, $|x| < 2M$. Therefore for $u < v$

$$V(\zeta_k, u + \Delta) = V(\zeta_k, u) + \frac{\partial V(\zeta_k, u)}{\partial u} + o(\Delta)$$

where the estimate is uniform in ω and u on Γ_{kt}^M. On the other hand

$$\begin{aligned} V(\zeta_{k+1}, u) &= V(\zeta_k + s_{k+1}, u) \\ &= V(\zeta_k, u) + s_{k+1}\frac{\partial V(\zeta_k, u)}{\partial x} + \frac{s_{k+1}^2}{2}\left(\frac{\partial^2 V(\zeta_k, u)}{\partial x^2} + \rho\right). \end{aligned}$$

Here

$$\rho = \rho(\zeta_k, s_{k+1}, u) = \frac{\partial^2 V(\zeta_k + \beta s_{k+1}, u)}{\partial x^2} - \frac{\partial^2 V(\zeta_k, u)}{\partial x^2}, \quad 0 \leqslant \beta \leqslant 1.$$

Hence, by (1.57), we find that on Ω_k^M

$$\mathbf{E}_{\mathfrak{M}_k} V(\zeta_k, u) = V(\zeta_k, u) + \frac{\partial V(\zeta_k, u)}{\partial x}(a(\zeta_k) + r_1)$$

$$+ \Delta \frac{\partial^2 V(\zeta_k, u)}{2\,\partial x^2}(b(\zeta_k) + r_2) + \mathbf{E}_{\mathfrak{M}_k} \frac{s_{k+1}^2}{2}\rho$$

where $|r_i(\omega)| < \varepsilon$,

$$|\mathbf{E}_{\mathfrak{M}_k} s_{k+1}^2 \rho| \leqslant \mathbf{E}_{\mathfrak{M}_k} s_{k+1}^2 \sup_{|x| < \delta_T} |\rho(\zeta_k, x, t)|$$

$$+ \mathbf{E}_{\mathfrak{M}_k}(s_{k+1}^2, |\rho| I(|s_{k+1}| > \delta_T)). \tag{1.59}$$

The boundedness of $\partial^2 V/\partial x^2$ implies that $|\rho| < c$, and its uniform continuity on $0 \leqslant u \leqslant v$, $|x| < 2M$ implies the uniform convergence.

$$\sup_{|x| < \delta_T} |\rho(\xi_k, x, t)| < \varepsilon_1 \to 0$$

for $\delta_T = o(1)$. We note now that since (PIII) is valid for any $\delta > 0$, then it will be valid for a sequence $\delta_T = o(1)$. If we choose such a δ_T in (1.59) then on Ω_k^M

$$\mathbf{E}_{\mathfrak{M}_k}(s_{k+1}^2 I(|s_{k+1}| > \delta_T)) \leqslant \varepsilon_2 \Delta$$

$$\mathbf{E}_{\mathfrak{M}_k} |s_{k+1}^2 \rho| \leqslant c(\varepsilon_1 \Delta + \varepsilon_2 \Delta) = o(\Delta).$$

Since $V(x, u)$ is a solution of (1.44), then this gives the required estimate for the difference (1.58).

1.6.4. We now prove (1.56). We have

$$\mathbf{E}_{\mathfrak{M}_0}(\phi(y(v)) = \mathbf{E}_{\mathfrak{M}_0}\phi(\zeta_n)$$

$$= \mathbf{E}_{\mathfrak{M}_0} V(\zeta_n, 0) I\left(\bigcap_{k=0}^{n-1} \Omega_k^M\right) + c(\omega)\mathbf{E}_{\mathfrak{M}_0} I\left(\bigcap_{k=0}^{n-1} \bar{\Omega}_k^M\right)$$

where $|c(\omega)| < c$ by the boundedness of V. Further, by (1.58),

$$\mathbf{E}_{\mathfrak{M}_0} V(\zeta_n, 0) I\left(\bigcap_{k=0}^{n-1} \Omega_k^M\right)$$

$$= I(\Omega_0^M)\mathbf{E}_{\mathfrak{M}_0} I(\Omega_1^M) \cdots I(\Omega_{n-1}^M)\mathbf{E}_{\mathfrak{M}_{n-1}} V(\zeta_n, 0)$$

$$= I(\Omega_0^M)\mathbf{E}_{\mathfrak{M}_0} I(\Omega_1^M) \cdots \mathbf{E}_{\mathfrak{M}_{n-2}} I(\Omega_{n-1}^M) V(\zeta_{n-1}, \Delta) + \varepsilon_{n-1}(\omega)\Delta$$

$$= (\text{since } I(\Omega_k^M)V(\zeta_k, u) = V(\zeta_k, u) - I(\bar{\Omega}_k^M)V(\zeta_k, u), \text{ then we can expand})$$

$$= I(\Omega_0^M)\mathbf{E}_{\mathfrak{M}_0} I(\Omega_1^M) \cdots \mathbf{E}_{\mathfrak{M}_{n-3}} I(\Omega_{n-2}^M) V(\zeta_{n-2}, 2\Delta)$$

$$- c_{n-1}(\omega)\mathbf{P}_{\mathfrak{M}_0}\left(\bigcap_{k=0}^{n-2} \Omega_k^M \Omega_{n-1}^M\right) + \varepsilon_{n-2}(\omega)\Delta + \varepsilon_{n-1}(\omega)\Delta$$

$$= \cdots$$

$$= I(\Omega_0^M)V(\zeta_0, n\Delta) - \sum_{k=1}^{n-1} c_{n-k}(\omega)\mathbf{P}_{\mathfrak{M}_0}\left(\bigcap_{i=0}^{n-k-1} \Omega_i^M \bar{\Omega}_{n-k}^M\right) + \Delta \sum_{k=1}^{n} \varepsilon_k(\omega)$$

where $|c_k(\omega)| < c$, $|\varepsilon_k(\omega)| < \varepsilon^*$. This means that

$$|\mathbf{M}_{\mathfrak{M}_0}\phi(y_T(v)) - I(\Omega_0^M)V(y_T(0), v)| < 2c\mathbf{P}_{\mathfrak{M}_0}\left(\bigcup_{k=0}^{n-1} \bar{\Omega}_0^M\right) + v\varepsilon^*$$

$$< 2c\mathbf{P}_{\mathfrak{M}_0}\left(\bigcup_{k=0}^{n-1} \bar{\Omega}_{kt,t}\right) + 2c\mathbf{P}_{\mathfrak{M}_0}\left(\max_{k\leqslant n-1} |y_T(k\Delta)| > M\sqrt{T}\right) + v\varepsilon^*.$$

By lemma 1, as $T \to \infty$, the right-hand side of this inequality can be made arbitrarily small through the choice of the number M. Since $I(\Omega_c^M) \xrightarrow[P]{} 1$ as $T \to \infty$, $M \geqslant r$ and $V(y(0), v) = \mathbf{E}\phi(w^{(y)}(v))|_{y=y(0)}$, then (1.56), and with it the convergence of the finite-dimensional distributions of $y(u)$, is proved. □

1.7. The proof of theorem 2 of section 1.5.

By theorem 1 of section 1.5 and theorem 1 of section 1.2, it remains for us to show that for any $\alpha > 0$

$$\lim_{\delta\to 0} \overline{\lim_{T\to\infty}} \mathbf{P}_T(\omega_\delta(y_T) > \alpha) = 0 \tag{1.60}$$

where

$$\omega_\delta(y) = \sup_{|u'-u''|\leqslant\delta} |y(u') - y(u'')|.$$

1.7.1. We begin by proving a lemma.

LEMMA 1. *Let $\tilde{y}(u)$ be a polygon constructed through the points $(k\Delta, y(k\Delta))$, $k = \overline{0, n}$. Then*

$$\rho_C(y, \tilde{y}) \xrightarrow[P]{} 0.$$

Proof. We recall the notation of section 1.5

$$\omega(r, s) = \bigcap_{r<u\leqslant s} \Omega_{u,t}; \quad h(r, s) = \bigcap_{r<u\leqslant s} H_u^\delta$$

(see (1.44)) and for brevity put $\Gamma_k = \Gamma_{kt}^M = \{|\xi(k, t)| < M\sqrt{T}\}$.

For arbitrary $\alpha > 0$

$$\mathbf{P}(\rho_C(y, \tilde{y}) > 4\alpha) \leqslant \mathbf{P}(\overline{\omega(0, U)}) + \mathbf{P}\left(\bigcup_{k=0}^{n} \bar{\Gamma}_k\right) + \mathbf{P}(\bar{h}(0, U))$$
$$+ \sum_{k=0}^{n-1} \mathbf{P}(|z_{kt,t}| > 2\alpha\sqrt{T}; \Omega_k^M)$$
$$+ \sum_{k=0}^{n-1} \mathbf{P}\left(\sup_{u \leqslant t} |z_{kt,u}| > 2\alpha\sqrt{T}; \omega(kt, t)h(kt, t)\Gamma_k\right). \tag{1.61}$$

Since we can estimate the first sum by the condition

$$\mathbf{P}_{\mathfrak{M}_k}(|z_{kt,t}| > 2\alpha\sqrt{T}; \Omega_k^M) \leqslant \frac{\varepsilon t}{4\alpha^2 T}$$

then the only difficulty that can arise is in the estimation of the last sum.

Denote $\omega(s) = \omega(kt, s)h(kt, s)\Gamma_k$, $\mathfrak{N}(s) = \mathfrak{M}(kt + s)$. Then, by (PIII) on $\omega(s)$, for $t \leqslant u \leqslant 2t$ we have

$$\mathbf{P}_{\mathfrak{N}(s)}(z_{kt+s,u} \leqslant -\alpha\sqrt{T}) \leqslant \frac{\varepsilon u}{\alpha^2 T} < \frac{1}{2} \tag{1.62}$$

for sufficiently large T. Since $\xi(u)$ (and, consequently, $z_{kt+s,u}$) is continuous from the right relative to u, then by a minor modification of lemma 2 of section 1.4 we can write

$$\mathbf{P}\left(\sup_{u \leqslant t} z_{kt,u} > 2\alpha\sqrt{T};\ \omega(t)\right) \leqslant 2\mathbf{P}(z_{kt,2t} > \alpha\sqrt{T};\ \omega(0)). \tag{1.63}$$

(The modification consists in replacing the first line of the proof of lemma 2 by the inequality

$$\mathbf{P}(z_{kt,2t} > x; \omega(0)) \geqslant \mathbf{E}\mathbf{E}_{\mathfrak{M}(\tau)} I(\tau \leqslant t) I(\omega(\tau)) I(z_{kt,2t} > x).$$

The remainder of the proof is the same and uses (1.62) only for $t \leqslant u \leqslant 2t$.)

Inequality (1.63) allows us to estimate the last sum in (1.61) as before. The lemma is proved. □

1.7.2. We now note that

$$\omega_\delta(y) \leqslant \omega_\delta(\tilde{y}) + 2\rho_c(y, \tilde{y})$$

and, consequently, it is enough to verify (1.60) for $\tilde{y}$. Since

$$\overline{\lim_{T\to\infty}}\ \mathbf{P}(\omega_{n-1}) = 0, \quad \overline{\lim_{T\to\infty}}\ \mathbf{P}\left(\bigcup_{k=1}^{n} \{|z_k| > \sqrt{\delta T}\}\right) = 0$$

and for suitable M the probability of $\gamma_{n-1} = \bigcap_{k=0}^{n-1} \Gamma_{kt}^M$ can be made arbitrarily close to 1, then it is sufficient for us to show that for any $\alpha > 0$

$$\overline{\lim_{T\to\infty}}\, \mathbf{P}(\omega_\delta(\tilde{y}) > 3\alpha; \omega_{n-1}\gamma_{n-1}b_n) \to 0$$

as $\delta \to 0$, where $b_k = \bigcap_{j=1}^{k} \{|z_j| < \sqrt{\delta T}\}$. By lemma 1, here we need only consider δ which are multiples of $\Delta = t/T$. We have $(L = \delta/\Delta)$

$$\mathbf{P}(\omega_\delta(\tilde{y}) > 3\alpha; \omega_{n-1}\gamma_{n-1}b_n) \leqslant \sum_{k=0}^{[v/\delta]} \mathbf{P}\left(\max_{l\leqslant L} |\xi_{kL+l} - \xi_{kL}| > \alpha\sqrt{T}; \omega_{n-1}\gamma_n b_n\right).$$

An estimate of this sum by the methods used in lemma 1 and lemma 2 of section 1.4 (and going back to Kolmogorov's inequality) easily leads to an estimate of

$$\sum_{k\leqslant v/\delta} \mathbf{P}\left(|\xi_{kL+L} - \xi_{kL}| > \frac{\alpha}{2}\sqrt{T}; \omega_{n-1}\gamma_{n-1}b_n\right). \tag{1.64}$$

1.7.3. Denoting

$$\omega_{k-1}\gamma_{k-1}b_k = e_k, \quad x_j = \frac{z_{kL+j}}{\sqrt{\beta t}}$$

where β is chosen later, we may write the terms of (1.64) in the form

$$\mathbf{P}\left(|x_1 + \cdots + x_L| > \frac{\alpha}{2}\sqrt{\frac{T}{\beta t}}; e_n\right) \tag{1.65}$$

where

$$|x_l| < \sqrt{\frac{\delta T}{\beta t}}$$

on the set $b_l \subset e_n$.

We make use of the following proposition (a generalization of the Bernshtein–Kolmogorov inequality).

LEMMA 2. *Let $\mu_1, \ldots, \mu_L$ be random variables and $\mathfrak{N}_1, \ldots, \mathfrak{N}_L$ a sequence of nested σ-algebras, $\mathfrak{N}_k \subset \mathfrak{N}_{k+1}$ such that $\mu_1, \ldots, \mu_k$ are measurable relative to $\mathfrak{N}_k$ and on certain ω-sets $e_l \in \mathfrak{N}_l$, $e_{l-1} \subset e_l$ we have*

$$|\mu_l| \leqslant Q; \quad l = \overline{1, L}.$$

In addition, let

$$\mathbf{E}_{\mathfrak{N}_{l-1}}(\mu_l; e_l) = 0, \quad \mathbf{E}_{\mathfrak{N}_{l-1}}(\mu_l^2; e_l) \leqslant 1.$$

Then, for $x > L/Q$

$$\mathbf{P}(|\mu_1 + \cdots + \mu_L| > x; e_n) \leqslant 2\,\mathrm{e}^{-(x/4Q)}. \tag{1.66}$$

The proof is based on a complete repetition of the arguments of the proof in the case $\mathbf{P}(e_l) = 1$ and μ_l independent (see, for example, Khinchin, 1936). Namely, we must use the inequalities

$$\mathbf{E}(\mathrm{e}^{\lambda\mu_l}; e_l) < \exp\left\{\frac{\lambda^2}{2}\left(1 + \frac{\lambda Q}{2}\right)\right\},$$

$$\mathbf{E}(\mathrm{e}^{\lambda(\mu_1 + \cdots \mu_L)}; e_n) < \exp\left\{\frac{\lambda^2 L}{2}\left(1 + \frac{\lambda Q}{2}\right)\right\}$$

$$\mathbf{P}(\mu_1 + \cdots + \mu_L > x; e_n) \leqslant \mathrm{e}^{-\lambda x}\mathbf{E}(\mathrm{e}^{\lambda(\mu_1 + \cdots + \mu_L)}; e_n)$$

in which we put $\lambda = 1/Q$ and replace L by the larger value xQ.

To make use of lemma 2 we put $\mathfrak{N}_l = \mathfrak{M}_{kL+l}$ and introduce the random variables

$$x_l^* = x_l - \mathbf{E}_{\mathfrak{N}_{l-1}}(x_l; e_l), \quad l = \overline{1, n}.$$

Since, by (PI), (PIII), and the property of the function a

$$|\mathbf{E}_{\mathfrak{N}_{l-1}}(x_l; e_l)| < \frac{ct}{\sqrt{T\beta t}}(1 + M),$$

then, on e_l, for sufficiently large T,

$$|x_l^*| \leqslant \sqrt{\frac{\delta T}{\beta t}} + \frac{ct(1 + M)}{\sqrt{\beta t T}} \leqslant \sqrt{\frac{\delta T}{2\beta t}} \equiv Q.$$

In addition, by (PII) and the property of the function b

$$\mathbf{E}_{\mathfrak{N}_{l-1}}[(x_l^*)^2; e_l] \leqslant \frac{c}{\beta}.$$

Choosing $\beta = c$ we can write (1.65) in the form

$$\mathbf{P}\left(|x_1 + \cdots + x_L| > \frac{\alpha}{2}\sqrt{\frac{T}{ct}}; e_n\right)$$

$$\leqslant \mathbf{P}\left(|x_1^* + \cdots + x_L^*| > \frac{\alpha}{2}\sqrt{\frac{T}{ct}} - \frac{cLt(1 + M)}{\sqrt{Tct}}; e_n\right) \tag{1.67}$$

where x_k^* satisfies the conditions of lemma 2. We now choose δ so that

$$\frac{cLt(1 + M)}{\sqrt{Tct}} = \sqrt{\frac{cT}{t}}\,\delta(1 + M) < \frac{\alpha}{4}\sqrt{\frac{T}{ct}}$$

and

$$x = \frac{\alpha}{4}\sqrt{\frac{T}{ct}} > \frac{L}{Q} = \frac{\delta T}{t}\sqrt{\frac{2ct}{\delta T}} = \sqrt{\frac{2c\delta T}{t}}.$$

Then, by lemma 2, the right-hand side of (1.67) does not exceed

$$\mathbf{P}\left(|x_1^* + \cdots + x_L^*| > \frac{\alpha}{4}\sqrt{\frac{T}{ct}}; e_n\right) \leqslant 2\exp\left\{-\frac{\alpha}{4}\sqrt{\frac{T}{ct}}\frac{1}{4}\sqrt{\frac{2ct}{\delta T}}\right\}$$
$$= 2\exp\left\{-\frac{\alpha}{8\sqrt{2\delta}}\right\}.$$

Returning to the estimate of $\omega_\delta(\tilde{y})$, we see that for sufficiently small δ (1.64) is estimated by the values

$$\frac{2}{\delta}e^{-(\alpha/8\sqrt{2\delta})} \to 0$$

as $\delta \to 0$. This means that

$$\overline{\lim_{T\to\infty}}\ \mathbf{P}(\omega_\delta(\tilde{y}) > 3\alpha) \to 0 \tag{1.68}$$

as $\delta \to 0$ and for any $\alpha > 0$. Theorem 2 is completely proved. □

Remark. From the proof it is clear that (PV) is only used in lemma 2. This means that under the conditions of theorem 1 (1.68) will hold and, consequently, the C-convergence of the distributions of $\tilde{y}_T$ to the distribution of w.

1.8. Conditions for 'mean' convergence to an unbounded diffusion

Let there exist a sequence $t = o(T)$ such that, as $T \to \infty$ and for all* u, $\delta > 0$, $\varepsilon > 0$ we have:

(MI) $\mathbf{E}|\mathbf{E}_{\mathfrak{M}(u)} z_{u,t} - \dfrac{t}{\sqrt{T}} a(\zeta(u))| < \varepsilon \dfrac{t}{\sqrt{T}}$.

(MII) $\mathbf{E}|\mathbf{E}_{\mathfrak{M}(u)} z_{u,t}^2 - tb(\zeta(u))| < \varepsilon t$.

(MIII) $\mathbf{E}(z_{u,t}^2;\ |z_{u,t}| > \delta\sqrt{T}) < \varepsilon t$.

Condition (MIV) *coincides with* (PIV) *but requires in addition that the derivatives* $\partial V/\partial u$, $\partial V/\partial x$, $\partial^2 V/\partial x^2$ *be uniformly continuous in* $0 \leqslant u \leqslant v$, $-\infty < x < \infty$.

For (MIV), as for (PIV), the conditions (A) and (B) mentioned in section 1.5 will be sufficient.

Remark 1. It must be noted that the convergence condition formulated in section 1.5 and the condition 'in the mean' mentioned above represent the two most natural approaches to the description of conditions for convergence to diffusion processes. It is

* As in the statement of theorem 1 of section 1.5, it is sufficient here to require only that (MI)–(MIII) hold for $u = kt$, $k = 0, \ldots, u/t = n$, and that $\mathfrak{M}(kt)$ denotes the σ-algebra generated by $\xi(0)$, $\xi(t)$, $\ldots$, $\xi(kt)$. These conditions are easily reformulated for the nonhomogeneous case (cf. the remarks concerning this in section 1.5).

clear that a necessary element of any conditions for convergence to a diffusion must be the closeness in some sense or other of the moments $\mathbf{E}_{\mathfrak{M}(u)}z_{u,t}$, $\mathbf{E}_{\mathfrak{M}(u)}z_{u,t}^2$ (or the 'truncated' moments) to the values $(t/\sqrt{T})a(\zeta(u))$, $tb(\zeta(u))$ respectively. But it is convenient to define smallness of the random variables

$$\gamma_1 = \mathbf{E}_{\mathfrak{M}(u)}z_{u,t} - \frac{t}{\sqrt{T}}a(\zeta(u)), \quad \gamma_2 = \mathbf{E}_{\mathfrak{M}(u)}z_{u,t}^2 - tb(\zeta(u))$$

either as smallness 'in probability' (which is done in section 1.5) or as smallness 'in the mean', that is, smallness of $\mathbf{E}|\gamma_1|^\alpha$, $\mathbf{E}|\gamma_2|^\alpha$ for some $\alpha > 0$. The latter method with $\alpha = 1$ forms forms the topic of this section. We consider $\alpha \neq 1$ as less convenient and $\alpha > 1$ as even unreasonable since it will entail very rigid requirements on the objects under discussion.

THEOREM 1. *Under conditions* (MI)–(MIV) *the finite-dimensional distributions of* y_T *converge to the distributions of* w.

We note that if (MI)–(MIII) remain valid for $\varepsilon = o(t/T)$, then they immediately imply (PI)–(PIII). For $u = 0, t, \ldots, (n-1)t$ (see the remark following the statement of theorem 1 of section 1.5) it is sufficient to take as $\Omega_{kt,t}$ the intersection of the sets

$$\Omega_k^{(1)} = \left\{\omega: \left|\mathbf{E}_{\mathfrak{M}(kt)}z_{kt,t} - \frac{t}{\sqrt{T}}a(\zeta(kt))\right| < \frac{\varepsilon_1 t}{T}\right\}$$

$$\Omega_k^{(2)} = \{\omega: |\mathbf{E}_{\mathfrak{M}(kt)}z_{kt,t}^2 - tb(\zeta(kt))| < \varepsilon_1 t\}$$

$$\Omega_k^{(3)} = \{\omega: \mathbf{E}_{\mathfrak{M}(kt)}(z_{kt,t}^2; |z_{kt,t}| > \delta\sqrt{T}) < \varepsilon_1 t\}$$

where $\varepsilon_1 = \sqrt{T\varepsilon/t}$. Then

$$\mathbf{P}\left(\bigcup_k \bar{\Omega}_{kt,t}\right) \leqslant \frac{3\varepsilon T}{\varepsilon_1 t} = o(1).$$

Proof of theorem 1. As we have seen in subsection 1.6.2, to prove convergence of finite-dimensional distributions it is sufficient to show that for any function $\phi \in C_2$

$$\mathbf{E}_{\mathfrak{M}(0)}\phi(y(v)) = M\phi(w^{(y)}(v))|_{y=y(0)} + r(\omega) \tag{1.69}$$

where $r(\omega) \underset{P}{\longrightarrow} 0$.

In the notation of subsections 1.6.3 and 1.6.4, we have

$$V(\zeta, u+\Delta) = V(\zeta_k, u) + \Delta V_t'(\zeta_k, u) + r_0(\omega)\Delta$$

where $|r_0(\omega)| < \varepsilon_0 = o(1)$ and ε_0 does not depend on u and ω,

$$V(\zeta_{k+1}, u) = V(\zeta_k, u) + \xi_{k+1}V_x'(\zeta_k, u)$$
$$+ \frac{\xi_{k+1}^2}{2}(V_{xx}''(\zeta_k, u) + \rho(\zeta_k, \xi_{k+1}, u))$$

$$\rho(\zeta_k, \xi_{k+1}, u) = V_{xx}''(\zeta_k + \beta\xi_{k+1}, u) - V_{xx}''(\zeta_k, u), \quad |\beta| < 1.$$

And, further (see (1.57)),

$$\mathbf{E}_{\mathfrak{M}_k} V(\zeta_{k+1}, u) - V(\zeta_k, u + \Delta)$$

$$= -r_0(\omega)\Delta + \Delta V_x'(\zeta_k, u) r_1(\omega) + \tfrac{1}{2}\Delta V_{xx}''(\zeta_k, u) r_2(\omega) + \mathbf{E}_{\mathfrak{M}_k} \frac{\xi_{k+1}^2 \rho}{2}$$

where, by (MI) and (MII),

$$\mathbf{E}|r_i(\omega)| < \varepsilon, \quad i = 1, 2.$$

Since V_{xx}'' is uniformly continuous and bounded, then

$$\begin{aligned}\mathbf{E}_{\mathfrak{M}_k} |\xi_{k+1}^2 \rho| &= \mathbf{E}_{\mathfrak{M}_k} (\xi_{k+1}^2 |\rho| I(|\xi_{k+1}| \leqslant \delta_T)) \\ &\quad + \mathbf{E}_{\mathfrak{M}_k} (\xi_{k+1}^2 |\rho| I(|\xi_{k+1}| > \delta_T)) \\ &\leqslant \varepsilon^* \Delta (b(\zeta_k) + r_2(\omega)) + r_3(\omega)\end{aligned}$$

where $\varepsilon^* = o(1)$, $\mathbf{E} r_3(\omega) < \varepsilon^{**} = o(1)$ and $\delta_T \to 0$ is some sequence for which (MIII) holds. Hence we obtain

$$\mathbf{E}_{\mathfrak{M}_k} V(\zeta_{k+1}, u) - V(\zeta_k, u + \Delta) = \Delta r(\omega).$$

Here $\mathbf{E}|r(\omega)| < \varepsilon_4 = o(1)$ and ε_4 does not depend on u and k. Therefore, with the obvious meaning for the notation $r^{(k)}(\omega)$,

$$\begin{aligned}\mathbf{E}_{\mathfrak{M}_0} \phi(y(v)) &= \mathbf{E}_{\mathfrak{M}_0} \mathbf{E}_{\mathfrak{M}_1} \cdots \mathbf{E}_{\mathfrak{M}_{n-1}} V(\zeta_n, 0) \\ &= \mathbf{E}_{\mathfrak{M}_0} \cdots \mathbf{E}_{\mathfrak{M}_{n-2}} V(\zeta_{n-1}, \Delta) + \Delta \mathbf{E}_{\mathfrak{M}_0} r^{(n)}(\omega) \\ &= \cdots \\ &= \mathbf{E}_{\mathfrak{M}_0} V(\zeta_1, (n-1)\Delta) + \mathbf{E}_{\mathfrak{M}_0} \sum_{k=1}^{n} r^{(k)}(\omega) \\ &= V(y(0), v) + \Delta \mathbf{E}_{\mathfrak{M}_0} \sum_{k=1}^{n} r^{(k)}(\omega).\end{aligned}$$

Since

$$V(y(0), v) = \mathbf{E}\phi(w^{(y)}(v))|_{y = y(0)}$$

and $\mathbf{E}|\Delta \mathbf{E}_{\mathfrak{M}_0} \sum_{k=1}^{n} r^{(k)}(\omega)| \leqslant v\varepsilon_4 = o(1)$, then (1.69) and, with it, theorem 1, is proved. □

1.9. Convergence to diffusion processes with reflection at a boundary

The results below, like theorems 1, 2 of section 1.5, use conditions 'in probability'.

Conditions (PRI)–(PRIII). *Let* $\xi_T(u)$ *be nonnegative and let* (PI)–(PIII) *be satisfied only in the domain* $\omega \in \{\zeta(u) = \xi_T(u)/\sqrt{T} > \varepsilon^*$ *for any* $\varepsilon^* > 0\}$.*

* In what follows we will suppose, as before, that $\varepsilon^* = o(1)$ as $T \to \infty$.

(PRIV) *There is a transition function* $\mathbf{P}(x, t, E)$ *of a Markov process* $w_R(t) \geqslant 0$ *such that*

$$V(x, u) = \int \phi(y)\mathbf{P}(x, t, \mathrm{d}y) = \mathbf{E}\phi(w_R^{(x)}(t)), \quad \phi \in C_2$$

in $0 \leqslant u \leqslant v, 0 \leqslant x < \infty$ *has continuous and bounded derivatives* $\partial^2 V/\partial x^2, \partial V/\partial x$, $\partial V/\partial u$ *and is a solution of the second boundary value problem*

$$\frac{\partial V}{\partial u} = a\frac{\partial V}{\partial x} + \frac{b}{2}\frac{\partial^2 V}{\partial x^2}$$

$$V(x, 0) = \phi(x), \quad x \geqslant 0, \quad \frac{\partial V(0, t)}{\partial x} = 0.$$

In addition, for some $c > 0$

$$|a(x)| < c(1 + |x|), \quad b(x) < c.$$

The process w_R in (PRIV) is, as is well known, a diffusion process on the half line with reflection at the boundary $\zeta = 0$ and with diffusion coefficient b, drift coefficient a. As before, $w_R^{(x)}(u)$ denotes the process with initial value $w_R^{(x)}(0) = x$.

It can be shown that conditions (A) and (B), sections 1.5 and 1.3, on a and b (defined on $[0, \infty)$) are sufficient for (PRIV).

Condition (PRV) *coincides with* (PV).

It remains to give a reflection condition which determines the behaviour of ξ near the origin.

(PRVI) *For any sufficiently small* ε^{**} *(or, for any sufficiently slowly decreasing sequence* ε^{**}*) there exist* $p > 0$, $Q > 0$ *such that for* $t_1 \sim (\varepsilon^{**})^2 TQ$ *in* $\Omega_{u,t} \cap \{|\zeta(u)| < \varepsilon^{**}\}$

$$\mathbf{P}_{\mathfrak{M}(u)}(z_{u,t_1} > \varepsilon^{**}\sqrt{T}) > p$$

holds.

THEOREM 1. *Let* $y_T(0)$ *have a proper limit distribution* p_0. *Then, under conditions* (PRI)–(PRVI) *the processes* $y_T(u) = \xi(uT)/\sqrt{T}$ *C-converge to a process* w_R *with initial distribution* p_0.

The meaning of (PRI)–(PRV) is fairly clear. Condition (PRVI) guarantees the lack of absorption in the domain $\zeta(u) < \varepsilon^{**}$ (in which, in general, (PI)–(PIII) will not be satisfied). A particle which hits this domain will, with positive probability, leave it in a time $o(T)$.

It is not difficult to see that all the conditions (PRI)–(PRVI) are essential for this theorem.

Proof. As already noted, throughout (PRI)–(PRIII), (PRV), (PRVI) we can regard the arbitrarily small numbers ε, δ, ε^*, ε^{**} as sequences converging to 0 as $T \to \infty$. Here, to simplify the notation subsequently we will, instead of several sequences $\varepsilon, \delta, \varepsilon^*, \varepsilon^{**}/2$, consider only one: $\delta = \delta_T$, defined for example by

$$\delta_T = \max(\varepsilon, \delta, \varepsilon^*, \varepsilon^{**}/2).$$

The general plan of the proof is as before. We will use the notations of sections 1.6 and 1.7 and suppose that $h = U/t$ is an integer.

1.9.1. LEMMA 1. *Under conditions* (PRI)–(PRVI)

$$\mathbf{P}\left(\max_{k \leqslant n} \xi_k > x\sqrt{T}\right) \to 0 \quad \text{as } x \to \infty, T \to \infty.$$

Proof. By transformations of the process we reduce the problem to lemma 1 of section 1.6. First, by the method of lemma 1 of section 1.6 with respect to $z_k = \xi_k - \xi_{k-1}$, we construct a sequence z_k^*. The elements of this sequence will satisfy (1.45) on $\Omega_{kt,t} \cap \{\zeta(kt) > \delta_T\}$. In addition, by the remark in section 1.7, $\mathbf{P}_{\mathfrak{M}_0}(\inf_{k \leqslant n} \xi_k^* > \gamma\sqrt{T})$, for fixed $\gamma > 0$, will converge to $\mathbf{P}(\inf_{u \leqslant v} w(u) > \gamma)$, where w is the corresponding diffusion process.

It suffices for us to show that

$$\mathbf{P}\left(\sup_{k \leqslant n} \xi_k^* > x\sqrt{T}\right) \to 0 \quad \text{as } x \to \infty, T \to \infty. \tag{1.70}$$

We introduce a sequence X_k as follows. Let $\gamma > 0$. Put

$$X_k = \xi_k^*, \quad k = \overline{0, n}, \quad \text{if } \inf_{k \leqslant n} \xi_k^* \geqslant \gamma\sqrt{T}.$$

If

$$k_1 = \min\{k: \xi_k^* < \gamma\sqrt{T}\} \leqslant n$$

then we put

$$X_k = \begin{cases} \xi_k^* & \text{for } k \leqslant k_1 \\ \xi_{k_1}^* & \text{for } k_1 \leqslant k \leqslant l_1 \end{cases}$$

where $l_1 = \min\{k > k_1: \xi_k^* \geqslant (1+\gamma)\sqrt{T}\}$. We will say for $l_1 \leqslant n$ that the first period ends at l_1 and call $D_1 = \xi_{l_1}^* - \xi_{k_1}^*$ the *defect of the first period*. Further

$$X_k = \xi_k^* - D_1 \quad \text{for } k \geqslant l_1, \quad \text{if } \inf_{l_1 \leqslant k \leqslant n} \xi_k^* \geqslant \gamma\sqrt{T}.$$

If

$$k_2 = \min\{k \geqslant l_1; \xi_k^* < \gamma\sqrt{T}\} \leqslant n$$

then again we put

$$X_k = \xi_{k_2}^* - D_1, \quad k = k_2, \ldots, l_2$$

where $l_2 = \min\{k > k_2: \xi_k^* \geqslant (1+\gamma)\sqrt{T}\}$. The second period ends at l_2. The defect of the second period $\xi_{l_2}^* - \xi_{k_2}^*$ is denoted D_2. For $k \geqslant l_2$ and up to the beginning of the third period we put

$$X_k = \xi_k^* - D_1 - D_2, \quad \text{etc.}$$

We now choose an integer K. If the beginning of the $(K+1)$st period is earlier than U: $k_{K+1} < n$, then we put

$$X_k = \xi_k^* - \sum_{j=1}^{K} D_j, \quad k = k_{K+1}, \ldots, n.$$

Let $x_k = X_k - X_{n-1}$, $m_k = \max_{i \leqslant k} |z_i^*|$. Then it is clear that

$$X_k \geqslant \xi_k^* - K(\sqrt{T} + 2m_k), \quad k = 0, \ldots, n$$

and everywhere on $\Omega_{kt,t}$

$$|\mathbf{E}_{\mathfrak{M}_k} x_{k+1}| \leqslant \frac{ct}{\sqrt{T}}\left(1 + \frac{X_k + K(\sqrt{T} + 2m_k)}{\sqrt{T}}\right)$$

$$\mathbf{E}_{\mathfrak{M}_k} x_{k+1}^2 < ct, \quad c = \text{const.} \tag{1.71}$$

$$\mathbf{E}_{\mathfrak{M}_k}(x_{k+1}^2, |x_{k+1}| > \delta_T \sqrt{T}) = r_3(\omega) < \varepsilon.$$

If, instead of $\Omega_{kt,t}$, we introduce the sets $\Omega_{kt,t}^* = \Omega_{kt,t} \cap \{m_k < \delta\sqrt{T}\} \in \mathfrak{M}_k$, concerning which, obviously, it can also be assumed that

$$\mathbf{P}\left(\bigcap_{k=0}^{n-1} \Omega_{kt,t}^*\right) \to 1$$

then the first inequalities in (1.71) may be written in the form

$$|\mathbf{E}_{\mathfrak{M}_k} x_{k+1}| < \frac{ct}{\sqrt{T}}\left(1 + \frac{X_k + K(1 + 2\delta_T)}{\sqrt{T}}\right). \tag{1.72}$$

We may now proceed to the estimate of (1.70). Denote by ν_n the number of complete periods in the interval $[0, n]$ and note that for some $q < 1$

$$\mathbf{P}(\nu_n > k) < q^k. \tag{1.73}$$

In fact as already noted $\mathbf{P}_{\mathfrak{M}_0}(\inf_{k \leqslant n} \xi_k^* > \gamma\sqrt{T})$, for $\xi_0^* \geqslant (1 + \gamma)\sqrt{T}$, converges as $T \to \infty$ to some positive limit. Therefore after the end of each period there is a uniformly positive probability that this period was the last. This implies (1.73).

Let $\alpha > 0$ be given. Choose K so that $\mathbf{P}(\nu_n > K) < \alpha/3$. Then, for sufficiently large $T(\omega_{n-1}^* = \bigcap_{k=0}^{n-1} \Omega_{kt,t}^*)$,

$$\begin{aligned}
\mathbf{P}\left(\sup_{k \leqslant n} \xi_k^* > x\sqrt{T}\right) &\leqslant \mathbf{P}(\nu_n > K) + \mathbf{P}(\bar{\omega}_{n-1}^*) \\
&\quad + \mathbf{P}(\sup \xi_k^* > x\sqrt{T}; \nu_n \leqslant K, \omega_{n-1}^*) \\
&\leqslant \frac{2\alpha}{3} + \mathbf{P}\left(\sup_{k \leqslant n} X_k > x\sqrt{T} - K(1 + 2\delta)\sqrt{T}\right).
\end{aligned}$$

By (1.71) and (1.72) an estimate of the last term can be found under the conditions of lemma 1 of section 1.6 (compare with (1.45)). Using this lemma with our chosen K we may, by choice of x, obtain that the term to be estimated does not exceed $\alpha/3$. The lemma is proved. □

In essence the result of lemma 1 is clear: we hit the domain $\zeta > x$, $x \to \infty$, as a rule, from the domain $\delta < \zeta < x$ (significant jumps are unlikely), in which there are the same laws as for the processes in the conditions of theorems 1 and 2 of section 1.5. Equally natural is another proposition.

1.9.2. LEMMA 2. *Let*

$$\nu = \min\{k \geqslant 0: \xi_k \leqslant \delta_T\sqrt{T}\}, \quad l \leqslant A\frac{\delta_T^2 T}{t}, \quad A = \text{const.} > 0.$$

Then on the set $\{\nu = k - l\} \in \mathfrak{M}_{k-l}$

$$\mathbf{E}_{\mathfrak{M}_{k-l}}(\xi_k^2; h((k-l)t, kt)) < c\delta_T^2 T, \quad c = \text{const.}$$

where the event $h(u, s)$ *is defined in (1.44).*

The proof of this lemma uses the transformation applied in lemma 1 (but to the levels $\delta_T\sqrt{T}$ and $3\delta_T\sqrt{T}$ instead of γ and $(1+\gamma)$) and the estimate of the variance in lemma 1 of section 1.6. Since this proof uses no new methods, we omit it.

1.9.3. We turn to the convergence of distributions. As before let

$$y_T(v) = \zeta_n = y_T(0) + s_1 + \cdots + s_n, \quad s_k = z_k/\sqrt{T}. \tag{1.74}$$

Our aim will be the proof of the convergence (see (1.56))

$$\mathbf{E}_{\mathfrak{M}_0}\phi(\zeta_n) - \mathbf{E}\phi(w_R^{(x)}(v))|_{x=\zeta_0} \to 0, \quad \phi \in C_2. \tag{1.75}$$

The terms s_k in (1.74), as earlier, satisfy (1.57), but only on the set $\Omega_k \cap \{\zeta_k > \delta_T\}$. Therefore in (1.58), which we reduced earlier to (1.56), there can arise 'discrepancies' near the boundary $\zeta = 0$. To compensate for these we must change the sequence $\zeta_0, \ldots, \zeta_n$ and the function V.

Let

$$\nu = \min\{k: \zeta_k \leqslant \delta_T\}, \quad L = \frac{4\delta_T^2 TQ}{t}.$$

Without loss of generality L can be assumed to be an integer. We construct a sequence $\lambda_0, \ldots, \lambda_n$ as follows (for brevity we omit the lower index T on δ_T):

$$\lambda_0 = \zeta_0, \ldots, \lambda_{\nu-1} = \zeta_{\nu-1}, \lambda_\nu = \delta$$

$$\lambda_{\nu+1} = \delta - \delta_1, \ldots, \lambda_{\nu+L} = \delta - \delta_1 L = 0$$

$$\lambda_{\nu+L+1} = \max(\delta, \zeta_{\nu+L+1}).$$

Here $\delta_1 = \delta/L$. The subsequent values $\lambda_{\nu+L+2}, \ldots$ are defined by the same rule—as if $\zeta_{\nu+L+1}$ were the first element of the sequence $\zeta_0, \zeta_1, \ldots$.

Let $\mathfrak{N}_k$ be the σ-algebra generated by the events

$$\{\mu_k = j, B\}, \quad j = k, k-1, \ldots, \max(0, k-L); \quad B \in \mathfrak{M}_j$$

where μ_k is the random variable

$$\mu_k = \begin{cases} k & \text{if } \lambda_k \geqslant \delta \\ k - l & \text{if } \lambda_k = \delta - l\delta_1, \quad l = 0, \ldots, L. \end{cases}$$

It is clear that λ_k is measurable relative to $\mathfrak{N}_k$, $\mathfrak{N}_0 = \mathfrak{M}_0 = \mathfrak{M}(0)$, $\mathfrak{N}_k \subset \mathfrak{N}_{k+1}$. The σ-algebra $\mathfrak{N}_k$ can be regarded as generated by events related to the trajectory of the process $\xi(u)$ up to the first hit of the sequence $\lambda_j, j = \max(0, k-L), \ldots, k$ in the domain $\lambda_j \leqslant \delta$.

Before going on to the proof of (1.75) we must calculate the conditional moments relative to $\mathfrak{N}_k$ of the increments

$$\kappa_{k+1} = \lambda_{k+1} - \lambda_k.$$

1.9.4. On different ω-sets these moments behave differently. Therefore we consider the cases individually:

(a) If $\omega \in \Omega_k \cap \{\lambda_k > 2\delta\}$, then, by (PIII), the behaviour of the first two moments of κ_{k+1} is the same as the behaviour of the moments of ξ_{k+1} and is defined by (1.57).

(b) If $\omega \in \Omega_k \cap \{\delta < \lambda_k \leqslant 2\delta\}$, then $\mathbf{E}_{\mathfrak{N}_k} \kappa_{k+1}^2 < ct$.

(c) If $\omega \in \Omega_k \cap \{\lambda_k = \delta - l\delta_1\}$, $l = 0, 1, \ldots, L-1$

$$\mathbf{E}_{\mathfrak{N}_k} f(\kappa_{k+1}) = f(-\delta_1)$$

for any measurable function f.

(d) If $\omega \in \Omega_k \cap \{\lambda_k = 0\}$, then, by (PRVI),

$$\mathbf{E}_{\mathfrak{N}_k} \kappa_{k+L} \geqslant 2\delta p + \delta(1-p) = \delta(1+p).$$

By lemma 2, on $\{\lambda_k = 0\}$

$$\mathbf{E}_{\mathfrak{N}_k}(\kappa_{k+1}^2; h((k-L)t, kt)) < c\delta^2.$$

1.9.5. Instead of $V(x, u)$, which is a solution of the second boundary value problem (see condition PRIV), we consider

$$V_\alpha(x, u) = V(x, u) + \alpha u + g(x)$$

where $g(x)$ has the properties

$$g(0) = 0$$

$$g'(x) = g_1\delta > 0 \qquad \text{for } x \in (0, \delta)$$

$$g'(x) = -g_2\delta < 0 \quad \text{for } x \in (\delta, 3\delta).$$

On $[\delta, \infty]$ $g(x)$ is chosen to be twice continuously differentiable

$$|g(x)| < g_2\delta, \quad |g'(x)| < g_2\delta, \quad |g''(x)| < g_2\delta.$$

It is not difficult to see that such a choice of g is always possible. The values g_1 and g_2 will be chosen later.

In order to prove (1.75) following the plan of section 1.6 it would be desirable to establish that for any $\alpha > 0$, $M > 0$, and for sufficiently large T on $\Omega_k^M = \Omega_{kt,t}\Gamma_{kt}^M$

$$\mathbf{E}_{\mathfrak{N}_k} V_\alpha(\lambda_{k+1}, u) - V_\alpha(\lambda_k, u + \Delta) \leqslant 0 \tag{1.76}$$

for all k, u.

1.9.6. (a) On $A_a = \Omega_k^M\{\lambda_k \geqslant 2\delta\}$ the verification of this inequality (see subsection 1.9.4(a)) is the same as the proof of (1.58). In addition to the difference in (1.58) here there is the term

$$\begin{aligned}\mathbf{E}_{\mathfrak{N}_k}&(g(\lambda_k + x_{k+1}) - g(\lambda_k)) - \alpha\Delta\\ &= g'(\lambda_k)\mathbf{E}_{\mathfrak{N}_k}\kappa_{k+1} + g''(\lambda_k + \beta\kappa_{k+1})\mathbf{E}_{\mathfrak{N}_k}\kappa_{k+1}^2 - \alpha\Delta\\ &= O(\delta\Delta) - \alpha\Delta, \quad |\beta| \leqslant 1\end{aligned}$$

which, obviously, will be the major part of the difference (1.76) and will be negative.

(b) On $A_b = \Omega_k^M\{\delta < \lambda_k \leqslant 2\delta\}$ we have

$$V_\alpha(\lambda_k, u+\Delta) = V(\lambda_k, u) + \alpha u + \alpha\Delta + g(\lambda_k) + \Delta V_u'(\lambda_k, u) + \Delta r_0(\omega)$$

$$\begin{aligned}V_\alpha(\lambda_k + \kappa_{k+1}, u) = {}& V(\lambda_k, u) + \alpha u + V_x'(\lambda_k, u)\kappa_{k+1}\\ &+ \tfrac{1}{2}V_{xx}''(\lambda_k + \beta\kappa_{k+1}, u)\kappa_{k+1}^2 + g(\lambda_k + \kappa_{k+1})\end{aligned}$$

where $|r_0(\omega)| < \varepsilon = o(1)$, $|\beta| < 1$.

Since on the ω-set A_b

$$\kappa_{k+1} = \begin{cases} s_{k+1} & \text{if } s_{k+1} \geqslant \delta - \lambda_k \\ \delta - \lambda_k & \text{if } s_{k+1} \leqslant \delta - \lambda_k \end{cases}$$

and $\mathbf{E}_{\mathfrak{N}_k}(\kappa_{k+1}^2; |\kappa_{k+1}| > \delta_T) = o(\Delta)$, then

$$\begin{aligned}\mathbf{E}_{\mathfrak{N}_k}&V_\alpha(\lambda_{k+1}, u) - V_\alpha(\lambda_k, u + \Delta)\\ &= -\alpha\Delta + o(\Delta) + V_x'(\lambda_k, u)\mathbf{E}_{\mathfrak{N}_k}[(\delta - \lambda_k - s_{k+1})I(s_{k+1} < \delta - \lambda_k)]\\ &\quad + V_{xx}''(\lambda_k, u)\mathbf{E}_{\mathfrak{N}_k}((\delta - \lambda_k)^2 - s_{k+1}^2)I(s_{k+1} < \delta - \lambda_k)\\ &\quad + \mathbf{E}_{\mathfrak{N}_k}(g(\lambda_{k+1}) - g(\lambda_k)).\end{aligned} \tag{1.77}$$

But on A_b we have

$$|V_x'(\lambda_k, u)| < 2\delta V'', \quad \text{where } V'' = \sup_{\substack{x \leqslant 1\\ u \leqslant v}} |V''(x, u)|$$

$$3\delta\mathbf{E}_{\mathfrak{N}_k}(\delta - \lambda_k - s_{k+1})I(s_{k+1} < \delta - \lambda_k)$$
$$\geqslant -\mathbf{E}_{\mathfrak{N}_k}((\delta - \lambda_k)^2 - s_{k+1}^2)I(s_{k+1} < \delta - \lambda_k) \geqslant 0$$

$$\mathbf{E}_{\mathfrak{N}_k} g(\lambda_k + \kappa_{k+1}) - g(\lambda_k)$$
$$= g_2\delta\mathbf{E}_{\mathfrak{N}_k}\kappa_{k+1} + o(\Delta)$$
$$= o(\Delta) - g_2\delta\mathbf{E}_{\mathfrak{N}_k}((\delta - \lambda_k - s_{k+1})I(s_{k+1} < \delta - \lambda_k)).$$

Here the estimate, as in (1.77), is uniform with respect to ω, u, and k. We now see that the difference (1.77) does not exceed

$$-\alpha\Delta + o(\Delta) + \delta\mathbf{E}_{\mathfrak{N}_k}((\delta - \lambda_k + s_{k+1})I(s_{k+1} < \delta - \lambda_k))[V'' + 3V'' - g_2] \leqslant 0$$

provided we put

$$g_2 \geqslant 4V''.$$

(c) On $A_c = \Omega_k^M\{\lambda_k = \delta - l\delta_1\}$, $l = \overline{0, L-1}$,

$$\mathbf{E}_{\mathfrak{N}_k} V_\alpha(\lambda_{k+1}, u) - V_\alpha(\lambda_k, u + \Delta)$$
$$= -\delta_1 V_x'(\lambda_k, u) + o(\delta_1^2) - \Delta V_u'(\lambda_k, u) - \alpha\Delta + o(\Delta) - \delta_1 g_1 \delta.$$

Since

$$\delta_1\delta = \frac{\delta^2}{L} = \frac{\Delta}{4Q}$$

then this value will also be negative for

$$g_1 > V'' + 4QV'$$

where $V' = \sup_{u \leqslant v} |V_u'(0, u)|$.

(d) Finally, on $A_d = \Omega_k^M\{\lambda_k = 0\}$, denoting temporarily, for brevity, $h((k - L)t, kt) = h$, we obtain

$$\mathbf{E}_{\mathfrak{N}_k}[V_\alpha(\kappa_{k+1}, u); h] - V_\alpha(0, u + \Delta)$$
$$= \mathbf{E}_{\mathfrak{M}_{k-L}}[V_{xx}''(\beta\kappa_{k+1}, u)\kappa_{k+1}^2; h]$$
$$- V_u'(0, u)\Delta + o(\Delta) - \alpha\Delta + \mathbf{E}_{\mathfrak{N}_k}[g(\kappa_{k+1}); h].$$

By subsection 1.9.4(d) this does not exceed

$$V''c\delta^2 + o(\delta^2) + g(\delta) + g'(\delta)\mathbf{E}_{\mathfrak{N}_k}[(\kappa_{k+1} - \delta); h]$$
$$+ \mathbf{E}_{\mathfrak{N}_k}[g''(\delta + \beta(\kappa_{k+1} - \delta))(\kappa_{k+1} - \delta)^2; h]$$
$$\leqslant V''c\delta^2 + o(\delta^2) + g_1\delta^2 - g_2\delta^2 p \leqslant 0$$

for $g_2 > (cV'' + g_1)/p$ and sufficiently large T. Here β denotes a number from $[0, 1]$.

Thus, we have established a relation somewhat different from (1.76). Namely, if we denote

$$B_k = h((k-L)t, kt) \cup \{\lambda_k \neq 0\}\bar{h}((k-L)t, kt)$$

then

$$\mathbf{E}_{\mathfrak{N}_k}(V_\alpha(\lambda_{k+1}, u), B_k) \leqslant V_\alpha(\lambda_k, u+\Delta).$$

Reasoning in a completely analogous way, we obtain, for $V_\alpha^* = V - \alpha u - g(x)$, the reverse inequalities for any $\alpha > 0$.

1.9.7. We can now pass on to the concluding part of the proof of convergence (1.77). We have

$$\mathbf{E}_{\mathfrak{N}_0}\phi(\lambda_n) = \mathbf{E}_{\mathfrak{N}_0}V_\alpha(\lambda_n, 0) - \mathbf{E}_{\mathfrak{N}_0}g(\lambda_n)$$

$$\mathbf{E}_{\mathfrak{N}_0}V_\alpha(\lambda_n, 0) = \mathbf{E}_{\mathfrak{N}_0}V_\alpha(\lambda_n, 0)I\left(\bigcap_{k=0}^{n-1}\Omega_k^M\right)I\left(\bigcap_{k=1}^{n}B_k\right) - \rho_0(\omega)$$

where

$$|\rho_0(\omega)| < c\mathbf{P}_{\mathfrak{M}_0}\left(\bigcup_{0}^{n-1}\bar{\Omega}_k^M\right) + c\mathbf{P}_{\mathfrak{M}_0}\left(\bigcup_{k=1}^{n}\bar{B}_k\right).$$

Since

$$I(\Omega_k^M)\mathbf{E}_{\mathfrak{N}_k}[V_\alpha(\lambda_{k+1}, u)I(B_k)] \leqslant I(\Omega_k^M)V_\alpha(\lambda_k, u+\Delta)$$
$$= V_\alpha(\lambda_k, u+\Delta)(1 - I(\bar{\Omega}_k^M))$$

then

$$\mathbf{E}_{\mathfrak{N}_0}V_\alpha(\lambda_n, 0)I\left(\bigcap_{k=0}^{n-1}\Omega_k^M\right)I\left(\bigcap_{k=1}^{n}B_k\right)$$
$$= I(\Omega_0^M)\mathbf{E}_{\mathfrak{N}_0}I(\Omega_1^M)I(B_1)\mathbf{E}_{\mathfrak{N}_1}\cdots I(\Omega_{n-1}^M)\mathbf{E}_{\mathfrak{N}_{n-1}}V_\alpha(\lambda_n, 0)I(B_n)$$
$$\leqslant I(\Omega_0^M)\mathbf{E}_{\mathfrak{N}_0}I(\Omega_1^M)I(B_1)\mathbf{E}_{\mathfrak{N}_1}\cdots$$
$$\cdots I(\Omega_{n-2}^M)\mathbf{E}_{\mathfrak{N}_{n-2}}V_\alpha(\lambda_{n-1}, \Delta)I(B_{n-1}) - \rho_n(\omega)$$
$$\leqslant \cdots$$
$$\leqslant V_\alpha(\lambda_0, v) - \sum_{k=1}^{n}\rho_k(\omega).$$

Here

$$\left|\sum_{k=1}^{n}\rho_k(\omega)\right| \leqslant c\mathbf{P}_{\mathfrak{N}_0}\left(\bigcup_{0}^{n-1}\bar{\Omega}_k^M\right).$$

Summarizing, we find that

$$\mathbf{E}_{\mathfrak{M}_0}\phi(\lambda_n) \leqslant V_\alpha(\lambda_0, v) + \rho^M = V(\lambda_0, v) + \alpha v + g(v) + \rho^M$$

where $g(v) \xrightarrow{P} 0$ as $T \to \infty$, and, using lemma 1 and the condition of the theorem, ρ^M can be made arbitrarily small by choice of M. If we use the similar reverse inequalities and the arbitrariness of α and M, then we may conclude that

$$\mathbf{E}_{\mathfrak{M}_0}\phi(\lambda_n) \underset{P}{\longrightarrow} V(\lambda_0, v) = \mathbf{E}\phi(w_R^{(y)}(v))|_{y=y(0)}.$$

$\mathbf{E}_{\mathfrak{M}_0}\phi(\zeta_n)$ also converges to this same limit, since

$$\mathbf{E}_{\mathfrak{M}_0}|\phi(\zeta_n) - \phi(\lambda_n)| < c\mathbf{E}_{\mathfrak{M}_0}|\zeta_n - \lambda_n|$$

$$\mathbf{E}_{\mathfrak{M}_0}|\zeta_n - \lambda_n| \leqslant \mathbf{E}_{\mathfrak{M}_0} \sum_{l=0}^{L} \mathbf{E}_{\mathfrak{N}_n} I(\lambda_n = \delta - l\delta_1)|\zeta_n - \lambda_n|$$

$$\leqslant \mathbf{E}_{\mathfrak{M}_0} \sum_{l=0}^{L} I(\lambda_n = \delta - l\delta_1)\mathbf{E}_{\mathfrak{M}_{n-l}}(\delta + |\zeta_n|).$$

By lemma 2 this sum does not exceed $c\delta$ for some $c > 0$. Convergence of the finite-dimensional distributions is proved.

1.9.8. The plan of the proof of C-convergence (the verification of the second condition of theorem 1 of section 1.2) also remains as before. In subsection 1.7.1 it is sufficient to consider the process $\xi^* = \max(\delta_T, \xi)$. In subsection 1.9.3 we must use the construction of lemma 1. A more detailed discussion is not necessary here since all the required elements of the proof are contained in the sections mentioned. □

Remark. The method used for the proof of the theorem in this section and consisting of the replacement of $V(x, u)$ by $V_\alpha(x, u)$ (see subsection 1.9.5), may also be successfully applied to the modification of theorems 1 and 2 of section 1.5 and the theorems of sections 1.9 and 1.10 in the following direction. It may be that the verification of conditions (I)–(III) of these theorems in a δ-neighbourhood of some point D is difficult or completely impossible. The modification is that if the fixed $\delta \geqslant 0$ may be chosen arbitrarily small, then in $|\zeta(u) - D| \leqslant \delta$ it is sufficient, with (I)–(III), to require that the 'reflection' condition (PRVI) is satisfied on both sides: for sufficiently small $\varepsilon^{**} > 0$ there are $p > 0, Q > 0$ such that for $t_1 \sim (\varepsilon^{**})^2 TQ$ in $\Omega_{u,t} \cap \{|\mathbf{f}(u) - D| < \varepsilon^*\}$ we have

$$\mathbf{P}_{\mathfrak{M}(u)}(z_{u,t_1} > \varepsilon^{**}\sqrt{T}) > p$$
$$\mathbf{P}_{\mathfrak{M}(u)}(z_{u,t_1} < -\varepsilon^{**}\sqrt{T}) > p.$$

Here it is assumed that condition (PV) on the absence of large jumps is satisfied.

Following the reasoning of subsection 1.9.5 of the above proof (naturally, another function $g(x)$ must be chosen), it is possible to show that such a violation of conditions (I)–(III) in a small domain $|\zeta(u) - D| \leqslant \delta = \varepsilon^{**}$ has influence neither on the convergence of processes nor on the nature of the limit distribution. It is also clear that this will all remain valid if a similar violation of (I)–(III) holds not at one but at several points.

1.10. Conditions for convergence to a diffusion with two reflecting boundaries

An acquaintance with the proof of theorem 1 of section 1.9 makes quite clear both the construction of conditions 'in probability' for convergence to a diffusion process with two reflecting boundaries and the proof of the theorem given below.

THEOREM 1. *Let the process* $\zeta(u) = \xi_T(u)/\sqrt{T}$ *take values in the strip* $0 \leqslant \zeta(u) \leqslant R$ *and let* (PRRI)–(PRRIII), *which consist of assuming that* (PI)–(PIII) *hold in the domain* $\omega \in \{R - \varepsilon^* \geqslant \zeta(u) \geqslant \varepsilon^*\}$ *for any* $\varepsilon^* > 0$, *be satisfied.*

Let the following condition also hold:

(PRRIV) *There is a transition function* $\mathbf{P}(x, t, E)$ *of some Markov process on* $[0, R]$ *such that the function*

$$V(x, u) = \int_0^R \phi(y)\mathbf{P}(x, u, \mathrm{d}y), \quad \phi \in C_2$$

has continuous and bounded derivatives $\partial^2 V/\partial x^2, \partial V/\partial u$ *for* $0 \leqslant u \leqslant 1, 0 < x < R$ *and is a solution of the boundary value problem*

$$\frac{\partial V}{\partial u} = a\frac{\partial V}{\partial x} + \frac{b\partial^2 V}{2\partial x^2}$$

$$V(x, 0) = \phi(x), \quad \frac{\partial V}{\partial x}(0, u) = 0, \quad \frac{\partial V}{\partial x}(R, u) = 0$$

The functions a and b are bounded on $[0, R]$.

The process w_{RR}, defined by this condition, is a diffusion process with reflection at the boundaries 0 and R. a and b are respectively its coefficients of drift and diffusion.

Condition (PRRV) *coincides with* (PV).

We will also assume condition (PRRVI) *which is that the processes* ζ *and* $R - \zeta$ *satisfy condition* (PRVI).

Under these conditions the distributions of $y_T(u) = \xi_T(uT)/\sqrt{T}$ *C-converge to the distribution of* w_{RR}.

The proof of this theorem is just a repetition of the proof of theorem 1 of section 1.9 and differs from it only by simplification, since, under the conditions of this section, $\sup_{u \leqslant v} y_T(u) \leqslant R$.

We note once again that the methods used above may be applied without significant changes to the study of convergence to a diffusion process, non-homogeneous in time, and to processes with reflection from smooth curvilinear boundaries. The case of curvilinear boundaries reduces to the rectilinear case by a change of variables.

1.11. Examples

The basic examples of use in queuing theory of the results of this chapter are in chapters 2 and 3. In this section we only consider fairly simple problems which illustrate the verification of the conditions in the theorems on convergence to degenerate and diffusion processes.

1.11.1. The generalized renewal process. Let $\tau_1 > 0, \tau_2 > 0, \ldots; \nu_1, \nu_2, \ldots$ be two independent of each other sequences of independent random variables. A wandering point jumps at epochs $t_0 = 0$, $t_1 = \tau_1$, $t_2 = \tau_1 + \tau_2, \ldots$ through distances $\nu_1, \nu_2, \ldots$. Our interest is in the position $X(t)$ of the point at time t, $t \in [0, U]$. If

$$\eta(t) = \min\{k : t_k = \tau_1 + \cdots + \tau_k > t\}$$

then

$$X(t) = \nu_1 + \cdots + \nu_{\eta(t)}.$$

For a series array the distributions of τ_j and ν_j and, consequently, the distribution of $X(t)$ depend on a parameter T. We denote

$$A = \frac{\mathbf{E}\nu}{\mathbf{E}\tau}$$

(we will sometimes omit the index j from τ_j and ν_j for brevity) and we find conditions under which the process

$$\xi(t) = \xi_T(t) = X(t) - At$$

will converge, with suitable normalization, to degenerate and diffusion processes.

We make use of the following well-known facts from renewal theory and the theory of summation of a random number of random variables.

1. Let

$$\mathbf{E}\tau > c > 0 \tag{1.78}$$

$$\mathbf{E}(\tau : \tau > N) \to 0, \quad \mathbf{E}(|\nu| : |\nu| > N) \to 0 \tag{1.79}$$

as $N \to \infty$, uniformly in T. (Condition (1.79), obviously, will always be satisfied if $\mathbf{E}\tau^{1+\gamma} <$ const., $\mathbf{E}|\nu|^{1+\gamma} <$ const. for some $\gamma > 0$.) Then

$$\left|\frac{\eta(t)}{t} - a\right| \xrightarrow[P]{} 0, \quad \left|\frac{\xi(t)}{t}\right| \xrightarrow[P]{} 0 \tag{1.80}$$

together with the first moments:

$$\mathbf{E}\left|\frac{\eta(t)}{t} - a\right| \to 0, \quad \mathbf{E}\left|\frac{\xi(t)}{t}\right| \to 0 \tag{1.81}$$

as $t = t(T) \to \infty$, where $a = (\mathbf{E}\tau)^{-1}$.

2. Let (1.78) be satisfied and let

$$\sigma_\tau^2 = \mathbf{D}\tau \geqslant c > 0, \quad \mathbf{E}(\tau^2; \tau \geqslant N) \to 0 \tag{1.82}$$

as $N \to \infty$, uniformly in T (for the latter condition it is sufficient that $\mathbf{E}\tau^{2+\gamma} <$

const. for some $\gamma > 0$). Then the distribution of

$$\eta^* = \frac{\eta(t) - at}{\sigma_\tau \sqrt{ta^3}}$$

converge to normal together with the second moments

$$\mathbf{E}(\eta^*)^2 \to 1.$$

Hence, in its turn, follows the same result on the convergence of the distribution of

$$\xi^* = \frac{\xi(t)}{\sigma \sqrt{t}} \tag{1.83}$$

provided

$$\sup_T \mathbf{E}(v^2 : |v| > N) \to 0 \tag{1.84}$$

as $N \to \infty$, and

$$\sigma^2 = a\mathbf{D}v + a^3\mathbf{D}\tau(\mathbf{E}v)^2 \geqslant c > 0. \tag{1.85}$$

We will now prove the following two fundamental results.

THEOREM 1. *If* (1.78) *and* (1.79) *are satisfied then*

$$\frac{\xi(tT)}{T} \underset{C}{\Rightarrow} 0.$$

THEOREM 2. *If* (1.78), (1.82), (1.84), *and* (1.85) *are satisfied then*

$$\frac{\xi(tT)}{\sigma\sqrt{T}}$$

C-converges to a standard Wiener process on $[0, v]$ $(U = Tv)$.

The proof of theorem 1 uses theorem 1 of section 1.4. As the σ-algebras $\mathfrak{M}(u)$ in that theorem we take the σ-algebras generated by the random variables

$$\eta(u); \quad \tau_1, \ldots, \tau_{\eta(u)}; \quad v_1, \ldots, v_{\eta(u)-1}$$

and as $\Omega_{u,\theta}$ we take

$$\Omega_{u,\theta} = \{\chi(u) \leqslant s\}, \quad \chi(u) = t_{\eta(u)} - u.$$

The increments

$$z_{u,t} = \xi(t + u) - \xi(u)$$

are, obviously, distributed as

$$-A\chi(u) + \xi^*(t - \chi(u))$$

where $\xi^*(\cdot)$ does not depend on $\chi(u)$ and has the same distribution as $\xi(\cdot)$. But for each v and t such that

$$0 \leqslant v \leqslant s, \quad s = o(\theta), \quad t \leqslant \theta$$

the random variable

$$z^* = -Av + \xi^*(t-v)(\xi^*(t-v) = 0 \text{ for } t - v \leqslant 0) \tag{1.86}$$

has, by (1.80) and (1.81), the properties

$$\frac{z^*}{\theta} \underset{P}{\longrightarrow} 0, \quad \mathbf{E}\left|\frac{z^*}{\theta}\right| \longrightarrow 0 \tag{1.87}$$

as $\theta \longrightarrow \infty$. Hence it follows that on $\Omega_{u,\theta}$

$$\mathbf{E}_{\mathfrak{M}(u)}|z_{u,t}| < \theta\varepsilon(\theta), \quad \varepsilon(\theta) = o(1).$$

Since, by weak convergence of the distribution of z^*/θ

$$\mathbf{E}\left(\left|\frac{z^*}{\theta}\right|; |z^*| \leqslant \varepsilon\theta\right) \longrightarrow 0$$

then it also follows from (1.87) that on $\Omega_{u,\theta}$

$$\mathbf{E}_{\mathfrak{M}(u)}(|z_{u,t}|; |z_{u,t}| > \varepsilon\theta) < \theta\varepsilon_1(\theta), \quad \varepsilon_1(\theta) = o(1).$$

To complete the verification of the conditions of theorem 1 of section 1.4 it remains to show that

$$\mathbf{P}\left(\max_{j \leqslant \eta(t)} |v_j| > \delta T\right) \longrightarrow 0$$

and that s and θ can be chosen so that

$$s = o(\theta), \quad \theta = o(T)$$

$$\mathbf{P}\left(\bigcap_k \Omega_{k\theta,\theta}\right) \geqslant \mathbf{P}\left(\bigcap_u \Omega_{u,\theta}\right) \geqslant \mathbf{P}\left(\max_{j \leqslant \eta(T)} \tau_j \leqslant s\right) \longrightarrow 1 \tag{1.88}$$

as $T \longrightarrow \infty$. Since $\eta(T)$, with probability close to 1, does not exceed cT for some c, then it is sufficient for us to show that

$$cT\mathbf{P}(\tau_1 > s) \longrightarrow 0, \quad cT\mathbf{P}(|v_1| > \delta T) \longrightarrow 0. \tag{1.89}$$

Since

$$T\mathbf{P}(\tau_1 > s) \leqslant \frac{T}{s}\gamma^2(s, T),$$

where $\gamma^2(s, T) = \mathbf{E}(\tau; \tau \geqslant s)$,

$$T\mathbf{P}(|v_1| > \delta T) \leqslant \frac{1}{\delta}\mathbf{E}(|v_1|; |v| > \delta T) \longrightarrow 0$$

then we obtain the required relations (1.88) and (1.89) if

$$\theta = T\sqrt{\gamma(s, T)}$$

and s is equal to the solution s_0 of the equation

$$s = T\gamma(s, T)$$

(to avoid trivial difficulties, we may suppose that $\gamma(s, T)$ is continuous in s and that $s_0 \to \infty$ as $T \to \infty$. The solution $s_0 = o(T)$ exists and is unique since $\gamma(s, T) \downarrow 0$ as $s \to \infty$).

Thus, the conditions of theorem 1 of section 1.4 are satisfied and theorem 1 is proved. □

The proof of theorem 2 is based on theorem 2 of section 1.5. Here the σ-algebra $\mathfrak{M}(u)$ and the set $\Omega_{u,\theta}$ will have their earlier meaning. By the Wald identity

$$\mathbf{E}\eta(t) = \frac{t + \mathbf{E}\chi(t)}{\mathbf{E}\tau}, \quad \mathbf{E}\xi(t) = \mathbf{E}\nu\mathbf{E}\eta(t) - At = A\mathbf{E}\chi(t)$$

where, for all t (see Mogul'skii, 1973),

$$\mathbf{E}\chi(t) \leqslant \frac{3}{2}\frac{\mathbf{E}\tau^2}{\mathbf{E}\tau} \leqslant c.$$

Hence and from (1.86) it follows that on $\Omega_{u,\theta}$

$$|\mathbf{E}_{\mathfrak{M}(u)} z_{u,\theta}| = O(s).$$

From the weak convergence of the distribution of (1.83) (together with the second moments) and from the representation (1.86) it follows that on $\Omega_{u,\theta}$, as $N = N(T) \to \infty$,

$$\mathbf{E}_{\mathfrak{M}(u)} z^2_{u,\theta} = O(s^2) + \theta$$

$$\mathbf{E}_{\mathfrak{M}(u)}(z^2_{u,\theta};\ |z_{u,\theta}| > N\sqrt{\theta}) = O(s^2) + o(\theta).$$

Thus, to ensure that the conditions of theorem 2 of section 1.5 hold, s and θ must satisfy

$$s = o\left(\frac{\theta}{\sqrt{T}}\right), \quad \theta = o(T). \tag{1.90}$$

In addition, we must have

$$\mathbf{P}\left(\bigcap_u \Omega_{u,\theta}\right) \geqslant \mathbf{P}\left(\max_{j \leqslant \eta(T)} \tau_j \leqslant s\right) \to 1$$

$$\mathbf{P}\left(\max_{j \leqslant \eta(T)} |\nu_j| < \delta\sqrt{T}\right) \to 1.$$

As before, the additional probabilities are estimated using the expressions

$$\mathbf{P}\left(\max_{j\leqslant cT}\tau_j > s\right) \leqslant cT\mathbf{P}(\tau_1 > s) \leqslant \frac{cT}{s^2}\,\mathbf{E}(\tau^2;\tau > s)$$
$$\mathbf{P}\left(\max_{j\leqslant cT}|\nu_j| > \delta\sqrt{T}\right) \leqslant \frac{c}{\delta^2}\,\mathbf{E}(|\nu^2|;|\nu| > \delta\sqrt{T}) \to 0. \tag{1.91}$$

To guarantee convergence to 0 in (1.91) and the satisfaction of (1.90), we put s equal to the solution of

$$s = \sqrt{T}\gamma(s, T), \quad \text{where } \gamma^3(s, T) = \mathbf{E}(\tau^2;\tau \geqslant s)$$

(under the same simplifying assumptions as before, which remove trivial difficulties). Then for $\theta = T\sqrt{\gamma(s, T)}$ (1.90) is obvious and the right side of (1.91) does not exceed $\gamma(s, T) \to 0$ as $s = s(T) \to \infty$. The theorem is proved. □

1.11.2. *A sum of generalized renewal processes.* In queuing problems we often have to deal with a 'sum' of some number of independent renewal processes $\xi^{(i)}(t)$, $i = 1, \ldots, m$, of the above form (say, for the description of the total work of m service channels, each controlled by sequences of independent variables, and with no stoppages).

It is not difficult to establish that the process

$$\xi(t) = \sum_{i=1}^{m} \xi^{(i)}(t), \quad \xi^{(i)}(t) = X^{(i)}(t) - A^{(i)}(t)$$

$$A^{(i)} = \frac{\mathbf{E}\nu^{(i)}}{\mathbf{E}\tau^{(i)}}$$

where $X^{(i)}(t)$ denotes the ith renewal process, will satisfy theorems 1 and 2 of the preceding part if the random variables $\tau^{(i)}$, $\nu^{(i)}$, which describe the ith process, satisfy the conditions of these theorems and if by σ^2 in theorem 2 we understand

$$\sigma^2 = \sum_{i=1}^{m}\left[\frac{\mathbf{D}\nu^{(i)}}{\mathbf{E}\tau^{(i)}} + \frac{\mathbf{E}^2\nu^{(i)}\mathbf{D}\tau^{(i)}}{\mathbf{E}^3\tau^{(i)}}\right] \geqslant c > 0. \tag{1.92}$$

Here the first of conditions (1.82) will be unnecessary.

The proof of this provides no difficulties and completely follows the path followed by the proofs of theorems 1 and 2. It is only necessary to take as $\mathfrak{M}(u)$ the product of the σ-algebras $\mathfrak{M}^{(i)}(u)$, corresponding to the terms $\xi^{(i)}(t)$, and put

$$\Omega_{u,\theta} = \bigcap_{i=1}^{m} \{\chi^{(i)}(u) \leqslant s\}$$

where $\chi^{(i)}(u)$ is the variable $\chi(u)$ for the process $\xi^{(i)}(t)$.

1.11.3. *Loaded autonomous service systems.* The results of the preceding two subsections allow us to immediately describe the behaviour of the queue length for loaded systems with so-called autonomous service (see Borovkov, 1976a).

We will assume that requests are received in the system in m_1 independent channels, in each of which through intervals of time $\tau_j^{(i)}$ requests proceed in groups of size $v_j^{(i)}$, $i = 1, \ldots, m_1; j = 1, 2, \ldots$. The service leads to m_2 channels, where in the ith channel (for notational convenience we will assume that i takes the values $m_1 + 1, \ldots, m_1 + m_2 = m$) the service begins only at epochs $0, \tau_1^{(i)}, \tau_1^{(i)} + \tau_2^{(i)}, \ldots$, independently of the input stream, and into service are received respectively $v_1^{(i)}$, $v_2^{(i)}, \ldots$ requests (or less, if the queue is not large enough; requests are assumed to form a queue pending service). If $\{\tau_j^{(i)}\}_{j=1}^{\infty}, \{v_j^{(i)}\}_{j=1}^{\infty}, i = 1, \ldots, m$, are made up of independent variables and are independent of each other, then the renewal process

$$X(t) = \sum_{i=1}^{m_1} X^{(i)}(t) - \sum_{i=m_1+1}^{m} X^{(i)}(t)$$

where $X^{(i)}(t)$ is generated by the sequence $\{\tau_j^{(i)}, v_j^{(i)}\}_{j=1}^{\infty}$ will be a process of the same type as considered in subsection 1.11.2.

If we assume for simplicity that the initial size of the queue $q(0)$ at time $t = 0$ is equal to 0 then, as shown in Borovkov (1976a), the size of the queue at time t, $q(t)$, will be

$$q(t) = X(t) - \inf_{[0,t]} X(u).$$

The loading of the system is characterized by the number

$$A = \sum_{i=1}^{m_1} \frac{\mathbf{E}v^{(i)}}{\mathbf{E}\tau^{(i)}} - \sum_{i=m_1+1}^{m} \frac{\mathbf{E}v^{(i)}}{\mathbf{E}\tau^{(i)}}$$

which, for large t and u, is approximately equal to

$$\frac{1}{u}[\mathbf{E}(X(t+u) - X(t))].$$

That is, obviously, the difference between the mean number of requests entering the system over a time u and the mean number of requests which the system can serve over this time under an infinite queue.

We will say that the system is in a *loaded state*, if in a series array (when all the controlling sequences $\{\tau_j^{(i)}\}, \{v_j^{(i)}\}$ depend on a parameter $T \to \infty$)

$$A = A(T) \to 0.$$

Consider the time interval $[0, U]$, $U = vT$, and suppose that $\{\tau_j^{(i)}, v_j^{(i)}\}$ satisfy the conditions of theorem 2 of subsection 1.11.1 (replacing the first condition of (1.82) by (1.92)). Then the process

$$y(t) = \frac{X(tT) - AtT}{\sigma\sqrt{T}}$$

will C-converge to a standard Wiener process. But

$$\frac{q(tT)}{\sigma\sqrt{T}} = \frac{1}{\sigma\sqrt{T}}\left[X(tT) - \inf_{[0,t]} X(uT)\right]$$

$$= y(t) + \frac{At\sqrt{T}}{\sigma} - \inf_{[0,t]}\left(y(u) - \frac{Au\sqrt{T}}{\sigma}\right).$$

Suppose now that as $T \to \infty$

$$A\sqrt{T} \to \alpha, \quad |\alpha| < \infty.$$

Then, since the functional $x(t) + at - \inf_{[0,t]}(x(u) + au)$ is continuous on C and measurable, we obtain that the distribution of $q(tC)/\sigma\sqrt{T}$ converges as $T \to \infty$ to the distribution of

$$w(t) + \frac{\alpha t}{\sigma} - \inf_{[0,t]}\left(w(u) - \frac{\alpha u}{\sigma}\right) \tag{1.92}$$

(or, what is the same thing, to the distribution of $\sup_{[0,t]}(w(u) + \alpha u/\sigma)$, if the question concerns only the one-dimensional distributions). Hence it is already easy to obtain an explicit formula for the distribution of $q(u)$ for large u (see Borovkov, 1976a; Skorokhod, 1961). In particular, if $A \to 0$ and $\alpha < 0$ and is large in absolute value, then, using $\mathbf{P}(w(u) < x + u, 0 < u < \infty) = 1 - \mathrm{e}^{-2x}$, we obtain

$$\mathbf{P}\left(q(T) > \frac{\sigma x}{|A|}\right) \approx \mathbf{P}\left(\frac{q(T)}{\sigma\sqrt{T}} > \frac{x}{\alpha}\right) \approx \mathbf{P}\left(\sup_{[0,1]}\left(w(u) + \frac{\alpha u}{\sigma}\right) > \frac{x}{\alpha}\right)$$

$$= \mathbf{P}\left(\sup_{[0,(\alpha^2/\sigma^2)]}(w(t) + t) > \frac{x}{\sigma}\right) \approx \mathrm{e}^{-(2x/\sigma)}.$$

(1.9.3) also holds for a system with the usual (not autonomous) service. However, the verification of the conditions of theorem 1 of section 1.5 in this case plays a much larger part (see, for example, section 3.6).

1.11.4. *The sum of renewal processes with increments depending on the renewal time.* Suppose in the example of subsection 1.11.2 that in the ith renewal process $v_j^{(i)}$ depends linearly on the intervals $\tau_j^{(i)}$. More precisely, let the process with index i be controlled by a sequence of independent vectors $(\tau_1^{(i)}, v_1^{(i)}), (\tau_2^{(i)}, v_2^{(i)}), \ldots$, where

$$v_j^{(i)} = b^{(i)}\tau_j^{(i)} + \Delta_j^{(i)}$$

$\Delta_j^{(i)}, j = 1, 2, \ldots$, are identically distributed and do not depend on $\tau_j^{(i)}$. The renewal processes $X^{(i)}(t)$, as before, are assumed to be independent of each other. Since

$$X^{(i)}(t) = v_1^{(i)} + \cdots + v_{(t)}^{(i)} = b^{(i)}(t + \chi(t) + \Delta_1^{(i)} + \cdots + \Delta_{(t)}^{(i)})$$

(here $\eta(t) = \min(k: \tau_1^{(i)} + \cdots + \tau_k^{(i)} > t)$, $\chi(t) = \tau_1^{(i)} + \cdots + \tau_{\eta(t)}^{(i)} - t)$, then it is easy to see that there is an analogue of theorems 1 and 2 of subsection 1.11.1 on C-convergence to 0 or to a diffusion process under corresponding conditions on $\{\tau_j^{(i)}\}$ and $\{\Delta_j^{(i)}\}$ (the condition on $v_j^{(i)}$ must be replaced by the same conditions on $\Delta_j^{(i)}$ and instead of (1.92) it is supposed that

$$\left(\sum\left[\frac{\mathbf{D}\Delta^{(i)}}{\mathbf{E}\tau^{(i)}} + \frac{\mathbf{E}^2\Delta^{(i)}\mathbf{D}\tau^{(i)}}{\mathbf{E}^3\tau^{(i)}}\right] \geqslant c > 0\right).$$

1.12. The connection between the conditions of theorem 1 of section 1.5 and strong mixing conditions

1.12.1. Stationary sequences. Let $\{\zeta_k\}_{-\infty}^{\infty}$ be a strictly stationary sequence,* $\mathbf{E}\zeta_k = 0$. satisfying the *uniformly strong mixing* (u.s.m.) condition

$$\sup_{\substack{A \in \mathfrak{M}_{-\infty}^k \\ B \in \mathfrak{M}_{k+n}^{\infty}}} |\mathbf{P}(B/A) - \mathbf{P}(B)| \leqslant \psi(n) \to 0 \tag{1.94}$$

as $n \to \infty$, where $\mathfrak{M}_k^l$, $l \geqslant k$, is the σ-algebra generated by $\zeta_k, \zeta_{k+1}, \ldots, \zeta_l$. This condition may also be written in the form

$$|\mathbf{P}_{\mathfrak{M}_{-\infty}^k}(B) - \mathbf{P}(B)| \leqslant \psi(n)$$

a.s. and for all $B \in \mathfrak{M}_{k+n}^{\infty}$.

We wish to show that u.s.m. together with the conditions

$$\mathbf{E}|\zeta_k|^{2+\delta} \leqslant c < \infty, \quad \delta > 0 \tag{1.95}$$

(to simplify the calculations we somewhat overestimate here the requirements on moments) and

$$\mathbf{E}Z_n^2 = n\sigma^2 + o(n), \quad 0 < \sigma < \infty, \quad Z_n = \zeta_1 + \cdots + \zeta_n \tag{1.96}$$

implies the conditions of theorem 1 of section 1.5 for $\xi(t) = Z_{[t]}$. The converse, naturally, is not valid since the conditions of theorem 1 of section 1.5 do not, in general, imply the existence of the moments of ζ_k and their sums. The weak dependence is required only in a very special form—for the first two moments of the increments. Thus, the condition of u.s.m. together with (1.95) and (1.96) is stronger than the condition of weak dependence in theorem 1 of section 1.5.

A similar picture obviously holds for processes $\zeta(t)$ with continuous time.

Let ξ be measurable relative to $\mathfrak{M}_{-\infty}^k$ and let η be measurable relative to $\mathfrak{M}_{k+n}^{\infty}$. Then, if

$$\mathbf{E}|\xi|^p < \infty, \quad \mathbf{E}|\eta|^q < \infty; \quad p, q > 1, \quad \frac{1}{p} + \frac{1}{q} = 1$$

* All of this discussion holds for a sequence $\{\zeta_k\}_0^{\infty}$.

it follows from (1.94) that (see Ibragimov and Linnik, 1965)

$$|\mathbf{E}\xi\eta - \mathbf{E}\xi\mathbf{E}\eta| \leqslant 2\psi^{1/p}(n)\mathbf{E}^{1/p}|\xi|^p\mathbf{E}^{1/q}|\eta|^q. \tag{1.97}$$

Hence, in particular, we find the following estimate for the correlation function $R(k) = \mathbf{E}\zeta_j\zeta_{j+k}$:

$$\text{if } \mathbf{E}|\zeta_j|^2 < \infty, \quad \text{then } |R(k)| \leqslant 2\sqrt{\psi(k)}\mathbf{E}|\zeta_j|^2.$$

Since (see Ibragimov and Linnik, 1965)

$$\mathbf{E}Z_n^2 = nR(0) + 2\sum_{j=1}^{n}(n-j)R(j) = \sigma^2 n(1 + o(1))$$

where

$$\sigma^2 = R(0) + 2\sum_{j=1}^{\infty} R(j)$$

we obtain that for (1.96) to hold it is sufficient that the series

$$\sum_{k=1}^{\infty} \psi^{\frac{1}{2}}(k)$$

converge and that σ be positive.

If (1.95) is satisfied then, putting $q = 2 + \delta$, $\xi = \zeta_j$, $\eta = \zeta_{j+k}$ in (1.97), we obtain as a sufficient condition for (1.96) the convergence of the series

$$\sum_{k=1}^{\infty} \psi^{1+\delta/2+\delta}(k) < \infty.$$

In the above result on the sufficiency of (1.94)–(1.96) for the weak dependence condition of theorem 1 of section 1.5 we may also suppose that a *series array* is considered, where the distribution of $\{\zeta_k\}_{-\infty}^{\infty}$ depends on some parameter (for example on n in the above theorem). The passage to a series array must be accompanied by a requirement of *uniformity* in conditions (1.94)–(1.96).

We denote

$$z_{k,m} = Z_{k+m} - Z_k, \quad \mathfrak{M}_{-\infty}^{k} = \mathfrak{M}(k).$$

THEOREM 1. *If conditions* (1.94)–(1.96) *are satisfied then there exist a sequence* $m = o(n)$ *and sets* $\Omega_k \in \mathfrak{M}(k)$ *satisfying*

$$\mathbf{P}\left(\bigcap_{0 \leqslant jm \leqslant n} \Omega_{jm}\right) \to 1$$

as $n \to \infty$, *on which*

(I) $\mathbf{E}_{\mathfrak{M}(k)} z_{k,m} = \dfrac{m}{\sqrt{n}} r_1(\omega)$

(II) $\mathbf{E}_{\mathfrak{M}(k)} z_{k,m}^2 = m\sigma^2(1 + r_2(\omega))$

(III) *for any* $N = N(n) \to \infty$

$$\mathbf{E}_{\mathfrak{M}(k)}(z_{k,m}^2; |z_{k,m}| > N\sqrt{m}) \leqslant \varepsilon m$$

where $|r_1(\omega)| < \varepsilon$, $|r_2(\omega)| < \varepsilon$, ε *an arbitrary fixed number.*

It is obvious that these conditions will imply the conditions of theorem 1 of section 1.5 for $a = 0$, $b = \sigma^2$.

It is easy to see that in our case the stationarity of $\{\zeta_k\}$ and the possibility of choosing m so that $M = n/m$ grows arbitrarily slowly permits some simplification of the form of conditions (I)–(III) by removing the two parameters m and n. Namely, these conditions will be satisfied if there is a set $\Omega = \Omega(n)$, with the property $\mathbf{P}(\Omega) \to 1$, on which

(I′) $|\mathbf{E}_{\mathfrak{M}(0)} Z_n| < \sqrt{n} r_1(\omega)$

(II′) $\mathbf{E}_{\mathfrak{M}(0)} Z_n^2 = n\sigma^2(1 + r_2(\omega))$

(III′) for any $N = N(n) \to \infty$

$$\mathbf{E}_{\mathfrak{M}(0)}(Z_n^2; |Z_n| > N\sqrt{n}) < \varepsilon n$$

where $r_i(\omega)$ and ε have their former meaning.

To obtain (I)–(III) from (I′)–(III′) we divide $[1, n]$ into a finite number $M = n/m$ of equal segments $[1, m], [m + 1, 2m], \ldots$ (assuming, for simplicity, that M is an integer). On each of these segments we construct a set Ω on which (I′)–(III′) hold (changing the initial reading from n to m). We then obtain that (I)–(III) will be satisfied for any finite $M = n/m$, and, consequently, for sufficiently slowly growing $n/m \to \infty$.

For the proof of (I′)–(III′) we will use the following result (Ibragimov and Linnik, 1965).

THEOREM 2. *Under conditions* (1.94)–(1.96) *the central limit theorem applies to* Z_n:

$$\lim_{n\to\infty} \mathbf{P}\left(\frac{Z_n}{\sigma\sqrt{n}} < x\right) = \frac{1}{\sqrt{2\pi}} \int_{-\infty}^{x} \mathrm{e}^{-u^2/2}\, du.$$

Here there is a constant c for which

$$\mathbf{E}|Z_n|^{2+\delta} < c^{2+\delta} n^{1+\delta/2}. \tag{1.98}$$

We also need the following inequality, which follows from (1.97). If η is measurable relative to $\mathfrak{M}_k^{\infty}$, $A \in \mathfrak{M}_{-\infty}^0$, $\mathbf{P}(A) > 0$, then

$$|\mathbf{E}(\eta/A) - \mathbf{E}\eta| \leqslant 2\mathbf{P}(A)^{-1/q}(\mathbf{E}|\eta|^q)^{1/q}\psi(k)^{1/p}. \tag{1.99}$$

We obtain this at once if we put $\xi = I(A)$ in (1.97) and divide both sides of (1.97) by $\mathbf{P}(A)$.

We prove initially the satisfaction of (I′). For $m < n$ we have

$$\mathbf{E}_{\mathfrak{M}(0)} Z_n = \mathbf{E}_{\mathfrak{M}(0)} Z_m + \mathbf{E}_{\mathfrak{M}(0)} z_{m,n-m}. \tag{1.100}$$

Here the first term, which we denote by ξ_1, has a moment of order $2 + \delta$:

$$\mathbf{E}|\xi_1|^{2+\delta} \leqslant \mathbf{E}|Z_m|^{2+\delta} \leqslant c\sigma^{2+\delta} m^{1+\delta/2}$$

therefore

$$\mathbf{P}(\xi_1 \geqslant \sqrt{n}\varepsilon(n)) \leqslant c^{2+\delta}\left(\frac{m}{n}\right)^{1+\delta/2} \varepsilon(n)^{-2-\delta}. \tag{1.101}$$

The function $\varepsilon(n) = o(1)$ will be chosen later.

For the estimate of the second term in (1.100), which we denote by η_1, we use (1.99) for $q = 2 + \delta$.

For any $A \in \mathfrak{M}(0)$, $\mathbf{P}(A) > 0$ we obtain

$$\begin{aligned} \left|\frac{1}{\mathbf{P}(A)} \int_A \eta_1 \, \mathbf{dP}(\omega)\right| &= |\mathbf{E}(\eta_1/\mathrm{A})| \\ &\leqslant 2\mathbf{P}(A)^{-1/(2+\delta)} (\mathbf{E}|Z_{n-m}^{2+\delta}|)^{1/(2+\delta)} \psi(m)^{(1+\delta)/(2+\delta)} \\ &\leqslant 2c^{1/(2+\delta)} \sigma \mathbf{P}(A)^{-1/(2+\delta)} \sqrt{n} \psi(m)^{(1+\delta)/(2+\delta)}. \end{aligned} \tag{1.102}$$

We take as A any set A_1 for which

$$\mathbf{P}(A_1) = \psi(m)^{\delta/2}$$

and on which the values of η_1 are greater than or equal to the values of η_1 on the complement $\bar{A}_1$ (without loss of generality we may suppose that such a set always exists). Since from (1.102) it follows that the mean value of η_1 on A_1 does not exceed

$$2c^{1/(2+\delta)} \sigma \sqrt{n\psi(m)} \equiv \gamma(n)$$

then this means that $\eta_1 \leqslant \gamma(n)$ on $\bar{A}_1$.

We now choose as $\Omega \in \mathfrak{M}(0)$ the set $\Omega_1 = \bar{A}_1 \cap \{\xi_1 < \sqrt{n\varepsilon(n)}\}$. Then by (1.101)

$$\mathbf{P}(\bar{\Omega}_1) \leqslant \psi(m)^{\delta/2} + c\sigma^{2+\delta}\left(\frac{m}{n}\right)^{1+\delta/2} \varepsilon(n)^{-2-\delta} \tag{1.103}$$

and for the proof of (I′) (that is, $\xi_1 + \eta_1 = o(\sqrt{n})$ on Ω and $\mathbf{P}(\bar{\Omega}) \to 0$) we need to show that the right-hand side of (1.103) converges to 0, $\varepsilon(n) \to 0$, $m \to \infty$, as $n \to \infty$. This can be achieved for any sequence $m = m(n) \to \infty$, $m = o(n)$, by putting

$$\varepsilon(n) = \left(\frac{m}{n}\right)^{\frac{1}{4}}.$$

We turn now to condition (II′). We have

$$\mathbf{E}_{\mathfrak{M}(0)} Z_n^2 = \mathbf{E}_{\mathfrak{M}(0)} Z_m^2 + 2\mathbf{E}_{\mathfrak{M}(0)} z_{m,n-m} Z_m + \mathbf{E}_{\mathfrak{M}(0)} z_{m,n-m}^2. \tag{1.104}$$

Here, as above, for the first term, which we denote by ξ_2, we obtain

$$\mathbf{E}\xi_2^{1+\delta/2} \leqslant \mathbf{E}|Z_m|^{2+\delta}$$

$$\mathbf{P}(\xi_2 > n\varepsilon(n)) \leqslant c\sigma^{2+\delta}\left(\frac{m}{n}\right)^{1+\delta/2} \varepsilon(n)^{-1-\delta/2}. \tag{1.105}$$

To estimate the last term in (1.104), we denote it by η_2 and consider the inequality (see (1.99) for $q = 1 + \delta/2$)

$$|\mathbf{E}(\eta_2/A) - \mathbf{E}\eta_2| \leqslant 2\mathbf{P}(A)^{-2/(2+\delta)}c^{2/(2+\delta)}\sigma^2 n\psi(m)^{\delta/(2+\delta)}.$$

The principle for the choice of A remains as before and its probability can be put equal to

$$\mathbf{P}(A) = \psi(m)^{\delta/4}. \tag{1.106}$$

We denote the set chosen by A_2. This will mean that on $\bar{A}_2$ the value of η_2 differs from $\mathbf{E}\eta_2 = \mathbf{E}z^2_{m,n-m} = \sigma^2(n-m)(1+o(1))$ by no more than

$$2c^{2/(2+\delta)}\sigma^2 n\psi(m)^{\delta/2(2+\delta)}.$$

Put $\Omega_2 = \bar{A}_2 \cap \{\xi_2 < n\varepsilon(u)\}$: we construct the required set since on it $\xi_2 + 2\sqrt{\xi_2\eta_2} = o(n)$, $\eta_2 = \sigma^2 n + o(n)$ and, consequently,

$$\mathbf{E}_{\mathfrak{M}(0)} Z_n^2 = \sigma^2 n(1 + o(1)).$$

It is necessary only that $m \longrightarrow \infty$, $m = o(n)$, and $\varepsilon(n) = o(1)$ are chosen so that the upper estimate for $\mathbf{P}(\bar{\Omega}_2)$ (see (1.105), (1.106))

$$\psi(m)^{\delta/4} + c^{2+\delta}\left(\frac{m}{n}\right)^{1+\delta/2} \varepsilon(n)^{-1-\delta/2} \longrightarrow 0.$$

For this it is sufficient to put

$$\varepsilon(n) = \left(\frac{m}{n}\right)^{\frac{1}{2}}.$$

It remains for us to verify condition (III′). This will follow easily from (II′) and the existence of a set Ω_3, $\mathbf{P}(\Omega_3) \longrightarrow 1$, on which

$$\lim_{n\to\infty} \mathbf{P}_{\mathfrak{M}(0)}\left(\frac{Z_n}{\sigma\sqrt{n}} < x\right) = \frac{1}{\sqrt{2\pi}} \int_{-\infty}^{x} \mathrm{e}^{-u^2/2}\,\mathrm{d}u \tag{1.107}$$

(then, obviously, for any N, on Ω_3

$$\mathbf{E}_{\mathfrak{M}(0)}\left(\frac{Z_n^2}{\sigma^2 n}; |Z_n| < \sigma N\sqrt{n}\right) \longrightarrow \frac{1}{\sqrt{2\pi}} \int_{-N}^{N} u^2\,\mathrm{e}^{-u^2/2}\,\mathrm{d}u\Bigg).$$

Together with (II′) this means that (III′) holds on $\Omega_3' = \Omega_2 \cap \Omega_3$.

The proof of (1.107) is similar to the preceding (with some simplifications), and we leave it to the reader.

If we now take as Ω the set

$$\Omega = \bigcap_{j=1}^{3} \Omega_j$$

then $\mathbf{P}(\Omega) \to 1$ and (I′)–(III′) will be satisfied on it. The theorem is proved. □

Remark 1. The condition of *strong mixing*, weaker than (1.94),

$$\sup_{\substack{A \in \mathfrak{M}_{-\infty}^{k} \\ B \in \mathfrak{M}_{k+n}^{\infty}}} |\mathbf{P}(AB) - \mathbf{P}(A)\mathbf{P}(B)| \leqslant \psi(n) \to 0$$

as $n \to \infty$, also implies the conditions of theorem 1 of section 1.5 if, in addition to (1.95) and (1.96), we require (or prove) (1.98) and the inequality $\psi(u) \leqslant cn^{-\gamma}$ for a suitable $\gamma > 0$ (see Ibragimov and Linnik, 1965).

Remark 2. The result on C-convergence of $\xi(tn)/\sigma\sqrt{n}$ to a Wiener process under conditions (1.94)–(1.96) is, apparently, simpler to prove not using theorem 2 of section 1.5, but by verification of a sufficient condition of the Kolmogorov–Prokhorov type (see (1.12)). In fact, in the fundamental Chebyshev inequality it is sufficient to show that for all $k \gg 1$

$$\mathbf{E}\left|\frac{1}{\sqrt{n}} Z_k\right|^{\alpha} \leqslant c\left(\frac{k}{n}\right)^{1+\beta}$$

for some $\alpha > 0, \beta > 0$ (we suppose $F(t) = t$). This inequality, by (1.98), will be satisfied if we put $\alpha = 2 + \delta$. Then

$$\mathbf{E}\left|\frac{1}{\sqrt{n}} Z_k\right|^{2+\delta} \leqslant n^{-1-\delta/2} c\delta^{2+\delta} k^{1+\delta/2} = c\sigma^{2+\delta}\left(\frac{k}{n}\right)^{1+\delta/2}.$$

The same type of remark could be made regarding the proof of convergence of finite-dimensional distributions (the investigation of this convergence using theorem 2 would be simpler direct rather than by proving theorem 1). However, our aim is the clarification of the relation between u.s.m. and the conditions of section 1.5.

1.12.2. *The renewal process generated by a sequence satisfying the uniform strong mixing condition.* The previous example can be considered as a generalized renewal process (see example 1 of section 1.11), for which $\tau_k = 1$ and v_k are dependent. Now let $v_k = 1$ and τ_k be dependent and satisfy the u.s.m. condition.

Namely, we consider a step-process $\eta(u)$ with stationary increments such that the intervals of constancy (or the intervals between the jumps of $\eta(u)$) have the same distribution as the variables τ_j from a sequence $\{\tau_j\}$ which satisfies the u.s.m. condition. To make this precise we introduce the jump variable $\chi(t)$, which is equal to the distance from t to the epoch of the first jump of $\eta(u)$ after t (if there is a jump at t then we put $\chi(t) = 0$). We denote the intervals of constancy subsequent to $t + \chi(t)$ by $\tau_1^*, \tau_2^*, \ldots$ *We will assume* that for any t the sequence* $\{\tau_j^*\}_{j=1}^{\infty}$ *is distributed the same as* $\{\tau_j\}_{j=1}^{\infty}$.

It is interesting to note that the increments of this process $\eta(u)$ will no longer, in general, satisfy u.s.m. This is the case even when the τ_k are independent. In fact, let, for example, $\mathbf{P}(\tau_k > x) = (1+x)^{-\alpha}$, $x \geqslant 0$, $\alpha > 1$. Then $\chi(t)$, for any t, has the distribution

$$\mathbf{P}(\chi(t) > x) = \int_x^{\infty} \frac{\mathbf{P}(\tau_1 > u)}{\mathbf{E}\tau_1}\,du = (1+x)^{-\alpha+1}$$

and $\eta(u)$, for $u > 0$, can be described in the form

$$\eta(u) = \min\{k: \chi(0) + \tau_1 + \cdots + \tau_k > u\}, \quad u \geqslant 0$$

where $\chi(0)$ and $\{\tau_j\}$ are independent. Put $A = \{\chi(0) > n\}$, $B = \{\eta(3n) - \eta(2n) = 0\}$, and denote, as in the previous subsection, by $\mathfrak{M}_a^i$ the σ-algebra generated by the $\eta(u)$ for $u \in [a, b]$. Then

$$A \in \mathfrak{M}_{-\infty}^n, \qquad B \in \mathfrak{M}_{2n}^{\infty}, \quad \mathbf{P}(B) = \mathbf{P}(\eta(n) = 0) \to 0$$

as $n \to \infty$, whereas

$$\mathbf{P}(B/A) \geqslant \mathbf{P}(\chi(0) > 3n/\chi(0) > n) = \frac{\mathbf{P}(\chi(0) > 3n)}{\mathbf{P}(\chi(0) > n)}$$

$$= \left(\frac{1+3n}{1+n}\right)^{-\alpha+1} \to 3^{-\alpha+1}.$$

This means that the difference $\mathbf{P}(B/A) - \mathbf{P}(B)$ does not converge to zero as $n \to \infty$ and condition (1.94) for $\eta(u)$ is not satisfied. Thus, we cannot use the results of subsection 1.12.1 to establish the conditions of theorem 1 of section 1.5 for the renewal process, described in the title of this section.

We will show that these conditions, nevertheless, are satisfied.

To characterize the weak dependence of $\eta(u)$ with the help of a sequence $\{\tau_j\}$ we change condition (1.94) in the following way. Let $\mathfrak{M}^*(t)$ be the σ-algebra generated by $\mathfrak{M}_{-\infty}^t$ and the variable $\chi(t)$, and let $\mathfrak{M}_n^*$ be the σ-algebra generated by $\tau_n^*, \tau_{n+1}^*, \ldots$; then, for $n \to \infty$

$$\sup_{\substack{A \in \mathfrak{M}^*(t) \\ B \in \mathfrak{M}_n^*}} |\mathbf{P}(B/A) - \mathbf{P}(B)| \leqslant \psi(n) \to 0. \tag{1.108}$$

In addition, we will suppose, as before, that τ_j (or τ_j^*) satisfy

$$\mathbf{E}\tau_j^{2+\delta} \leqslant c < \infty, \quad \mathbf{D}(\tau_1 + \cdots + \tau_n) = n\sigma^2(1 + o(1)) \tag{1.109}$$

where $0 < \sigma < \infty$.

* It is not clear for which stationary sequences $\{\tau_j\}$ with finite $\mathbf{E}\tau_j$ (apart from sequences with independent elements) there is a process $\eta(u)$ with such properties.

THEOREM 3. *Under conditions* (1.108), (1.109) *the process* $\eta(u)$ *satisfies the conditions of theorem 1 of section 1.5.*

This means that $\eta(u)$ in a large interval of time will converge, after normalization, to a Wiener process.

Remark. If we consider a series array then we must impose additional conditions on $\{\tau_j\}$ which guarantee the uniform convergence to 0 as $N \to \infty$ of the probability

$$\mathbf{P}(\chi(t) > N).$$

It is possible, for example, to give the following conditions (see Borovkov, 1967b) which guarantee such decrease.

1. There exist constants $\alpha > 0$ and $\beta < 1$ such that almost everywhere

$$\mathbf{P}_{\mathfrak{M}(0)}(\tau_1 < \alpha) < \beta.$$

2. There is a function $P(x)$, which does not depend on the index of the series, for which

$$\int_0^\infty P(x)\,\mathrm{d}x < \infty, \quad \mathbf{P}_{\mathfrak{M}(0)}(\tau_1 > x) \leqslant P(x).$$

Here $\mathfrak{M}(0)$ is the σ-algebra generated by $\tau_0, \tau_{-1}, \ldots$.

Proof of theorem 3. By the stationarity of the increments of $\eta(u)$ and the remarks made in subsection 1.12.1 (see conditions (I′)–(III′)) it is sufficient for us to show that there is a set $\Omega = \Omega(n)$, $\mathbf{P}(\Omega) \to 1$, on which

$$\text{(I'')} \quad \left|\mathbf{E}_{\mathfrak{M}^*(0)}\left(\eta(t) - \frac{t}{a}\right)\right| < \sqrt{t r_1(\omega)}$$

$$\text{(II'')} \quad \mathbf{E}_{\mathfrak{M}^*(0)}\left(\eta(t) - \frac{t}{a}\right)^2 = \frac{t\sigma^2}{a^3}(1 + r_2(\omega))$$

$$\text{(III'')} \quad \mathbf{E}_{\mathfrak{M}^*(0)}\left(\eta(t) - \frac{t}{a}\right)^{2+\delta'} < ct^{1+\delta'/2}, \quad \delta' > 0$$

where $a = \mathbf{E}\tau_j$, $c = \text{const.}$, $|r_i(\omega)| < \varepsilon$ as $t \to \infty$ for any preassigned $\varepsilon > 0$.

But, as is easy to see by repeating the arguments of subsection 1.12.1, there is a set Ω^* in $\mathfrak{M}^*(0)$, $\mathbf{P}(\Omega^*) \to 1$, on which the conditional distribution of $(1/\sigma\sqrt{n})Z_n$ for $Z_n = \tau_1^* + \cdots + \tau_n^* - an$ will converge as $n \to \infty$ to normal together with the moments of order $2 + \delta'$, $\delta' < \delta$, $\delta' > 0$ (see subsection 1.12.1 and theorem 2).

We now use the following well-known method of investigation of the distribution of $\eta(u)$: consider the equality

$$\mathbf{P}_{\mathfrak{M}^*(0)}\left(\eta(t) - \frac{t}{a} < \frac{x\sigma\sqrt{t}}{a^{3/2}}\right) = \mathbf{P}_{\mathfrak{M}^*(0)}(\chi(0) + Z_n + an > t) \tag{1.110}$$

where

$$n = \left[\frac{t}{a} + \frac{x\sigma\sqrt{t}}{a^{3/2}}\right].$$

Here $[\alpha]$ denotes the integer part of α. Take $\Omega' = \{\omega: \chi(0) \leqslant y(n)\} \in \mathfrak{M}^*(0)$, where $y(n) \to \infty$ as $n \to \infty$, $y(n) = o(\sqrt{n})$. Then, obviously, $\mathbf{P}(\Omega') \to 1$ and on $\Omega^* \cap \Omega'$ the probability (1.110) is equal to

$$\mathbf{P}_{\mathfrak{M}^*(0)}(Z_n > -x\sigma\sqrt{n} + o(\sqrt{n})) \to 1 - \Phi(-x) = \Phi(x) = \frac{1}{\sqrt{2\pi}} \int_{-\infty}^{x} e^{-u^2/2}\, du.$$

In addition, the convergence of the left-hand side of (1.110) to normal on $\Omega^* \cap \Omega'$ will hold, together with the moments of order $2 + \delta'$ for some $\delta' > 0$, $\delta' < \delta$. In fact, consider the sums

$$\begin{aligned} &\sum_{k \geqslant 1} k^{1+\delta'} \mathbf{P}_{\mathfrak{M}^*(0)}\left(\eta(t) - \frac{t}{a} > k \right) \\ &\sum_{1 \leqslant k \leqslant t/a} k^{1+\delta'} \mathbf{P}_{\mathfrak{M}^*(0)}\left(\eta(t) - \frac{t}{a} < -k \right) \end{aligned} \tag{1.111}$$

which estimate at the same time the conditional moment of order $2 + \delta'$. By relations of the form (1.110)

$$\sum_{k \geqslant 1} k^{1+\delta'} \mathbf{P}_{\mathfrak{M}^*(0)}\left(\eta(t) - \frac{t}{a} > k \right) = \sum_{k \geqslant 1} k^{1+\delta'} \mathbf{P}_{\mathfrak{M}^*(0)}(Z_n < -ak + o(\sqrt{n}))$$

where $n = [t/a + k]$. By Chebyshev's inequality, the part of the sum for values $k \leqslant c_1 t$, for large enough t, does not exceed

$$2 \sum_{1 \leqslant k \leqslant c_1 t} k^{1+\delta'} \frac{\sigma^{2+\delta}(t/a + k)^{1+\delta/2}}{(ak)^{2+\delta}} \leqslant c t^{1+\delta/2}. \tag{1.112}$$

To estimate $\sum_{k > c_1 t}$, we construct a new renewal process $\tilde{\eta}(u)$, for which the intervals $\tilde{\tau}_j$ between jumps are $\min(A, \tau_j^*)$. It is clear that $\tilde{\eta}(u) \geqslant \eta(u)$ and that $\tilde{\tau}_j$ has finite moments of any order, in particular of order $4 + 2\delta$. Therefore, proceeding as in (1.112), for the remainder of the sum we obtain the majorant

$$c_2 \sum_{k > c_1 t} k^{1+\delta'} \frac{(t + k)^{2+\delta}}{k^{4+2\delta}} \leqslant c_3.$$

In a similar, but somewhat simpler, way we estimate the second sum in (1.111). Thus condition (III′) is proved.

Conditions (I′) and (II′) are now proved very simply with the help of (III′) and the convergence of (1.110) on $\Omega = \Omega^* \cap \Omega'$ to normal. In fact, on the constructed set

$$\begin{aligned} \mathbf{E}_{\mathfrak{M}^*(0)}\left(\eta(t) - \frac{t}{a} \right)^2 = {} & \mathbf{E}_{\mathfrak{M}^*(0)}\left[\left(\eta(t) - \frac{t}{a} \right)^2 ; \left| \eta(t) - \frac{t}{a} \right| < N\sqrt{t} \right] \\ & + \mathbf{E}_{\mathfrak{M}^*(0)}\left[\left(\eta(t) - \frac{t}{a} \right)^2 ; \left| \eta(t) - \frac{t}{a} \right| \geqslant N\sqrt{t} \right]. \end{aligned}$$

The first term here has the form $(t\sigma^2/a^3)(1 + r(\omega))$, where $|r(\omega)|$, by choice of N, can be made less than any preassigned $\varepsilon > 0$. The second term does not exceed

$$\frac{1}{(N\sqrt{t})^{\delta'}} \cdot \mathbf{E}_{\mathfrak{M}^*(0)}\left(\eta(t) - \frac{t}{a}\right)^{2+\delta'} \leqslant \frac{ct}{N^{\delta'}}$$

and also, by the choice of N, can be made less than εt.

Condition (I″) is verified in a similar way. The theorem is proved. □

It is easy to verify that theorem 3 remains valid for a *nonstationary* renewal process defined by

$$\eta(u) = \min\{k\colon \tau_1 + \cdots + \tau_k > u\}.$$

1.12.3. In the previous two subsections we have considered stationary processes with the u.s.m. condition and renewal processes generated by sequences with the u.s.m. condition. However, other conditions can be given, connected, say, only with the moments of the initial random variables, which are sufficient to imply the conditions of theorem 1 of section 1.5.

It is not difficult to see, for example, that if the sequence of random variables $\{\zeta_j\}_{j=-\infty}^{\infty}$ is stationary and has the properties

(I) $|\mathbf{E}_{\mathfrak{M}(0)}\zeta_k| < \phi_1(k), \quad \sum_{k=1}^{n} \phi_1(k) = o(\sqrt{n})$

(II) $|\mathbf{E}_{\mathfrak{M}(0)}\zeta_k\zeta_{k+j}| < \phi_2(j), \quad \phi_2 = \sum_{j=1}^{\infty} \phi_2(j) < \infty$

(III) $|\mathbf{E}_{\mathfrak{M}(0)}\zeta_{j-l}\zeta_j\zeta_k\zeta_{k+m}| \leqslant c \min(\phi_3(l), \phi_3(m))$

$$\phi_3 = \sum_{l=1}^{\infty} l\phi(l) < \infty$$

then the conditions of theorem 1 of section 1.5 will be satisfied (from (II) and (III) a direct calculation shows (cf. Borovkov, 1976a), that for $Z_n = \zeta_1 + \cdots + \zeta_n$

$$\mathbf{E}_{\mathfrak{M}(0)}Z_n^2 \sim n(\phi_2(0) + 2\phi_2)$$

$$\mathbf{E}_{\mathfrak{M}(0)}Z_n^4 \leqslant 3cn^2(8\phi_3 + 1)).$$

In all the above examples we have dealt with convergence to a spatially homogeneous diffusion (the coefficients $a(x)$ and $b(x)$ are constant). Examples of convergence to a nonhomogeneous diffusion are obtained in sections 3.4–3.6.

In this section we have not considered the connection between the u.s.m. condition and the conditions of theorem 1 of section 1.4 on C-convergence to a degenerate process. We leave this to the reader who may find the necessary conditions in terms of u.s.m. by comparison of the discussion of sections 1.4 and 1.12.

However, it must be noted that for a sum of stationarily connected variables $\zeta_1, \zeta_2, \ldots; \mathbf{E}\zeta_n = a$, C-convergence of

$$y_n(t) = \frac{Z[nt]}{n}, \quad 0 \leqslant t \leqslant v, \quad Z_k = \zeta_1 + \cdots + \zeta_k$$

to the function at follows immediately from the ergodicity of the sequence $[\zeta_j]$ (see the remarks at the beginning of section 1.4).

CHAPTER 2

Limit Theorems for Systems with Intensive Input Stream and a Large Number of Service Channels

We pass on now directly to the study of service processes. As already noted in the introduction, there are two possible approaches to the definition of service processes each with its own directions of research. The first of these is based on the traditional ideas; it is explicitly or implicitly used in almost all the articles and books on queuing theory (for some remarks on this see Borovkov, 1976a). In the definition of the service process here appear the so-called controlling sequences (or processes), composed of random variables, and a service algorithm. Such a definition will inevitably be mainly descriptive, since, under these conditions, a complete formal construction of a random process, uniquely describing the evolution of the system, would require the introduction of a complicated phase space and that is not speaking of the difficulties of effectively assigning the distributions of the process itself.

Nevertheless, this approach to the definition and study of queuing systems must be acknowledged as fundamental and we will use it for the description of many real systems.

However, there is another approach to the definition of a service process, which turns out to be much more convenient in clarifying the general limit laws which are inherent in a service system. In this approach the service process is considered as a vector-valued probability process with nondecreasing step-components

$$\mathbf{S}(t) = (e(t), r(t), s(t))$$

which characterize respectively *the input process, the number of calls which are refused, and the number of calls which are served up to the epoch t*. It turns out that, if the components of $\mathbf{S}(t)$ satisfy very broad assumptions, which are simple enough to verify in concrete problems, then the existence of interesting asymptotic laws for the service systems may be established. This is possible when some characteristics of the system (the queue length, the number of busy lines, the intensity of the input stream, etc.) are unboundedly increasing. The limit laws obtained under these conditions can be used effectively in the study of very complex service systems, including systems with control. For example, for the so-called busy systems (see Barrer, 1957; Blomqvist, 1973; Borovkov, 1966; Borovkov, 1967a, 1972e, 1973, 1980; Cohen, 1973; Gaver, 1968; Iglehart, 1965, 1973; Iglehart and Whitt, 1970; Kac, 1947; Krupin, 1976; Skorokhod, 1961, 1964; Viskov and Prokhorov, 1964) the distribution of the queue length, suitably normalized, is, under very broad assumptions, well approximated by the distribution of a diffusion process.

In subsequent sections it will often be convenient to use precisely this simpler and more general definition of service process which is unconnected with many of the traditional assumptions and ideas and which carry, as a rule, a partial character. Here we restrict the discussion to the 'one step' system, where requests are indistinguishable and are served once.

Thus, *by a service process on a time interval* $[0, U]$ *we will mean an arbitrary three-dimensional step process*

$$\mathbf{S}(t) = \{e(t), r(t), s(t); 0 \leqslant t \leqslant U\}; \quad r(0) = s(0) = 0,$$

whose components do not decrease, and have the property

$$q(t) = e(t) - r(t) - s(t) \geqslant 0.$$

$e(t)$ describes the input stream so that the difference $e(t) - e(u)$ is equal to the number of requests joining the system between the time t and u. Similarly, $r(t)$ describes the number of requests which are refused and $s(t)$ the number of requests served by the system up to t. $q(t)$ is the '*occupation*' or '*queue*' of the system. The notations e, r, s, and q are formed from the first letters of the English words *entrance*, *refusal*, *service*, and *queue*. It is obvious that for ordinary queuing systems $r(t) \equiv 0$, and for systems with refusals $q(t) = e(t) - r(t) - s(t)$ is the number of busy service lines.

The process $\mathbf{S}(t)$ may be considered as given on the space $D^3(0, U)$ with the σ-algebra generated by the cylinder sets. Where necessary we will suppose that the components of $\mathbf{S}(t)$ are continuous from the right. All characteristics of the system which are usually studied may be considered as measurable functionals of $\mathbf{S}(t)$. We note that the basic ones are the queue process $\{q(t)\}$ and the 'refusal process' $\{p(t) = r(t)/e(t)\}$.

In concrete systems the joint distribution of the components e, r, and s is usually given by local properties (the distribution of service time, time between incoming requests, and so on), and also by algorithms which describe the nature of the queuing process.

We have already remarked that asymptotic regularity, as a rule, has a collective nature. This concerns not only the nature of the controlling processes but also the algorithms defining the type of the system. As an illustration we note the following. It is well known that one of the sources of the variety of service systems, and the complications arising in their study, is the presence of a large number of distinct algorithms which determine the behaviour of the systems near the 'boundary' $e - r - s = 0$. But, it is to just this behaviour 'close to zero', as we shall see in this chapter and in chapter 3, that the asymptotic results are not sensitive, and are almost independent of it. This is one of the important circumstances which allow us to develop *general* methods of asymptotic analysis in service theory, that is, methods which are applicable for very different types of systems and controlling sequences.

Thus we will study *asymptotic laws* for service processes. Therefore, in all that follows, in discussing $\mathbf{S}(t)$ we will have in mind a *series array*, that is, a family (sequence) of processes $\{\mathbf{S}_T(t)\}$, depending on a parameter T, which can be regarded as unboundedly increasing and identified with some initial characteristic of the system (for example, with the intensity of the input stream in the conditions of section 2.1). Sometimes instead of T it is convenient to discuss a small parameter δ characterizing, say, the capacity of the system, as in the examples of section 1.11.

The present chapter comprises the study of multichannel systems with intensive input stream. In section 2.1 we give a classification of these systems according to the degree of loading (the latter being defined as the ratio of the intensity of the input stream to the number of service channels) and establish rough theorems (of the type of laws of large numbers) on the asymptotic behaviour of the number of busy channels for all types of loading. In section 2.2 we establish more precise results, describing the behaviour over a finite time interval of the number of busy channels for 'underloaded' systems (when some of the channels remain free). In section 2.3 similar results are obtained for convergence to a stationary process connected with the consideration of the processes in receding intervals of time. In section 2.4 the connection between multichannel systems with intensive input streams and branching processes with intensive immigration is established. This connection allows the description of limit processes for the number of particles of a subcritical process with intensive immigration. Section 2.5 is devoted to the discussion of problems connected with the behaviour of the number of busy lines when the system is in a 'transitional' regime of loading. In section 2.6 the basic results of sections 2.2 and 2.3 are generalized to the case of dependent waiting times. Sections 2.4–2.6, in view of their technical complexity, are marked with asterisks and may be omitted on a first reading (similarly the proof of theorem 3 in section 2.1). Finally in section 2.7 we obtain the proper limit distribution for the non-normalized number of free channels in 'overloaded' systems with refusals (that is, for systems in which the intensity of the input stream is more than the mean capacity of the system).

The contents of this chapter follow Borovkov (1967c, 1977a). The results of section 2.7 are significantly generalized in Borisov and Borovkov (1980).

2.1. Description of the systems. Rough theorems for the number of busy lines and for the probability of refusal

2.1.1. In this chapter we will discuss systems with the so-called intensive input streams, when the number of requests joining the system in unit time is large. The assumptions on the nature of the input stream will be very general—satisfied in practically all concrete problems of any interest. The service system will be assumed to have a very definite form: it consists of a finite or infinite number n of service channels; the service times $\tau_j = \tau_j^s$ (τ_j^s is equal to the service time of the jth request; the index s will be omitted) form a stationary sequence $\{\tau_j\}$ satisfying some sort of weak dependence condition. Requests are directed to any of the free channels. For definiteness we may suppose the channels to be numbered and the channels with the lowest numbers are occupied first. If all the channels are busy then the request is either refused or takes its place in a queue depending on the type of system considered.

Intensity of the input stream means that in the series array the increments

$$e(1) - e(0) = e_T(1) - e_T(0)$$

are unboundedly increasing as $T \to \infty$. We may suppose, for example, that

$$\mathbf{E}(e(t) - e(0)) = Tm(t)$$

where $m(t)$ is some nondecreasing function not depending on T. It is clear that an intensive stream $e_T(t)$ may be obtained by the usual route of compressing time by T.

We denote by $q(t) = q_T(t)$ the number of busy lines in the system at time t and put $e(0) = 0$, $q(0) = 0$,

$$F(t) = \mathbf{P}(\tau_j \leqslant t), \quad G(t) = 1 - F(t)$$

$$Q(t) = \int_0^t G(t-u)\, \mathrm{d}m(u).$$

(In this chapter it is convenient for the distribution function of τ_j to be continuous from the right.)

First we will obtain some rough theorems (of the type of the law of large numbers) on the asymptotic behaviour of the number of busy lines $q(t)$ and the 'probability of refusal' $p(t)$ (the second, of course, only for systems with refusals) over a finite time interval $[0, t_0]$, for some $t_0 > 0$.

2.1.2. We will assume in the first part of this section that the following conditions are satisfied:

(I) *The sequence of processes $e(u)/T$ C-converges* to the function $m(u)$ on* $[0, t_0]$.

(II) *The sequence $\{\tau_j\}$ is stationary and satisfies the strong mixing condition (s.m.), that is,*

$$\sup_{A \in \mathfrak{S}_1^k, B \in \mathfrak{S}_{k+s}^\infty} |\mathbf{P}(AB) - \mathbf{P}(A)\mathbf{P}(B)| = \phi(s) \to 0$$

* For the definition of C-convergence see section 1.2.

as $s \to \infty$, *where* $\mathfrak{S}_k^l = \sigma(\tau_k, \ldots, \tau_l)$ *is the* σ*-algebra generated by* $\tau_k, \ldots, \tau_l$.

(III) *The number of channels n varies with T so that as* $T \to \infty$

$$\frac{n}{T} \to a > 0.$$

To simplify the exposition here we will take $m(u) \equiv u$. This makes the notation and results clearer, which is essential in the first stage of the discussion. From section 2.2, where $m(u)$ is an arbitrary nondecreasing function, it is clear that this assumption is minor. For $m(u) \equiv u$ we will have

$$Q(t) = \int_0^t G(u)\,\mathrm{d}u, \quad Q(t) \to \mathbf{E}\tau_1 \quad \text{as } t \to \infty.$$

According to the size of a we consider queuing systems in three essentially different situations:

(I) *Underloaded systems* when $a > Q(t_0)$.

(II) Systems in a 'transitional' state when $a = Q(t_0)$. Such systems could simply be called *loaded.*

(III) *Overloaded systems* when $a < Q(t_0)$.

Later, in sections 2.2, 2.3, and 2.5, this classification will be made more precise. The following results indicate that the terminology is natural: for underloaded systems a subset of the channels, comparable in size with n, remain free with probability close to 1. For loaded and overloaded systems almost all channels are busy.

We begin with the consideration of underloaded systems.

THEOREM 1. *Let* $a > Q(t_0)$. *Then under conditions* (I)–(III) *the sequence of processes* $q(u)/T$ *C-converges on* $[0, t_0]$ *to the function* $Q(u)$:

$$\frac{q(u)}{T} \underset{C}{\to} Q(u), \quad u \in [0, t_0].$$

Proof. Suppose first that $n = \infty$. We find the form of $q(t)$ (as before, for brevity, we will frequently omit the index T). We denote by $v_1 < v_2 < \ldots$ the jump points of the input process $e(t)$, which we will take as continuous from the right, and by $\nu_1, \nu_2, \ldots$ the sizes of these jumps. In addition let

$$\nu(t) = \min\{k \geqslant 1 : v_{k+1} > t\}$$

be the index of the last jump on $[0, t]$ of $e(t)$. Then

$$q(t) = \sum_{k=1}^{\nu(t)} \sum_{j=1}^{\nu_k} \chi_k^j(t) \tag{2.1}$$

where

$$\chi_k^j(t) = \begin{cases} 1 & \text{if } v_k + \tau > t \\ 0 & \text{otherwise} \end{cases}$$

$k = 1, \ldots, v(t)$; τ is the service time of the request numbered $\nu_1 + \cdots + \nu_{k-1} + j$. In what follows it is often convenient for us to deal with representations of $q(t)$ of the form (2.1). To abbreviate this description we will suppose that $\nu_k \equiv 1$ (requests join one by one) and, consequently, $\nu(t) = e(t)$. This assumption, as the reader may easily verify (see, for example, the remark after (2.28)), has no influence on the proof of the theorem—it neither simplifies not complicates it. However, the brevity in writing unquestionably helps in the reading and, therefore, we will use this assumption; that is, we will suppose that

$$q(t) = \sum_{k=1}^{e(t)} \chi_k(t)$$

$$\chi_k(t) = \begin{cases} 1 & \text{if } v_k + \tau_k > t \\ 0 & \text{otherwise.} \end{cases}$$

We denote by $\mathfrak{E}$ the algebra generated by the input stream, that is, events of the form $\{e(u) = l\}$, $u \leqslant t_0$. Then

$$\mathbf{E}_{\mathfrak{E}} q(t) = \sum_{k=1}^{e(t)} \mathbf{P}(\tau_k > t - v_k) = \int_0^{e(t)} G(t - v)\, de(v)$$

$$V(t) = q(t) - \mathbf{E}_{\mathfrak{E}} q(t) = \sum_{k=1}^{e(t)} x_k$$

where x_k can also be written in the form

$$x_k = \chi(\tau_j > b_j), \quad b_j = t - v_j$$

$\chi(A)$ is the indicator of the event A.

Since

$$\frac{1}{T}\mathbf{E}_{\mathfrak{E}} q(t) = \frac{1}{T}\int_0^t G(t - v)\, de(v) = \frac{1}{T}\left[e(t) - \int_0^t e(v)\, dG(t - v)\right] \underset{C}{\rightarrow} Q(t) \tag{2.2}$$

it is sufficient for us to show that on $[0, t_0]$

$$\frac{1}{T} V(u) \underset{C}{\rightarrow} 0.$$

Fix a trajectory $e(u)$, $u \leqslant t_0$, assuming that it belongs to the set

$$B_e = \left\{ \sup_{u \leqslant t} \left| \frac{e(u)}{T} - u \right| \leqslant \varepsilon_e(T), \right.$$

$$\left. \sup_{u \leqslant t} \left| \frac{1}{T} \int_0^u G(u-v)\, de(v) - Q(v) \right| \leqslant \varepsilon_e(T) \right\} \tag{2.3}$$

where $\varepsilon_e(T) \to 0$ is chosen so that $\mathbf{P}(B_e) \to 1$. By (2.2) and condition (I) such a function ε_e exists. Further,

$$\mathbf{E}_{\mathfrak{E}} x_k = 0, \quad |\mathbf{E}_{\mathfrak{E}} x_k x_j| \leqslant \phi(|j-k|) \to 0$$

as $|j-k| \to \infty$. Hence it easily follows that

$$\mathbf{E}_{\mathfrak{E}} V(u) = 0$$

$$\mathbf{E}_{\mathfrak{E}} V^2(u) \leqslant e(u) + 2 \sum_{k=1}^{e(t)} (e(u) - k)\phi(k) = o(e^2(u)).$$

This means that $\mathbf{E}_{\mathfrak{E}} V^2(u) = o(T^2)$ on B_e. Therefore as $T \to \infty$, $e \in B_e$

$$\mathbf{P}_{\mathfrak{E}}(|V(u)| > \varepsilon T) \leqslant \frac{o(T^2)}{\varepsilon^2 T^2} \to 0. \tag{2.4}$$

Now consider the partition of $[0, t_0]$ into intervals of length Δ. Since $(\Delta y(u) = y(u+\Delta) - y(u);\ e \in B_e)$

$$\Delta V(u) \leqslant \Delta q(u) + \Delta Q(u) + 2\varepsilon_e(T) \leqslant \Delta e(u) + \Delta + 2\varepsilon_e(T) \leqslant 2\Delta + 4\varepsilon_e(T)$$

then for arbitrary $\varepsilon > 0$

$$\mathbf{P}_{\mathfrak{E}}\left(\sup_{u \leqslant t_0} |V(u)| > 2\varepsilon T \right)$$

$$\leqslant \mathbf{P}_{\mathfrak{E}}\left(\max_k |V(k\Delta)| > \varepsilon T \right) + \mathbf{P}_{\mathfrak{E}}(2\Delta + 4\varepsilon_e(T) > \varepsilon). \tag{2.5}$$

The second term is equal to zero for $\Delta \leqslant \varepsilon/3$ and sufficiently large T. The first term in (2.5) for the chosen and fixed Δ converges to 0 by (2.4).

Thus, on B_e, $\mathbf{P}(B_e) \to 1$,

$$\mathbf{P}_{\mathfrak{E}}\left(\sup_{u \leqslant t_0} |V(u)| > 2\varepsilon T \right) \to 0$$

as $T \to \infty$. The theorem is proved for the case $n = \infty$.

The result for finite $n \to \infty$ in such a way that $n/T \to a > Q(t_0)$ follows since the probability that the process $Tq(t)$, constructed for $n = \infty$, will reach the level $n \sim aT$, $a > Q(t_0)$, converges to zero as $T \to \infty$. But below this level the processes $q(t)$ for finite and infinite n behave the same. Therefore the systems for these finite values of n behave asymptotically as the systems in the case $n = \infty$. The theorem is proved. □

We pass on to the consideration of *overloaded* systems when $a < Q(t_0)$. Of most interest in this connection are the *systems with refusals.* For a remark concerning overloaded *systems with waiting* (there $q(t) = n$ from some time) see after theorem 3. Thus, let n be such that

$$\frac{n}{T} \longrightarrow a < Q(t_0). \tag{2.6}$$

We denote

$$Q_1(u) = \min\,[a, Q(u)], \quad u \in [0, t_0].$$

THEOREM 2. *Under the conditions* (I), (II), *and* (2.6)

$$\frac{q(u)}{T} \underset{C}{\longrightarrow} Q_1(u), \quad u \in [0, t_0].$$

In other words, up to order $o(T)$ the trajectory $q(u)$ coincides with the curve $\min\,(TQ(u), n)$.

Proof. Let t_1 be a solution of the equation

$$Q(u) = a.$$

Then, by theorem 1, it is sufficient to prove the C-convergence

$$\frac{q(u)}{T} \longrightarrow a$$

on the segment $t \in [t_1, t_0]$ (C-convergence to $Q(u)$ on $[0, t_1 - \varepsilon]$, for any $\varepsilon > 0$, follows immediately from theorem 1).

Suppose that simultaneously with our system there is another system, governed by the same processes $\{e(u)\}$ and $\{\tau_j\}$, but with an infinite number of channels. For this system we denote the number of busy channels by $\tilde{q}(u)$ and introduce the event

$$B_\Delta = \left\{\frac{1}{T} \sup_{u \leqslant t_0} |\Delta\tilde{q}(u) - T\Delta Q(u)| \leqslant \Delta\varepsilon_\Delta(T)\right\}$$

where $\varepsilon_\Delta(T) \to 0$ as $T \to \infty$, and Δ is fixed. By theorem 1, $\varepsilon_\Delta(T)$ can be chosen so that $\mathbf{P}(B_\Delta) \to 1$.

We now use the following theorem on C-convergence to a degenerate process (see section 1.4). Let there be a sequence of nonnegative processes $\{\xi(u) = \xi_T(u) \geqslant 0;\ t_1 \leqslant u \leqslant t_0\}$, continuous from the right, and depending on a parameter T. Further, let $\mathfrak{M}(u)$, $u \in [t_1, t_0]$, be a collection of nested σ-algebras $\mathfrak{M}(u_1) \subset \mathfrak{M}(u_2)$ for $u_1 \leqslant u_2$, such that $\xi(t)$ is measurable with respect to $\mathfrak{M}(u)$ for $t \leqslant u$. Δ_T will denote a sequence $\Delta_T \to 0$ as $T \to \infty$ and ε and c (with or without indices) will denote, respectively, arbitrarily small numbers and constants.

Finally, we put

$$\Gamma_u = \{\varepsilon_1 \leqslant \xi(u) < c_1\}.$$

THEOREM A. *Suppose there exist a sequence Δ_T and a system of sets $\Omega_{u,T} \in \mathfrak{M}(u)$ such that*

$$\Omega(u) = \bigcap_{t_1 \leqslant s \leqslant u} \Omega_{u,T} \in \mathfrak{M}(u), \quad \mathbf{P}(\Omega(t_0)) \to 1$$

and on $\Omega_{u,T} \cap \Gamma_u$ for all $\Delta \leqslant \Delta_T$ and each fixed ε, ε_1, c_1 we have:

(I) $\mathbf{E}_{\mathfrak{M}(u)} \Delta\xi(u) \leqslant \varepsilon \Delta_T$

where $\Delta\xi(u)$ denotes the increment $\Delta\xi(u) = \xi(u + \Delta) - \xi(u)$.

(II) *For any fixed $\Delta > 0$ and $T \to \infty$*

$$\mathbf{E}_{\mathfrak{M}(u)}(|\Delta\xi(u)|; |\Delta\xi(u)| > \delta\sqrt{\Delta_T}) < \varepsilon\Delta_T.$$

(III) *The probability that at least one jump of $\xi(u)$ on $[t_1, t_2]$ will exceed δ, converges to 0 as $T \to \infty$.*

If, in addition, $\mathbf{P}(\xi(t_1) > \delta) \to 0$ as $T \to \infty$, then

$$\xi(u) \underset{C}{\to} 0, \quad u \in [t_1, t_0].$$

We return to the question of C-convergence to 0 of

$$\frac{1}{T}[q(u) - a], \quad u \in [t_1, t_0].$$

Using (2.6) we may discuss the equivalent problem of C-convergence to 0 of

$$\frac{1}{T}[q(u) - n] \leqslant 0, \quad u \in [t_1, t_0].$$

For fixed trajectories of $e(u)$ and $\{\tau_j\}$ on $B_e \cap B_\Delta \cap \{q(u) \leqslant n - \varepsilon_1 T\}$ we will have (by definition, $\delta(s) = 1$ for $s < n$ and $\delta(s) = 0$ for $s = n$),

$$\Delta q(u) = q(u + \Delta) - q(u) = -\sum_{j=1}^{e(u)} \chi(u < \tau_j + v_j \leqslant u + \Delta)\delta(q(v_j)) + \sum_{j=e(u)+1}^{e(u+\Delta)} \chi(\tau_j + v_j \leqslant u + \Delta) \tag{2.7}$$

provided $\Delta \leqslant \varepsilon_1/2$, since, in this case, no more than $\varepsilon_1 T$ requests join the system in the interval $(u, u + \Delta)$ and, consequently, there will be no refusals ($\delta(q(v_j)) = 1$ for $j \in [e(u) + 1, e(u + \Delta)]$). This means

$$\Delta q(u) \geqslant -\sum_{j=1}^{e(u)} \chi(u<\tau_j+v_j \leqslant u+\Delta) + \sum_{j=e(u)+1}^{e(u+\Delta)} \chi(\tau_j+v_j \leqslant u+\Delta)$$
$$= \tilde{q}(u+\Delta) - \tilde{q}(u) \geqslant T\Delta Q(u) - T\varepsilon_\Delta(T) \geqslant -T\varepsilon_\Delta(T). \tag{2.8}$$

This inequality, the sense of which is very simple, means that condition (I) of theorem A is satisfied ($\mathfrak{M}(u) = \mathfrak{E} \times \mathfrak{S}$, where $\mathfrak{E}$ and $\mathfrak{S}$ are, respectively, generated by $\{e(u)\}$ and $\{\tau_j\}$, and $\Omega_{u,r} = B_e B_\Delta \in \mathfrak{M}(u)$. Since the increments of $q(u)$ are measurable relative to $\mathfrak{E} \times \mathfrak{S}$, then the conditional expectation $\mathbf{E}_{\mathfrak{M}(u)}$ in theorem A becomes superfluous).

Condition (II) follows because on $B_e B_\Delta$

$$\frac{1}{T}|\Delta q(u)| \leqslant \frac{1}{T}(\Delta e(u) + |\Delta \tilde{q}(u)|) \leqslant 2\Delta + \Delta Q(u) \tag{2.9}$$

and, consequently, the conditional expectation in condition (II) is equal to 0. From (2.9) it also follows that $(1/T)q(t_1)$ is close to $Q(t_1) = a$.

Condition (III) of theorem A, on jumps, follows from the C-convergence of the processes $e(u)$ and $\tilde{q}(u)$.

Therefore, by that theorem,

$$\frac{1}{T}[q(u) - n] \underset{C}{\longrightarrow} 0, \quad u \in [t_1, t_0].$$

The theorem is proved. □

We now consider a queuing system with refusals in the transitional regime (loaded systems), when

$$\frac{n}{T} \longrightarrow a = Q(t_0). \tag{2.10}$$

For such systems, in the obvious way, we can construct processes $q_1(u)$ and $q_2(u)$ corresponding to the same control, but with, respectively, n_1 and n_2 channels, $n_1 < n < n_2$, for which

$$q_1(u) < q(u) < q_2(u).$$

Using such majorants, we easily obtain from theorems 1 and 2:

COROLLARY 1. *Under conditions* (I), (II), (2.10) *theorem 2 still holds.*

A similar result for systems with queues in the case $Q(t_0) = a$ will follow from comparison of theorems 1, 2, and 3.

COROLLARY 2. *Let* $f(u) = \max_{v \leqslant u} q(u)$ *be the number of lines which are 'operational' at some time in* $[0, u]$ *and let* $y(u) = f(u) - q(u)$ *be the number of free* *lines amongst those which have been operational. Then, from the monotonicity of $Q_1(u)$ and theorems 1 and 2 it follows that on $[0, t_0]$*

$$\frac{q(u)}{f(u)} \xrightarrow[c]{} 1, \quad \frac{y(u)}{f(u)} \xrightarrow[c]{} 0.$$

We have seen here that despite the naturalness and transparency of theorem 2 its proof still requires some effort. To a large extent this remark reduces to the following result.

Let $\tilde{s}(u)$ denote the number of requests accepted for service in the system up to time u. 'The probability of refusal' is the quantity

$$p(u) = \frac{e(u) - \tilde{s}(u)}{e(u)}.$$

We note that if $s(u)$ denotes the number of requests served by the system up to time u, then obviously

$$\tilde{s}(u) = s(u) + q(u)$$

and, consequently, $p(u) = 1 - (s(u) + q(u))/e(u)$.

THEOREM 3. *Let conditions* (I), (III) *hold for* $a \leqslant Q(t_0)$ *and let the sequence* $\{\tau_j\}$ *satisfy the uniform strong mixing condition (u.s.m.): as* $s \to \infty$

$$\sup_{A \in \mathfrak{S}_1^k, B \in \mathfrak{S}_{k+s}^\infty} |\mathbf{P}(B/A) - \mathbf{P}(B)| = \phi(s) \to 0.$$

Then, on $[0, t_0]$

$$\frac{s(u)}{T} \to \sigma(u) = \int_0^a H(u - Q^{-1}(v))\, dv$$

$$p(u) \to \pi(u) = 1 - \frac{1}{u}\left[Q_1(u) + \int_0^a H(u - Q^{-1}(v))\, dv\right]$$

where $Q^{-1}(v)$ is the function inverse to $Q(v)$, $H(v)$ is the renewal function of the random variable τ_1: $H(v) = \sum_{k=1}^\infty \mathbf{P}(\tau + \cdots + \tau^k \leqslant v)$, where the τ^j are distributed as τ_1 and are independent.

Obviously, if we consider *systems with waiting* in the case $a \leqslant Q(t_0)$, then for them, with high probability, $q(t) \sim TQ(t)$ for $t < t_1$ and $q(t) = n$ for $t > t_1$ (t_1, as before, is the solution of the equation $Q(t_1) = a$). For such systems our interest is in the 'occupation' of the system $q_1(t)$, being the sum of the number of busy lines, and the number of requests awaiting service. In our notation

$$q_1(t) = e(t) - s(t).$$

It is quite clear that the asymptotic behaviour of $s(t)$ both for systems with waiting and systems with refusals for $a < Q(t_0)$ will be identical (up to t_1 the processes $s(t)$ behave the same. After t_1 they also behave essentially the same since, in both cases, with probability close to 1, almost all channels are busy). Therefore, using the first assertion of theorem 3, we obtain for systems with waiting that

$$\frac{q_1(u)}{T} \underset{C}{\longrightarrow} u - \sigma(u) = u - \int_0^a H(u - Q^{-1}(v))\,dv.$$

A rigorous proof of this result differs from the proof of theorem 3 only by simplifications. Therefore we omit it.

As we shall see later (see corollary 2 of theorem 3) $H(u - Q^{-1}(v)) \sim u/\mathbf{E}\tau_1$ for large u and $v < \mathbf{E}\tau_1$. Consequently, for such u,

$$\frac{q_1(u)}{T} \approx u\left(1 - \frac{a}{\mathbf{E}\tau_1}\right).$$

We return to theorem 3. To simplify the arguments we prove this theorem under the additional assumption that

$$\mathbf{P}(\tau_j \geqslant h) = 1$$

for some $h > 0$. We will make some remarks in connection with this assumption after the proof of the theorem.

In addition, in order not to complicate the statement and notations we will suppose that condition (I) on the C-convergence of $e(u)/T$ is satisfied on $[0, t_0 + 1]$. We note, in general, that it is natural to suppose that condition (I) holds for any fixed t_0. Since in condition (II) the 'whole' stationary sequence $\{\tau_j\}$ is given, then this allows us to consider $q(u)$, $s(u)$, and others also in any fixed segment $[0, t_0]$.

Proof. Obviously it is sufficient to consider C-convergence of the processes

$$\frac{s(u)}{T}, \quad u \leqslant t_0.$$

Let $\lambda_{n,i}$ denote the total idle time of the ith channel in $[0, t_0]$ from the instant u_i at which it becomes operational (u_i is the solution of the equation $f(u) = i$), and let

$$\lambda_E = \max_{i \in E} \lambda_{n,i}$$

where E denotes the integers.

LEMMA 1. *Under conditions* (I) *and* (II), *for any* $p > 0$ *and* $\Delta > 0$ *there is a set* E *of indices of service channels, with not less than* $n(1 - p)$ *elements, for which*

$$\mathbf{P}(\lambda_E > \Delta) \to 0$$

as $T \to \infty$.

Proof. Suppose that there exist np channels in each of which the idle time in $[0, t_0]$ from the instant of becoming operational is greater than Δ. This means that there is a $t^* \leqslant t_0$ such that the number of free channels $z(t^*)$ at this time will be greater than $np\Delta/t_0$. But the probability of such an event, by corollary 2 of theorem 2, converges to zero. The lemma is proved. □

LEMMA 2. *For any* $p > 0$ *it is possible to remove* np *channels from the* n *channels in such a way that for any channel* j *from the remaining set* E_p *and for any* $\Delta > 0$

$$p_j = \mathbf{P}(\mathscr{E}_\Delta \text{ contains } j) \to 1 \tag{2.11}$$

where $\mathscr{E}_\Delta$ *is a random set, consisting of channels for which* $\lambda_{n,j} \leqslant \Delta$.

Proof. Order the channels relative to the size of $\mathbf{P}(j \in \mathscr{E}_\Delta)$ and form the set E_p of $n(1-p)$ channels for which the probabilities are maximal.

Assuming that (2.11) is not satisfied for all $j \in E_p$ we can give $\varepsilon > 0$ such that the number $N(\mathscr{E}_\Delta)$ of elements in $\mathscr{E}_\Delta$ satisfies

$$\mathbf{E}N(\mathscr{E}_\Delta) = \sum_{j=1}^{n} p_j < n(1-p) + np(1-\varepsilon) = n - np\varepsilon. \tag{2.12}$$

We now apply lemma 1, choosing as p the number $p_1 = \frac{1}{2}\varepsilon p$, where p is the number of lemma 2. We then obtain $\mathbf{P}(N(\mathscr{E}_\Delta) > n(1-p_1)) \to 1$ and, consequently, $\mathbf{E}N(\mathscr{E}_\Delta) > n(1-p_1)(1+o(1)) > n - np\varepsilon$. This contradicts (2.12). The lemma is proved. □

We can now construct an estimate for $s(u)$. Since $\{e(u)\}$ and $\{\tau_j\}$ are given for all $u \geqslant 0$ and $j \geqslant 1$, then $s(u)$ is defined on the whole semiaxis $u \geqslant 0$. It is obvious that

$$s(u) = \sum_{i=1}^{n} v_i(u)$$

where $v_i(u)$ is the number of requests served in the ith channel up to time u.

Let $i_1, i_2, \ldots$ be the numbers of the requests served in the ith channel and let u_i be the time of the first job in the ith channel, so that i_k and u_i are random variables. Then, if $\eta_i(v)$ denotes the renewal process for the sequence $\{\tau_{i_k}\}_{k=1}^{\infty}$:

$$\{\eta_i(v) = l\} = \{\tau_{i_1} + \cdots + \tau_{i_l} \leqslant v, \tau_{i_1} + \cdots + \tau_{i_{l+1}} > v\}$$

then

$$v_i(u) \leqslant \eta_i(u - u_i) \tag{2.13}$$

and for $i \in \mathscr{E}_\Delta$

$$v_i(u) \geqslant \eta_i(u - u_i - \Delta). \tag{2.14}$$

To make use of these inequalities we require

LEMMA 3. *Let* $\mathbf{P}(\tau_j \geqslant h) = 1$, $h > 0$ *and let* $\{\tau_j\}$ *satisfy the u.s.m. condition. Then for any fixed L the elements of* $\{\tau_{i_l}\}_{l=1}^{L}$ *are asymptotically independent and distributed like* τ_1. *More precisely, for any Borel sets* $\{A_l\}_{l=1}^{L}$, *having the property* $Q^{-1}(\alpha_i) + \sum_{l=1}^{L-1} x_l < t_0$ *for any* $x_l \in A_l$ *we have*

$$\mathbf{P}\left(\bigcap_{l=1}^{L} \{\tau_{i_l} \in A_l\}\right) - \prod_{l=1}^{L} \mathbf{P}(A_l) \to 0 \tag{2.15}$$

where $\alpha_i = \lim_{T\to\infty} i/T$, $\mathbf{P}(A) = \mathbf{P}(\tau_1 \in A)$.

Let $p > 0$ *be arbitrary. If* $i \in E_p$, $j \in E_p$ (E_p *is defined in lemma 2) and j does not belong to some* $E(i)$, *depending only on i and p and containing not more than np elements, then the sequences* $\{\tau_{i_l}\}_{l=1}^{L}$ *and* $\{\tau_{j_l}\}_{l=1}^{M}$ (j_l *is the index of the lth request served in the jth channel) are asymptotically independent on the set*

$$Q^{-1}(\alpha_i) + \sum_{l=1}^{L-1} \tau_{i_l} < t_0, \quad Q^{-1}(\alpha_j) + \sum_{l=1}^{M-1} \tau_{j_l} < t_0$$

that is, for arbitrary $A_1, \ldots, A_L$; $C_1, \ldots, C_M$, *having the property* $Q^{-1}(\alpha_i) + \sum_{l=1}^{L-1} x_l < t_0$, $Q^{-1}(\alpha_j) + \sum_{l=1}^{M-1} \tau_{j_l} < t_0$ *for any* $x_l \in A_l$, $y_l \in C_l$, *we have*

$$\mathbf{P}\left(\bigcap_{l=1}^{L} \{\tau_{i_l} \in A_l\} \bigcap_{l=1}^{M} \{\tau_j \in C_l\}\right) - \prod_{l=1}^{L} \mathbf{P}(A_l) \prod_{l=1}^{M} \mathbf{P}(C_l) \to 0.$$

We note that if we do not require the existence of $\lim_{T\to\infty} i/T$ and $\lim_{T\to\infty} j/T$, then, as will be clear from the proof, the lemma still holds. In this case the condition, for example, on $\{A_l\}_{l=1}^{L}$ should be described in the form $Q^{-1}(i/T) + \sum_{l=1}^{L-1} x_l < t_0$ for any $x \in A_l$. But since, in general, i depends on T, then A_l would also be assumed dependent on T.

Proof. Again fix a trajectory $e(t)$ from the set B_e defined in (2.3). Then in an interval of time h, for sufficiently large T, not less than $hT/2$ requests will join the system.

Further, since $\tau_j \geqslant h$, then the values τ_j will have no influence on the trajectories of $q(u)$ and $s(u)$ from the instant v_j, of appearance of the jth request, up to $v_j + h$. From here and the above remarks it follows that

$$\{i_k = l\} \in \mathfrak{F}_{l-\frac{1}{2}hT} \tag{2.16}$$

where $\mathfrak{F}_k = \sigma(\tau_1, \ldots, \tau_k)$ is the σ-algebra generated by $\tau_1, \ldots, \tau_k$; for simplicity we take $\frac{1}{2}hT$ to be an integer.

Now consider the probability

$$\mathbf{P}\left(\bigcap_{k=1}^{L} \{\tau_{i_k} \in A_k\}\right) = \sum_{l} \mathbf{P}\left(\bigcap_{k=1}^{L-1} \{\tau_{i_k} \in A_k\}, i_L = l\right) \mathbf{P}(\tau_l \in A_L/D_l)$$

where

$$D_l = \bigcap_{k=1}^{L-1} \{\tau_{i_k} \in A_k\}\{i_L = l\}.$$

Since $i_k \leqslant i_L - \frac{1}{2}hT$ for $k \leqslant L-1$, then alongside (2.16) $D_i \in \mathfrak{F}_{l-\frac{1}{2}hT}$. Hence it follows that

$$|\mathbf{P}(\tau_l \in A_L/D_l) - \mathbf{P}(\tau_l \in A_L)| < \phi(\tfrac{1}{2}hT) \to 0$$

$$\mathbf{P}\left(\bigcap_{k=1}^{L} \{\tau_{i_k} \in A_k\}\right) - \mathbf{P}(A_L)\mathbf{P}\left(\bigcap_{k=1}^{L-1} \{\tau_{i_k} \in A_k\}\right) \to 0$$

$$\mathbf{P}\left(\bigcap_{k=1}^{L} \{\tau_{i_k} \in A_k\}\right) - \prod_{k=1}^{L} \mathbf{P}(A_k) \to 0.$$

Since $\mathbf{P}(B_e) \to 1$, then (2.15) is proved.

We now prove the second part of the lemma. Consider the event

$$B_q = \left\{\sup_{u \leqslant t_0} \left|\frac{\tilde{q}(u)}{T} - Q(u)\right| \leqslant \varepsilon_q(T)\right\}$$

$$\varepsilon_q(T) = o(1), \quad \mathbf{P}(B_q) \to 1.$$

Since the time u_i of the first job in the ith channel is the least solution of the equation

$$f(u) = i$$

(we suppose that the ith channel is occupied only when all the previous $i-1$ channels are busy), then, by corollary 2 of theorem 2, on B_q

$$u_i \in \left[Q^{-1}\left(\frac{i}{T}\right) - c\varepsilon_q, Q^{-1}\left(\frac{i}{T}\right) + c\varepsilon_q\right], \quad c = \text{const.} \tag{2.17}$$

Since $i_1 = e(u_i)$, then on B_q

$$i_1 \in \left[TQ^{-1}\left(\frac{i}{T}\right) - T\varepsilon^*, TQ^{-1}\left(\frac{i}{T}\right) + T\varepsilon^*\right] \tag{2.18}$$

where $\varepsilon^* = \varepsilon_e + c\varepsilon_q = o(1)$.

For simplicity we first consider the three-dimensional distributions

$$P(\tau_{i_1} \in A_i, \tau_{i_2} \in A_2, \tau_{j_1} \in C_1)$$

and choose as the set $E(i)$, in the statement of the lemma, the set of numbers j for which $|j-i| > \Delta T$.

The basic circumstance, used later, is that under these conditions the indices i_1, i_2, j_1, with probability close to 1, will be separated by large intervals of time.

Denote by D the event

$$\{\mathscr{E}_\Delta \ni i, \mathscr{E}_\Delta \ni j\}$$

where the $\mathscr{E}_\Delta$ correspond to the interval $[0, t_0 + 1]$, and suppose that the distribution of τ_1 is continuous, $j > i$.

Since $i \in E_p$, $j \in E_p$, then, by lemma 2, $\mathbf{P}(D) \to 1$,

$$\mathbf{P}(\tau_{i_1} \in A_1, \tau_{i_2} \in A_2, \tau_{j_1} \in C_1)$$

$$= o(1) + \varepsilon(\Delta) + \sum_{\lambda=1}^{2} \sum_{k,l,s} \mathbf{P}(D, B_q; i_1 = k, \tau_k \in A_1^\lambda;$$

$$i_2 = l, \tau_l \in A_2; j_1 = s, \tau_s \in C_1) \quad (2.19)$$

where $\varepsilon(\Delta) \to 0$ as $\Delta \to 0$, the sums over k and s, by (2.18), are bounded by the limits $TQ^{-1}(i/T) \pm T\varepsilon^*$ and $TQ^{-1}(j/T) \pm T\varepsilon^*$ respectively. The sets A_1^λ, $\lambda = 1, 2$ are defined by

$$A_1^1 = A_1 \cap [0, Q^{-1}(\alpha_j) - Q^{-1}(\alpha_i) - 2\Delta]$$

$$A_1^2 = A_1 \cap [Q^{-1}(\alpha_j) - Q^{-1}(\alpha_i) + 2\Delta, \infty].$$

Consider the sum in (2.19) for $\lambda = 1$. Since $i \in \mathscr{E}_\Delta$ and $\tau_{i_1} \in A_1^1$, then the service of the i_2th request begins before the time $u^* = TQ^{-1}(j/T) - \Delta T + o(T)$. This means that the summation with respect to i_2 occurs in the domain

$$i_2 < e(u^*) < TQ^{-1}\left(\frac{j}{T}\right) - \frac{\Delta T}{2}. \quad (2.20)$$

At the same time, as already noted,

$$j_1 > TQ^{-1}\left(\frac{j}{T}\right) - \varepsilon^* T$$

and this domain is at least a distance $\frac{1}{3}\Delta T$ from (2.20). Consequently, if $\frac{1}{3}\Delta < \frac{1}{2}h$ (see the proof of the first part of the lemma), then in the first sum

$$A = \{i_1 = k, \tau_k \in A_1^1; \tau_2 = l, \tau_l \in A_2; j_1 = s\} \in \mathfrak{F}_{s - \frac{1}{3}\Delta T}$$

and therefore

$$|\mathbf{P}(\tau_s \in C_1/A) - \mathbf{P}(\tau_1 \in C_1)| < \phi(\tfrac{1}{3}\Delta T) \to 0.$$

From here and the first part of the lemma it is already easy to obtain that the sum in (2.19), corresponding to $\lambda = 1$, converges to

$$\mathbf{P}(A_1^1)\mathbf{P}(A_2)\mathbf{P}(C_1)$$

(after this selection of indices of summation the events D and B_q should be removed from under the probability sign. This changes the general sum by no more than $o(1)$).

We now consider the sum for $\lambda = 2$. Here the maximal index will be i_2, which, as in the first part of the lemma, allows us to distinguish the multiplier $\mathbf{P}(A_2)$. In the remaining sums the domains of summation with respect to i_1 and j_1 differ by an interval of length at least $\frac{1}{2}\Delta T$. Therefore, proceeding as before, we obtain

$$\mathbf{P}(\tau_{i_1} \in A_1^2, \tau_{j_1} \in C_1) - \mathbf{P}(A_1^2)\mathbf{P}(C_1) \to 0.$$

Since Δ is arbitrary, then we have established that the left-hand side of (2.19) converges to $\mathbf{P}(A_1)\mathbf{P}(A_2)\mathbf{P}(C_1)$.

If the distribution of τ_1 has a discrete component, for example a jump at a, then to $E(i)$ it is necessary to also add the ΔT-neighbourhood of x, for which

$$Q^{-1}\left(\frac{i}{T}\right) + a = Q^{-1}\left(\frac{x}{T}\right).$$

This excludes the proximity (with significant probability) of u_j and the initial service time of the i_2th request (and, thus, the proximity of j_1 and j_2). In the same way, by enlarging $E(i)$, we can remove other possibilities for the proximity of j_1 and j_2 (with significant probability), connected with the existence of other points of discontinuity of the distribution of τ_1. Since the number of discontinuities to be considered is finite, then, obviously, the number of points $N(E(i))$ can be made less than np.

The joint distributions of $\tau_{i_1}, \tau_{i_2}, \ldots$ and $\tau_{j_1}, \tau_{j_2}, \ldots$ of larger dimension are discussed in a similar way. The lemma is proved. □

We now return to the inequalities (2.13) and (2.14). From lemma 3 we obtain immediately that if $\eta(v)$ is the renewal process for τ_1 ($\eta(v) = k$, if $\tau^1 + \cdots + \tau^k \leqslant v$, $\tau^1 + \cdots + \tau^{k+1} > v$), where τ^j are distributed as τ_1 and are independent), then

$$\mathbf{P}(\eta_i(v) = k) \to \mathbf{P}(\eta(v) = k). \tag{2.21}$$

In addition, if $i \in E_p$, $j \in E_p E(i)$, then

$$\mathbf{P}(\eta_i(v) = k, \eta_j(u) = l) \to \mathbf{P}(\eta(v) = k)\mathbf{P}(\eta(u) = l). \tag{2.22}$$

By the first inequality of (2.13), on B_q (see (2.17))

$$s(u) \leqslant \sum_{i=1}^{n} \eta_i(u - u_i) \leqslant \sum_{i=1}^{n} \eta_i\left(u - Q^{-1}\left(\frac{i}{T}\right) + c\varepsilon_q\right).$$

The terms $\eta_i^* = \eta_i(b_i)$, where $b_i = u - Q^{-1}(i/T) + c\varepsilon_q$, by (2.21) and (2.22) and the fact that $\eta_i(v) \leqslant v/h$, have for $u < t_0$ the properties ($H(v) = \mathbf{E}\eta(v)$, $H(v) = 0$ for $v < h$)

$$\mathbf{E}\eta_i^* - H(b_i) \to 0$$

and for $i \in E_p$, $j \in E_p E(i)$

$$\mathbf{E}\eta_i^*\eta_j^* - H(b_i)H(b_j) \to 0. \tag{2.23}$$

Since the proportion of indices i and j for which (2.23) holds (and for which, consequently, η_i^* and η_j^* are asymptotically uncorrelated), can be made arbitrarily close to 1, then

$$\mathbf{E}\sum_{i=1}^{n}\eta_i^* = \sum_{i=1}^{n} H(b_i) + o(n)$$

$$\mathbf{D}\sum_{i=1}^{n}\eta_i^* = \sum_{i,j=1}^{n} \mathbf{E}(\eta_i^* - \mathbf{E}\eta_i^*)(\eta_j^* - \mathbf{E}\eta_j^*) = o(n^2).$$

Therefore, by Chebyshev's inequality, with probability converging to 1,

$$s(u) \leqslant \sum_{i=1}^{n}\eta_i(b_i) = \sum_{i=1}^{n} H\left(u - Q^{-1}\left(\frac{i}{T}\right) + c\varepsilon_q\right) + o(T)$$

$$= T\int_0^a H(u - Q^{-1}\,|\,v)\,\mathrm{d}v + o(T).$$

Similarly, with the help of (2.14), we obtain, with probability converging to 1, that for arbitrary $\Delta > 0$ and $p > 0$

$$s(u) \geqslant T\int_0^a H(u - Q^{-1}(v) - \Delta)\,\mathrm{d}v - cT(\Delta + p), \quad c = \text{const}.$$

From these inequalities it follows that

$$\frac{s(u)}{T} \underset{P}{\rightarrow} \int_0^a H(u - Q^{-1}(v))\,\mathrm{d}v.$$

Together with the monotonicity of $s(u)$ this means that on $[0, t_0]$

$$\frac{s(u)}{T} \underset{C}{\rightarrow} \int_0^a H(u - Q^{-1}(v))\,\mathrm{d}v.$$

The theorem is proved. □

COROLLARY 1. *The limit 'probability of refusal'* $\pi(u) \equiv 0$ *for* $a = Q(t_0)$.

In fact, from the identity

$$\int_0^u (1 + H(u - z))G(z)\,\mathrm{d}z = u$$

which follows from the properties of the renewal function, by the substitution $z = Q^{-1}(v)$ we obtain

$$\int_0^{Q(u)} H(u - Q^{-1}(v))\,\mathrm{d}v = u - Q(u).$$

Hence, and from theorem 3, it follows that $\pi(u) \equiv 0$.

It is easy to see that the formula for $\pi(u)$ in theorem 3 gives $\pi(u) = 0$ for $a > Q(t_0)$. This means that theorem 3 is valid for any a.

COROLLARY 2. *If* $n/T \to a$, $r \equiv a/\mathbf{E}\tau_1 < 1$, *then*

$$p(t) \underset{P}{\to} 1 - r_1.$$

In fact, by the renewal theorem, $H(t - Q^{-1}(v)) \to t/\mathbf{E}\tau_1$ as $t \to \infty$ since $Q^{-1}(v)$ is bounded for all $v < \mathbf{E}\tau_1$. Consequently, for large t

$$\int_0^a H(t - Q^{-1}(v))\,\mathrm{d}v = \frac{t}{\mathbf{E}\tau_1} a + o(t) = tr + o(t).$$

It remains only to use theorem 3 and $Q_1(t) \leqslant \mathbf{E}\tau_1$.

COROLLARY 3. *If* $r < 1$ *and the distribution of* τ_1 *is nonlattice-like then the 'probability of refusal'* $\pi_\Delta(t)$ *on the interval* $[t, t + \Delta]$, $\Delta > 0$ *converges to* $1 - r$ *as* $t \to \infty$.

We have, as $t \to \infty$

$$\frac{\tilde{s}(t+\Delta)-\tilde{s}(t)}{e(t+\Delta)-e(t)} \underset{P}{\to} \frac{1}{\Delta}\int_0^a [H(t+\Delta-Q^{-1}(v))-H(t-Q^{-1}(v))]\,\mathrm{d}v$$
$$\overset{\text{def}}{=} 1 - \pi_\Delta(t).$$

By the local renewal theorem, as $t \to \infty$

$$\pi_\Delta(t) = 1 - \frac{1}{\Delta} a \frac{\Delta}{\mathbf{E}\tau_1} + o(1) = 1 - r + o(1).$$

Remark 1. The condition $\mathbf{P}(\tau_1 \geqslant h) = 1$ was introduced to reduce the technical complexity. The reader may obtain lemma 3 in the general case if, alongside the sets D, B_e, and B_q we also use the set B_τ, which bounds from below the values $\tau_{i_k}, \tau_{j_k}, k \leqslant L$, by Δ. Since $\mathbf{P}(B_\tau)$ can be made arbitrarily close to 1 by the choice of $\Delta > 0$, then the proof of the lemma still holds if h is replaced by Δ and B_τ is introduced where necessary under the probability sign, as was done with D and B_q in (2.19).

Remark 2. The u.s.m. condition in the proof of lemma 3, evidently, is also not essential and may be replaced by the s.m. condition. However, implementing this change leads to some technical problems.

Remark 3. The regular behaviour of $s(u)$ and 'the probability of refusal' $p(u)$ significantly condition the regular method of occupation of channels which ensured the C-convergence $e(u)/T \underset{C}{\to} u$. Under other regimes of occupation of channels, or for arbitrary initial conditions $(v_1, \ldots, v_n)$, where v_1 is the time at which the ith channel may begin to serve requests, the behaviour of these processes may be essentially different. If, for example, the v_i are independent random variables with density $G(v)/\mathbf{E}\tau_1$, then the distribution of $q(u)$ will at once be close to stationary.

Remark 4. The process $\sigma(u)$ (the limit of $s(u)/T$) is easily described using an integral equation. In fact, on $\Delta v = [v, v+\Delta]$, for $v < t_1$, approximately $T\Delta$ requests join the system and of these, up to time t, approximately $F(t-v)T\Delta$ ($F(u) = \mathbf{P}(\tau_1 \leqslant u)$) will be served. For $a < Q(t_0)$ and $u < t_1$, on Δv, the number of requests joining will be approximately the same as the number of channels freed (i.e. $\Delta s(v)$), and of these, up to time t, approximately $F(t-v)\Delta s(v)$ will be served. Consequently, for the limit process $\sigma(t)$ we may write

$$\sigma(t) = \int_0^t F(t-v)\,\mathrm{d}\sigma, \quad \text{for } t \leqslant t_1$$

$$\sigma(t) = \int_0^{t_1} F(t-v)\,\mathrm{d}v + \int_{t_1}^t F(t-v)\,\mathrm{d}\sigma(v) \quad \text{for } t \geqslant t_1.$$

This equation is easily rewritten in the form

$$\sigma(t) = \int_0^t F(t-v)\,\mathrm{d}\sigma(v) + c(t)$$

where

$$c(t) = \int_0^{min(t_1,t)} F(t-v)\,\mathrm{d}(v-\sigma(v)) = \int_0^{\min(t_1,t)} F(t-v)G(v)\,\mathrm{d}v$$

is a known function.

Taking Fourier–Stieltjes transforms

$$\tilde{\sigma}(\lambda) = \int \mathrm{e}^{\mathrm{i}\lambda t}\,\mathrm{d}\sigma(t), \quad \tilde{c}(\lambda) = \int \mathrm{e}^{\mathrm{i}\lambda t}\,\mathrm{d}c(t)$$

we obtain

$$\tilde{\sigma}(\lambda) = \frac{c(\lambda)}{1-f(\lambda)}, \quad \sigma(t) = \int_0^t H(t-v)\,\mathrm{d}c(v)$$

where

$$f(\lambda) = \mathbf{E}\,\mathrm{e}^{\mathrm{i}\lambda\tau_1}.$$

Remark 5. A result similar to theorem 3 may also be obtained in the case when service in each of the n channels proceeds via its 'own' stationary sequence not depending on the others (see Borisov and Borovkov, 1980).

We now pass on to a more precise description of the asymptotic behaviour of the processes $q(t)$.

2.2. Limit theorems for the number of busy lines for underloaded systems

In this section we will call a system *underloaded* if

$$\frac{n - TQ(t_0)}{A(T)} \to \infty$$

for some function $A(T) \to \infty$, $A(T) = o(T)$, which will be made more precise later. This means that here we are partly concerned even with loaded systems (from the point of view of the classification of section 2.1).

In section 2.1 we have seen that for such systems $q(t)$ is close to $TQ(t)$. In this section we will clarify the behaviour of the error $q(t) - TQ(t)$.

The function $m(u)$ in the relation $\mathbf{E}(e(u) - e(0)) = Tm(u)$ will now be arbitrary since under the conditions of this section $Q(t)$ will be defined by the equality (see the beginning of section 2.1)

$$Q(t) = \int_0^t G(t-u)\, dm(u), \quad G(t) = 1 - F(t), \quad F(t) = \mathbf{P}(\tau_j \leqslant t).$$

We will also need the notation

$$P(t) = \int_0^t F(t-u)\, dm(u).$$

Concerning the sequence $\{\tau_j\}$ we consider firstly the simplest case when this sequence is composed of independent random variables.

THEOREM 1. *Let the following conditions hold.*

(1) *There exist $t_0 > 0$, a nondecreasing function $m(t)$, and a function $B(T) \to \infty$ as $T \to \infty$ such that the sequence*

$$x(t) = \frac{e(t) - Tm(t)}{B(T)} \tag{2.24}$$

C-converges on $[0, t_0]$ to some process $\{\xi(t); 0 \leqslant tj \leqslant t_0\}$.*

(2) *The τ_j are independent and $P(t)$ satisfies a Hölder condition (for this it is enough that at least one of F and m satisfy this condition).*

(3) *The number $n = n_T$ is such that*

$$\frac{n - TQ(t_0)}{B(T)} \to \infty$$

as $T \to \infty$, where $A(T) = \max(B(T), \sqrt{T})$. *In addition†*

$$\frac{B(T)}{A(T)} \to B \geqslant 0, \quad \frac{\sqrt{T}}{A(T)} \to K \geqslant 0 \quad \textit{as } T \to \infty \ (\textit{so that } \max(B, K) = 1).$$

Under these conditions the process

$$z(t) = \frac{q(t) - TQ(t)}{A(T)}$$

* For the definition of C-convergence see section 1.2.

† It is actually possible to consider only three pairs of values of B and K: (1, 1), (0, 1), and (1, 0). However, a detailed discussion of these three cases would take up a great deal of space.

C-converges on $[0, t_0]$ *to*

$$\zeta(t) = B \int_0^t G(t-u)\,\mathrm{d}\xi(u) + K\theta(t) \tag{2.25}$$

where $\theta(t)$ *is a centred Gaussian process, not depending on* $\xi(t)$, *with covariance function*

$$\begin{aligned} R(t, t+u) &= \mathbf{M}\theta(t)\theta(t+u) \\ &= \int_0^t G(t+u-v)F(t-v)\,\mathrm{d}m(v), \quad u \geqslant 0. \end{aligned}$$

By the stochastic integral in (2.25) we mean (the result of integration by parts; the trajectory of $\xi(t)$ being almost surely continuous)

$$\int_0^t G(t-u)\,\mathrm{d}\xi(u) = \xi(t) - \int_0^t \xi(t-u)\,\mathrm{d}F(u).$$

Some remarks about this theorem will be made after its proof.

Proof. We first assume that $n = \infty$. As in section 2.1 we establish that

$$q(t) = \sum_{k=1}^{e(t)} \chi_k(t) \tag{2.26}$$

$$\chi_k(t) = \begin{cases} 1 & \text{if } v_k + \tau_k > t \\ 0 & \text{otherwise.} \end{cases}$$

We denote, as before, by $\mathfrak{E}$ the σ-algebra generated by the input stream, that is, events of the form $\{e(u) = l\}$, $u \leqslant t_0$, and consider the characteristic function of the joint distribution of $q(t)$ and $q(t+u)$, $u \geqslant 0$:

$$\begin{aligned} &\mathbf{E}\exp\{\mathrm{i}\lambda q(t) + \mathrm{i}\mu q(t+u)\} \\ &\quad = \mathbf{E}\mathbf{E}_{\mathfrak{E}} \exp\left\{\sum_{k=1}^{e(t)} (\lambda\chi_k(t) + \mu\chi_k(t+u)) + \sum_{k=e(t)+1}^{e(t+u)} \mu\chi_k(t+u)\right\}. \end{aligned} \tag{2.27}$$

But $e(t)$ and $e(t+u)$ are measurable with respect to $\mathfrak{E}$ and the random variables τ in (2.26), associated with different requests, are independent. Therefore, denoting the conditional expectation in (2.27) by $\mathbf{E}_{\mathfrak{E}}$, we obtain

$$\mathbf{E}_{\mathfrak{E}} = \prod_{k=1}^{e(t)} \mathbf{E}_{\mathfrak{E}} \exp \mathrm{i}(\lambda\chi_k(t) + \mu\chi_k(t+u)) \prod_{k=e(t)+1}^{e(t+u)} \mathbf{E}_{\mathfrak{E}} \exp \mathrm{i}\mu\chi_k(t+u).$$

The random variables $\lambda\chi_k(t) + \mu\chi_k(t+u)$ take the values $\lambda + \mu, \lambda, 0$ respectively on the sets $\tau_k > t + u - v_k$, $t + u - v_k \geqslant \tau_k > t - v_k$, $\tau_k \leqslant t - v_k$. Therefore, denoting, for brevity, the probabilities of these sets by

$$G_k = G(t+u-v_k), \quad f_k = F(t+u-v_k) - F(t-v_k), \quad F_k = F(t-v_k)$$

we obtain

$$\begin{aligned}
\log \mathbf{E}_{\mathfrak{E}} &= \sum_{k=1}^{e(t)} \log\left[F_k + f_k e^{i\lambda} + G_k e^{i(\lambda+\mu)}\right] \\
&\quad + \sum_{k=e(t)+1}^{e(t+u)} \log\left[1 - G_k(1 - e^{i\mu})\right] \\
&= \sum_{k=1}^{e(t)} \log\left[1 + f_k\left(i\lambda - \frac{\lambda^2}{2}\right) + G_k\left(i(\lambda+\mu) - \frac{(\lambda+\mu)^2}{2}\right)\right. \\
&\qquad\qquad \left. + (f_k + G_k)O(|\lambda|^3 + |\mu|^3)\right] \\
&\quad + \sum_{k=e(t)+1}^{e(t+u)} \log\left[1 + G_k\left(i\mu - \frac{\mu^2}{2}\right) + G_k O(|\mu|^3)\right].
\end{aligned}$$

Here the estimate is uniform with respect to T, t, u, and the space of elementary events. If we use, for sufficiently small λ and μ, the expansion of log as a series, then, after elementary transformations, we obtain

$$\begin{aligned}
\log \mathbf{E}_{\mathfrak{E}} = i\lambda &\sum_{k=1}^{e(t)} G(t-v_k) + i\mu \sum_{k=1}^{e(t+u)} G(t+u-v_k) \\
&- \frac{\lambda^2}{2} \sum_{k=1}^{e(t)} G(t-v_k)F(t-v_k) \\
&+ \frac{\mu^2}{2} \sum_{k=1}^{e(t+u)} G(t+u-v_k)F(t+u-v_k) \\
&- \lambda\mu \sum_{k=1}^{e(t)} G(t+u-v_k)F(t-v_k) \\
&+ O\left(\sum_{k=1}^{e(t)} G(t-v_k)(|\lambda|^3 + |\mu|^3)\right) \\
&+ O\left(\sum_{k=e(t)+1}^{e(t+u)} G(t+u-v_k)|\mu|^3\right)
\end{aligned}$$

where the estimate, as before, is uniform. It is easy to see that this expression may be written as

$$\begin{aligned}
\log \mathbf{E}_{\mathfrak{E}} = i\lambda &\int_0^t G(t-v)\,de(v) + i\mu \int_0^{t+u} G(t+u-v)\,de(v) \\
&- \frac{\lambda^2}{2} \int_0^t G(t-v)F(t-v)\,de(v)
\end{aligned}$$

$$- \frac{\mu^2}{2} \int_0^{t+u} G(t+u-v)F(t+u-v)\,\mathrm{d}e(v)$$
$$- \lambda\mu \int_0^t G(t+u-v)F(t-v)\,\mathrm{d}e(v)$$
$$+ \left(\int_0^t G(t-v)\,\mathrm{d}e(v) + \int_t^{t+u} G(t+u-v)\,\mathrm{d}e(v) \right)$$
$$O(|\lambda|^3 + |\mu|^3). \tag{2.28}$$

We note that we have a representation for $\mathbf{E}_{\mathfrak{C}}$ purely in terms of $e(t)$. It is easy to see that this will be just the same if the assumption $v_k \equiv 1$ is not used.

We recall now that, by (2.24),

$$e(t) = Tm(t) + B(T)x(t) \tag{2.29}$$

where $x(t)$ C-converges to $\xi(t)$. If we substitute (2.29) into (2.28) and use the notations $Q(t)$ and $R(t, t+u)$, which were introduced earlier, then we obtain

$$\log \mathbf{E}_{\mathfrak{C}} = \mathrm{i}\lambda TQ(T) + \mathrm{i}\mu TQ(t+u) + \mathrm{i}\lambda B(T)X(t) + \mathrm{i}\mu B(T)X(t+u)$$
$$- \frac{\lambda^2 T}{2} R(t,t) - \frac{\mu^2 T}{2} R(t+u, t+u) - \lambda\mu TR(t, t+u)$$
$$+ T(Q(t) + Q(t+u))O(|\lambda|^3 + |\mu|^3)$$
$$+ B(T)(|X(t)| + |X(t+u)| + X(t,u))O(\lambda^2 + \mu^2) \tag{2.30}$$

where the estimate O is uniform in all variables,

$$X(t) = \int_0^t G(t-v)\,\mathrm{d}x(v)$$
$$X(t,u) = \left| \int_0^t G(t-v)F(t-v)\,\mathrm{d}x(v) \right|$$
$$+ \left| \int_0^t G(t+u-v)F(t-v)\,\mathrm{d}x(v) \right|$$
$$+ \left| \int_0^{t+u} G(t+u-v)F(t+u-v)\,\mathrm{d}x(v) \right|.$$

This equality for $\mathbf{E}_{\mathfrak{C}}$ shows how $q(t)$ must be normalized to obtain the proper limit distribution. Namely, we must consider the joint distributions of

$$\frac{q(t) - TQ(T)}{A(T)}.$$

Then, on the basis of (2.30), we find

$$\log \mathbf{E}_{\mathfrak{C}} \exp \left\{ \mathrm{i}\lambda \frac{q(t) - TQ(T)}{A(T)} + \mathrm{i}\mu \frac{q(t+u) - TQ(t+u)}{A(T)} \right\}$$

$$= \mathrm{i}\lambda \frac{B(T)}{A(T)} X(t) + \mathrm{i}\mu \frac{B(T)}{A(T)} X(t+u) - \frac{\lambda^2}{2} \frac{T}{A^2(T)} R(t,t)$$

$$- \frac{\mu^2}{2} \frac{T}{A^2(T)} R(t+u, t+u) - \lambda\mu \frac{T}{A^2(T)} R(t, t+u)$$

$$+ [TA^{-3}(T)(Q(t) + Q(t+u)) + B(T)A^{-2}(T)(1 \times (t)$$

$$+ |X(t+u)| + X(t,u))]O(1)$$

where the estimate O is uniform in T and t, u, ω for fixed λ and μ. But the functionals

$$X(t) = x(t) - \int_0^t x(t-v)\,\mathrm{d}F(v)$$

as $X(t, u)$, are continuous in the uniform metric. Consequently, under the conditions of the theorem, the distribution of

$$\mathbf{E}_{\mathfrak{C}} \exp \left\{ \mathrm{i}\lambda \frac{q(t) - TQ(t)}{A(T)} + \mathrm{i}\mu \frac{q(t+u) - TQ(t+u)}{A(T)} \right\} \tag{2.31}$$

converges, as $T \to \infty$, to the distribution* of

$$\exp \left\{ \mathrm{i}\lambda B\Xi(t) + \mathrm{i}\mu B\Xi(t+u) - \frac{K^2}{2} r(\lambda, \mu) \right\} \tag{2.32}$$

where

$$\Xi(t) = \int_0^t G(t-v)\,\mathrm{d}\xi(v)$$

$$r(\lambda, \mu) = \lambda^2 R(t,t) + 2\lambda\mu R(t, t+u) + \mu^2 R(t+u, t+u).$$

The quantity (2.31) is uniformly bounded for real λ and μ. Therefore this also means the convergence of the expectations of (2.31) and (2.32). This means, for $T \to \infty$

$$\mathbf{E} \exp \{\mathrm{i}\lambda z(t) + \mathrm{i}\mu z(t+u)\}$$

$$\to \exp\left\{ -\frac{K^2}{2} r(\lambda, \mu) \right\} \mathbf{E} \exp \{\mathrm{i}\lambda B\Xi(t) + \mathrm{i}\mu B\Xi(t+u)\}.$$

* This is the only place where the C-convergence of x is used in the proof of convergence of the finite-dimensional distributions. Obviously it is sufficient here to require the convergence of the distributions of $X(t)$ and $X(t, u)$.

But the joint distribution of $\zeta(t)$ and $\zeta(t+u)$ (see (2.25)) has the same characteristic function.

The passage to arbitrary finite-dimensional distributions, obviously, brings in no real changes in these arguments.

The convergence of the finite-dimensional distributions for the case $n = \infty$ is proved. We will now prove the C-convergence of the distributions of $z(t)$ and $\xi(t)$ on $[0, t_0]$. We have

$$z(t) = \frac{q(t) - TQ(t)}{A(T)} = z_2(t) - z_1(t)$$

where

$$z_1(t) = A^{-1}(T) \sum_{k=1}^{e(t)} (G(t - v_k) - \chi_k(t))$$

$$z_2(t) = A^{-1}(T)\left[\sum_{k=1}^{e(t)} G(t - v_k) - TQ(t)\right]$$

$$= A^{-1}(T)\left[\int_0^t G(t - v)\,\mathrm{d}e(v) - TQ(t)\right] = A^{-1}(T)B(T)X(t). \tag{2.33}$$

Let

$$\omega_\Delta^C(z) = \sup_{\substack{|t-u| \leqslant \Delta \\ 0 \leqslant t,u \leqslant t_0}} |z(t) - z(u)|$$

be the 'modulus' of continuity in $C(0, t_0)$. Obviously

$$\omega_\Delta^C(z) \leqslant \omega_\Delta^C(z_1) + \omega_\Delta^C(z_2).$$

Therefore, if we manage to prove a compactness condition for the distribution of the z_i in $C(0, t_0)$:

$$\lim_{\Delta \to 0} \limsup_{T \to \infty} \mathbf{P}(\omega_\Delta^C(z_i) > \delta) = 0 \tag{2.34}$$

for the components z_i, then we obtain the required C-convergence of the processes z (see theorem 1 of section 1.2). For z_2 this condition is satisfied. In fact,

$$\Delta X(t) = \Delta x(t) - \int_{-\Delta}^{t} \Delta x(v)\,\mathrm{d}G(t - v)$$

(here we take $\Delta y(t) = y(t + \Delta) - y(t)$ and $x(v) = 0$ for $v < 0$). Hence it follows that

$$\omega_\Delta^C(X) \leqslant 2\omega_\Delta^C(x) \tag{2.35}$$

and, by the conditions of the theorem on C-convergence of the processes x, condition (2.34) is satisfied for z_2.

Now consider $z_1(t)$. We have

$$\Delta z_1(t) = A^{-1}(T)\left[\sum_{k=1}^{e(t+\Delta)} (\chi_k(t) - G(t - v_k)) - \sum_{k=1}^{e(t+\Delta)} (\chi_k(t+\Delta) - G(t + \Delta - v_k))\right].$$

In the first sum here $\chi_k(t) = G(t - v_k) = 1$ for $k > e(t)$, therefore we may write

$$\Delta z_1(t) = A^{-1}(T) \sum_{k=1}^{e(t+\Delta)} (\xi_k - p_k) \tag{2.36}$$

where $\xi_k = \chi_k(t) - \chi_k(t + \Delta)$ are random variables independent over $\mathfrak{E}$ and taking the value 1 with probability

$$p_k = -\Delta G(t - v_k) = \Delta F(t - v_k)$$

and the value 0 with the complementary probability $q_k = 1 - p_k$. If we denote

$$E(t) = \int_0^t F(t - v)\, de(v) \tag{2.37}$$

then

$$\sum_{k=1}^{e(t+\Delta)} p_k = \Delta \sum_{k=1}^{e(t)} F(t - v_k) = \Delta E(t).$$

We note now that

$$A^4(T)\mathbf{E}_{\mathfrak{E}}(\Delta z_1(t))^4 \leqslant \Delta E(t) + 3(\Delta E(t))^2. \tag{2.38}$$

In fact

$$\begin{aligned} A^4(T)\mathbf{E}_{\mathfrak{E}}(\Delta z_1(t))^4 &= \sum_k \mathbf{E}_{\mathfrak{E}}(\xi_k - p_k)^4 + 6\sum_{k<j} \mathbf{E}_{\mathfrak{E}}(\xi_k - p_k)^2(\xi_j - p_j)^2 \\ &= \sum_k p_k q_k (p_k^3 + q_k^3) + 6 \sum_{k<j} p_k q_k p_j q_j \\ &\leqslant \sum_k p_k q_k + 3\left(\sum_k p_k q_k\right)^2 \end{aligned}$$

from which (2.38) follows.

So we return to the proof of (2.34).

Suppose, without loss of generality, that $t_0 = 1$, $\Delta = 2^{-l}$, and put

$$\omega(t, t + \Delta) = \sup_{u \in [t, t+\Delta]} |z_1(t) - z_1(u)|$$

$$\omega_\Delta^{[N]} = \max_{|k/2^n - j/2^n| \leqslant \Delta} \left| z_1\left(\frac{k}{2^N}\right) - z_1\left(\frac{j}{2^N}\right)\right|.$$

Then, taking $N > l$, we obtain

$$\omega_\Delta^C(z_1) \leqslant \omega_\Delta^{[N]} + 2 \max_{0 \leqslant k \leqslant 2^N} \omega\left(\frac{k}{2^N}, \frac{k+1}{2^N}\right)$$

$$\mathbf{P}(\omega_\Delta^C(z_1) \geqslant 3\delta) \leqslant \mathbf{P}(\omega_\Delta^{[N]} \geqslant \delta) + \mathbf{P}\left(\bigcup_{k=0}^{2^N-1} \left\{\omega\left(\frac{k}{2^N}, \frac{k+1}{2^N}\right) \geqslant \delta\right\}\right). \quad (2.39)$$

Consider the first term. It is easy to see that

$$\bigcap_{r=l}^{N} \bigcap_{k=1}^{2^N} \left\{\left| z_1\left(\frac{k}{2^r}\right) - z_1\left(\frac{k-1}{2^r}\right)\right| < \frac{\delta}{r^2}\right\}$$

contains $\{\omega_\Delta^{[N]} < \delta\}$. Since the inverse inclusion holds for the complementary events we find (for $l \geqslant 3$)

$$\mathbf{P}(\omega_\Delta^{[N]} \geqslant \delta) \leqslant \mathbf{P}\left(\bigcup_{r=l}^{N} \bigcup_{k=1}^{2^r} \left\{\left| z_1\left(\frac{k}{2^r}\right) - z_1\left(\frac{k-1}{2^r}\right)\right| \geqslant \frac{\delta}{r^2}\right\}\right). \quad (2.40)$$

Using an inequality of the Chebyshev type for conditional probabilities (Doob, 1953, p. 35), and denoting

$$\Delta_k^r y = y\left(\frac{k}{2^r}\right) - y\left(\frac{k-1}{2^r}\right)$$

we obtain from (2.38)

$$\mathbf{P}_{\mathfrak{E}}\left(|\Delta_k^r z_1| \geqslant \frac{\delta}{r^2}\right) \leqslant [\Delta_k^r E + 3(\Delta_k^r E)^2] A^{-4}(T)\delta^{-4} r^8$$

$$\mathbf{P}_{\mathfrak{E}}\left(\bigcup_{k=1}^{2^r} \left\{|\Delta_k^r z_1| \geqslant \frac{\delta}{r^2}\right\}\right) \leqslant \left[E(1)\left(1 + 3 \max_{k \leqslant 2^r} \Delta_k^r E\right)\right] A^{-4}(T)\delta^{-4} r^8. \quad (2.41)$$

But, from the definition of $E(t)$ (see (2.37)), it follows that

$$E(t) = TP(t) + B(T)(x(t) - X(t))$$
$$\Delta_k^r E = T\Delta_k^r P + B(T)\Delta_k^r(x - X). \quad (2.42)$$

Since, by the Hölder condition on P,

$$\Delta_k^r P \leqslant c2^{-r\alpha} \quad (2.43)$$

for some $\alpha > 0$ and since

$$\Delta_k^r(x - X) \leqslant \omega_{2^{-r}}^C(x) \leqslant \omega_{2^{-l}}^C(x) \quad (2.44)$$

then on

$$A = \{\omega_{2^{-l}}^C(x) < 1, |x(1) - X(1)| < B(T)\}$$

the right-hand side of (2.41) does not exceed

$$[TP(1) + B^2(T)][1 + 3Tc2^{-r\alpha} + 3B(T)]A^{-4}(T)\delta^{-4}r^8$$
$$\leqslant c\delta^{-4}r^8(2^{-r\alpha} + A^{-1}(T)).$$

Consequently (2.40) implies

$$\begin{aligned}\mathbf{P}(\omega_\Delta^{[N]} \geqslant \delta) &\leqslant 1 - \mathbf{P}(A) + \sum_{r=l}^{N} c\delta^{-4}r^8(2^{-r\alpha} + A^{-1}(T)) \\ &\leqslant 1 - \mathbf{P}(A) + c\delta^{-4}(l^8 2^{-l\alpha} + N^9 A^{-1}(T)).\end{aligned}$$

If we take $N \leqslant T^{1/20}$ then, obviously,

$$\lim_{t\to\infty} \limsup_{T\to\infty} \mathbf{P}(\omega_\Delta^{[N]} \geqslant \delta) = 0.$$

We pass on to the second term in (2.39). By (2.36) we have

$$\begin{aligned}\omega(t, t + \Delta) &\leqslant A^{-1}(T) \max\left(\sum_{k=1}^{e(t+\Delta)} \xi_k, \Delta E(t)\right) \\ &\leqslant 2A^{-1}(T) \max\left(\sum_{k=1}^{e(t+\Delta)} \xi_k - \Delta E(t), \Delta E(t)\right) \\ &= 2 \max(\Delta z_1(t), A^{-1}(T)\Delta E(t)).\end{aligned}$$

Consequently

$$\begin{aligned}&\mathbf{P}\left(\bigcup_{k=1}^{2^N} \left\{\omega\left(\frac{k-1}{2^N}, \frac{k}{2^N}\right) \geqslant \delta\right\}\right) \\ &\qquad\leqslant \mathbf{P}\left(\bigcup_{k=1}^{2^N} \left\{|\Delta_k^N z_1| \geqslant \frac{\delta}{2}\right\}\right) + \mathbf{P}\left(\bigcup_{k=1}^{2^N} \left\{|\Delta_k^N E| \geqslant \delta\frac{A(T)}{2}\right\}\right). \quad (2.45)\end{aligned}$$

We need not estimate the first probability here since the probabilities just estimated in (2.40) and (2.41) are of broader events. In the second probability, by (2.42)–(2.44), we have

$$\Delta_k^N E \leqslant cT2^{-N\alpha} + B(T)\omega_{2^{-N}}^C(x).$$

Therefore, for sufficiently large T and $N \geqslant (2/\alpha)\log(T)$ (recall, the only restriction on N is $N \leqslant T^{1/20}$), the probability to be estimated in (2.45) does not exceed

$$\mathbf{P}\left(\omega_{2^{-N}}^C(x) \geqslant \frac{\delta}{4}\right) \to 0$$

as $T \to \infty$. Thus (2.34), and together with it the C-convergence of the processes $z(t)$ for $n = \infty$, is proved.

For finite n the result follows because the probability that z, constructed for

$n = \infty$, will reach the level n converges to 0 as $T \to \infty$. But below these levels z, for finite or infinite n, behaves asymptotically as for $n = \infty$. □

In the following remarks, for simplicity, we will only consider the most interesting case, corresponding to a homogeneous input stream, where the function $m(t)$ is linear:

$$m(t) = mu, \quad m > 0.$$

Remark 1. The origin of the two components in the representation of the limit process for

$$\frac{q(t) - TQ(t)}{A(T)}, \quad Q(t) = m \int_0^t G(u)\,du$$

is easily explained. From the number of requests $\Delta e(u) = Tm(\Delta) + B(T)\Delta x(u)$, which arrive in the interval $(u, u + \Delta)$, it follows from the De Moivre–Laplace theorem that approximately

$$\Delta e(u)G(t-u) + \sqrt{\Delta e(u)G(t-u)F(t-u)}\,w \tag{2.46}$$

of them will remain in the system up to time t, where w is some normally distributed variable. Since for different Δu the variables w are independent, then $\sqrt{\Delta}w$ may be represented as the increment $\Delta w(u)$ of a Wiener process and for (2.46) we obtain

$$TmG(t-u)\Delta + B(T)G(t-u)\Delta x(u) + \sqrt{TmG(t-u)F(t-u)}\,\Delta w(u). \tag{2.47}$$

Summing over Δu we arrive at the following approximate representation for $q(t)$:

$$Tm \int_0^t G(t-u)\,du + B(T) \int_0^t G(t-u)\,d\xi(u)$$
$$+ \sqrt{Tm} \int_0^t \sqrt{G(t-u)F(t-u)}\,dw(u).$$

If we subtract $Tm \int_0^t G(t-u)\,du$ and divide by $A(T)$ then, as $T \to \infty$, we obtain

$$B \int_0^t G(t-u)\,d\xi(u) + K \int_0^t \sqrt{mG(t-u)F(t-u)}\,dw(u). \tag{2.48}$$

It must be noted, however, that this expression is valid only for the *one-dimensional* distributions of $q(t)$. A representation for $q(t)$, as for *processes* not having such simple expressions, requires more precise discussions. To understand this it is sufficient to remark that the first components in (2.25) and (2.48) coincide, but the correlation function of the second (Gaussian) components are, in general, different.

Remark 2. Theorem 1, as also the subsequent results, makes it possible to study the joint limit distribution of $e(t)$ and $q(t)$. In particular, it is easy to write out the form of the limit distribution for the number of requests which are served.

2.3. Convergence to a stationary process

Clearly, in order that the distributions of $z(t) = (q(t) - TQ(t))/AT$ converge, as

$T \to \infty, t \to \infty$, to stationary it is necessary to impose further conditions on the characteristics of the system.

For example, we must require

$$\int_0^\infty G(t)\,\mathrm{d}t = \mathbf{E}\tau_j < \infty$$

(see the general theorem on existence of stationary distributions for $q(t)$ as $t \to \infty$ (Borovkov, 1976a; Miller and Sentilles, 1975). The necessity of this condition will be clear and immediate from what follows).

In addition, we must require that the input stream $e(t)$ becomes more and more homogeneous for large t.

For example, if $m(t) = mt$, the τ_j are bounded by some constant c, and $\xi(t)$ in the condition of theorem 1 of section 2.2 is a process with stationary increments, then stationarity of the limit distribution $\zeta(t)$ already holds for $t > c$. This follows immediately from theorem 1 of section 2.2.

In the case of unbounded τ_j or $m(t) \neq mt$, the following result holds.

Let, as before,

$$x(t) = \frac{e(t) - Tm(t)}{B(T)}, \quad G(t) = \mathbf{P}(\tau_j > t)$$

$$Q(t) = \int_0^t G(t - v)\,\mathrm{d}m(v).$$

THEOREM 1. *Suppose that the following conditions are satisfied.*

(1) *In each finite interval of values of t the processes $x(t_0 + t) - x(t_0)$ C-converge as $T \to \infty$, $t_0 = t_0(T) \to \infty$, to a process $\xi(t)$ with increments which are strictly stationary. $\xi(t)$ may be regarded as being given on the whole axis.*

In addition, uniformly in T and u*

$$\begin{aligned} \mathbf{E}|x(u + t) - x(u)| &\leqslant c \max(1, |t|) \\ \mathbf{E}|\xi(u + t) - \xi(u)| &\leqslant c \max(1, |t|). \end{aligned} \tag{2.49}$$

* Instead of (2.49) we could have required

$$\mathbf{P}\Big(\bigcup_{|t|>N} \{|x(u + t) - x(u)| > c|t|\}\Big) \to 0$$

$$\mathbf{P}\Big(\bigcup_{|t|>N} \{|\xi(u + t) - \xi(u)| > c|t|\}\Big) \to 0$$

as $N \to \infty$ uniformly with respect to T. This condition, or condition (2.49), together with C-convergence on finite intervals, will guarantee the defined form of convergence of processes given on the whole axis, considered in Borovkov (1972d) and Sakhanenko (1974) (see convergence of the so-called Λ-continuous functionals).

The function $m(t)$ takes the form

$$m(t) = \int_0^t (m + \varepsilon(u))\,du, \quad |\varepsilon(u)| < cu^{-\alpha} \tag{2.50}$$

for some $\alpha > 0$.

(2) *The random variables τ_j are independent*†

$$G(t) \leqslant ct^{-\alpha-1}. \tag{2.51}$$

(3) *The number of channels n is such that*

$$\frac{n - Tm\mathbf{E}\tau_j}{A(T)} \to \infty$$

as $T \to \infty$, where, as before, $A(T) = \max(B(T), \sqrt{T})$,

$$\frac{B(T)}{A(T)} \to B \geqslant 0, \quad \frac{\sqrt{T}}{A(T)} \to K \geqslant 0.$$

Under these conditions, for t_0 so large that $t_0^{2\alpha_T - 1} \to \infty$, the process

$$z(t) = \frac{q(t_0 + t) - Tm\mathbf{E}\tau_j}{A(T)}$$

C-converges on each finite interval of values of t to a stationary process

$$\zeta(t) = K\theta(t) + B\int_{-\infty}^t G(t - v)\,d\xi(v)$$

where $\theta(t)$ is a stationary centred Gaussian process, not depending on $\xi(t)$, with covariance function

$$R(u) = \mathbf{E}\theta(t)\theta(t + u) = m\int_0^\infty F(v)G(v + u)\,dv.$$

Thus, entry to a stationary regime occurs, beginning with values $t > T^{1/(2\alpha)}$. It will be clear from the proof that for small values of t the distribution of $z(t)$, in general, does not converge to stationary.

Proof. For simplicity we restrict to the case $n = \infty$. As in the theorem of section 2.2, we begin with the proof of convergence of the finite-dimensional distributions. Since (2.26) still holds we can use the ready-made formula (2.30) for the conditional characteristic function $\mathbf{E}_{\mathfrak{E}}$, relative to $\mathfrak{E} = \sigma(e(t), t \geqslant 0)$, of the

† Apparently, instead of (2.50) and (2.51) it is enough to require

$$Q(t) = m\mathbf{E}\tau_j + \int_0^t \varepsilon(v)\,dv, \quad |\varepsilon(v)| < cv^{-\alpha-1}.$$

joint distribution of $q(t_0 + t)$ and $q(t_0 + t + u)$:

$$\mathbf{E}_{\mathfrak{E}} = \mathbf{E}_{\mathfrak{E}}\, e^{i\lambda q(t_0+t)+i\mu q(t_0+t+u)}.$$

By this formula (put $t_0 + t = s$ for brevity)

$$\begin{aligned}\log \mathbf{E}_{\mathfrak{E}} &= i\lambda TQ(s) + i\mu TQ(s+u) + i\lambda B(T)X(s) + i\mu B(T)X(s+u)\\ &\quad - \tfrac{1}{2}\lambda^2 TR(s,s) - \tfrac{1}{2}\mu^2 TR(s+u, s+u) - \lambda\mu TR(s, s+u)\\ &\quad + T(Q(s) + Q(s+u))O(|\lambda|^3 + |\mu|^3)\\ &\quad + B(T)(|X(s)| + |X(s+u)| + X(s, s+u))O(\lambda^2 + \mu^2)\end{aligned} \tag{2.52}$$

where the estimate O is uniform in all variables.

Consider the terms in (2.52). We have

$$\begin{aligned}Q(s) &= m\int_0^s G(s-v)\,dv + \int_0^s G(s-v)\varepsilon(v)\,dv\\ &= m\mathbf{E}\tau_j - m\int_s^\infty G(v)\,dv + \int_0^s G(s-v)\varepsilon(v)\,dv.\end{aligned}$$

Decomposing the last integral into two parts $\int_0^{s/2} + \int_{s/2}^s$, we easily obtain (see conditions (2.50) and (2.51)) the estimate $O(s^{-\alpha})$ for it. Since, in addition, it is obvious that $\int_s^\infty G(v)\,dv = O(s^{-\alpha})$ then, for fixed t and u, we find

$$Q(s) = m\mathbf{E}\tau_j + O(t_0^{-\alpha}), \quad Q(s+u) = m\mathbf{E}\tau_j + O(t_0^{-\alpha}).$$

We now consider the process

$$X(s) = \int_0^s G(s-v)\,dx(v) = \int_0^{s-N} + \int_{s-N}^s. \tag{2.53}$$

The second integral here, equal to $\int_0^N G(u)\,dx(s-u)$, converges for fixed N, by the C-convergence of $x(t_0 + v) - x(t_0)$, to

$$\int_0^N G(u)\,d\xi(t-u) = \int_{t-N}^t G(t-v)\,d\xi(v).$$

The first integral in (2.53), by the choice of N, can be made arbitrarily small (in probability). In fact,

$$\int_0^{s-N} G(s-v)\,dx(v) = x(s-N)G(s) + \int_0^{s-N} (x(s-N) - x(v))\,dG(s-v).$$

By condition (2.49), as $s \to \infty$

$$\begin{aligned}\mathbf{E}\left|\int_0^{s-N} G(s-v)\,dx(v)\right| &\leqslant c(s-N)G(s) + c\int_0^{s-N} \max(1, s-N-v)\,dG(s-v)\\ &= o(1) + cNG(N) + c\mathbf{E}(\tau_j; |\tau_j| \geqslant N).\end{aligned}$$

This expression, by choice of N, can be made arbitrarily small uniformly with respect to all sufficiently large T.

Since a similar conclusion holds for

$$\mathbf{E}\left|\int_{-\infty}^{t-N} G(t-v)\,\mathrm{d}\xi(v)\right|$$

then we obtain that the distribution of $X(s)$ converges to the distribution of

$$\Xi(t) = \int_{-\infty}^{t} G(t-v)\,\mathrm{d}\xi(v).$$

In exactly the same way we show the convergence of the distributions of $\mathrm{i}\lambda X(s) + \mathrm{i}\mu X(s+u)$ and $X(s, u)$.

We turn, finally, to the function $R(s, s+u)$ which appears in (2.52). We have (compare with the estimate of $Q(s)$)

$$R(s, s+u) = m\int_0^s G(s+u-v)F(s-v)\,\mathrm{d}v + O(s^{-\alpha})$$

$$= m\int_0^\infty G(v+u)F(v)\,\mathrm{d}v + O(t_0^{-\alpha}).$$

Substituting these results into (2.52) we find that the distribution of

$$\mathbf{E}_{\mathfrak{C}} \exp\{\mathrm{i}\lambda z(t) + \mathrm{i}\mu z(t+u)\}$$

converges as $T \to \infty$ to the distribution of

$$\exp\{\mathrm{i}\lambda B\Xi(t) + \mathrm{i}\mu B\Xi(t+u) - \tfrac{1}{2}K^2[\lambda^2 R(0) + 2\lambda\mu R(u) + \mu^2 R(0)]\}$$

provided that $t_0^{-\alpha}T = o(A(T))$. Since this is satisfied for $t_0^{-2\alpha} \gg T$ then, reasoning as in the proof of the theorem of section 2.2, we arrive at the convergence of the two-dimensional distributions of $z(t)$. The convergence of arbitrary finite-dimensional distributions, obviously, can be proved in precisely the same way.

We now prove C-convergence. Here, as in the theorem of section 2.2, the process* $z(t)$ can be represented (with precision up to the variable $TA^{-1}(T)(Q(s) - m\mathbf{E}\tau_j) = o(1)$, $s = t_0 + t$) in the form of a sum of two components $z_1(t)$ and $z_2(t)$, defined in (2.33) (replacing t by $s = t_0 + t$ on the right-hand sides). $z_2(t)$ is equal to

$$z_2(t) = \frac{B(T)}{A(T)} X(s).$$

The increment $\Delta z_2(t)$ is given by

$$\Delta z_2(t) = \frac{B(T)}{A(T)}\left[\Delta x(s) - \int_{-\Delta}^{s} \Delta x(v)\,\mathrm{d}G(s-v)\right].$$

* The notation $z(t)$ here and in section 2.2 have different meanings.

Since

$$\mathbf{E}\left|\int_{-\Delta}^{s-N} \Delta x(v)\, dG(s-v)\right| \leqslant c \int_{-\Delta}^{s-N} dG(s-v) \leqslant cG(N)$$

by the choice of N, can be made arbitrarily small, and

$$\left|\int_{s-N}^{s} \Delta x(v)\, dG(s-v)\right| \leqslant \omega_{\Delta}^{C,N}(x_0)$$

where $\omega_{\Delta}^{C,N}(x_0)$ is the modulus of continuity of $x_0(t) = x(t_0+t) - x(t_0)$ on a segment of length N, then

$$\limsup_{T\to\infty} \mathbf{P}(\omega_{\Delta}^{C}(z_2) > \delta) \leqslant \limsup_{T\to\infty} \left[\mathbf{P}(\omega_{\Delta}^{C,N}(x_0) > \delta/2) + \mathbf{P}\left(\left|\int_{-\Delta}^{s-N} x(v)\, dG(s-v)\right| > \delta/2\right)\right]. \quad (2.54)$$

Given $\varepsilon > 0$ we first choose N so that the second term is less than $\varepsilon/2$, and then Δ so that the first term is less than $\varepsilon/2$. Since the left-hand side of (2.54) does not depend on N, then this will mean compactness in C of the distributions of $z_2(t)$.

Now consider

$$z_1(t) = A^{-1}(T) \sum_{k=1}^{e(s)} (\chi_k(s) - G(s - v_k))$$

and follow the same method of proof as used in section 2.2. Without loss of generality, we need only estimate the modulus of continuity $\omega_{\Delta}^{C}(z_1)$ on the unit interval $t \in [0, 1]$ $(t_0 \leqslant s \leqslant t_0 + 1)$. Introducing again the notations $\omega(t, t+\Delta)$, $\omega_{\Delta}^{[N]}$, we obtain the inequalities (2.39)–(2.41), but with E replaced by $E_0(t) = E(t_0 + t) - E(t_0)$. Before using (2.42) we note that by our condition $m(v)$ has a derivative $m'(v) = m + \varepsilon(v) \to m$ as $v \to \infty$. Consequently, $P(v)$ has a derivative

$$P'(v) = \int_0^v (m + \varepsilon(u))\, dF(v - u) \to m$$

as $v \to \infty$. Consequently, for sufficiently large t_0,

$$\Delta P_0(t) \leqslant 2m\Delta$$

where the index 0 on a function $y(t)$ means that we are considering the difference $y_0(t) = y(t_0 + t) - y(t_0)$.

Therefore, by (2.44), with the obvious changes, we obtain all the inequalities which held in section 2.2 (all of the changes are connected with the replacement of E, P, x, and X by E_0, P_0, x_0, and X_0). As a result, for $N \leqslant T^{1/20}$, we have

$$\lim_{t\to\infty} \limsup_{T\to\infty} \mathbf{P}(\omega_{\Delta}^{[N]} \geqslant \delta) = 0$$

and it remains for us to obtain the same relation for

$$\mathbf{P}\left(\bigcup_{k=1}^{2^N}\left\{\omega\left(\frac{k-1}{2^N},\frac{k}{2^N}\right)\geqslant\delta\right\}\right).$$

But here the reasoning is as in the proof in section 2.2. This completes the proof of C-convergence to a stationary process and the proof of the theorem. □

Remark 1. From the proof it is easy to see that with exponential decrease of $G(t)$ and $\varepsilon(t)$ the entry of the system to a stationary regime would occur with a value of t of the order of $\log T$.

If $\xi(t)$ is a Wiener process, then the stationary process

$$\zeta(t)=K\theta(t)+B\int_{-\infty}^{t}G(t-u)\,\mathrm{d}\xi(u)$$

together with $\theta(t)$ will be Gaussian. Its correlation function is

$$R(u)=mK^2\int_0^\infty G(v+u)F(u)\,\mathrm{d}v+B^2\int_0^\infty G(v)G(v+u)\,\mathrm{d}v$$

$$=\int_0^\infty [K^2mF(v)+B^2G(v)]G(u+v)\,\mathrm{d}v.$$

In the special case when τ_j is exponentially distributed

$$G(t)=\mathrm{e}^{-\beta t},\quad \beta>0$$

we obtain

$$R(u)=\frac{K^2m+B^2}{2\beta}\,\mathrm{e}^{-\beta u}.$$

This means that $\zeta(t)$ is a Markovian stationary Gaussian process taking the form

$$\zeta(t)=\sqrt{K^2m+B^2}\int_{-\infty}^{t}\mathrm{e}^{-\beta(t-v)}\,\mathrm{d}w(v)$$

where $w(t)$ is a stationary Wiener process. We will obtain the same family of processes as limits in chapter 3 and will describe them as diffusion processes with constant coefficient of diffusion and with coefficient of drift $a(x)=-ax$, $a>0$ (see section 1.3).

To conclude this section we make the following remark. Consider the limit distribution of the number of served requests $s(t)=e(t)-q(t)$ (suppose $n=\infty$). Since

$$y(t)=\frac{s(t)-TP(t)}{A(T)}=\frac{B(T)}{A(T)}x(t)-\frac{q(t)-TQ(t)}{A(T)}$$

then from the above proof it is easy to obtain a result on the convergence of $y(t_0 + t)$ to a process

$$-K\theta(t) + B\left(\xi(t) - \int_{-\infty}^{t} G(t-v)\,\mathrm{d}\xi(v)\right)$$

$$= -K\theta(t) + B\int_{-\infty}^{t} \xi(v)\,\mathrm{d}G(t-v) \quad (2.55)$$

with stationary increments (instead of $-K\theta(t)$ here we can write $K\theta(t)$, that is, the distributions of these processes coincide).

In addition, $s(t)$, considered as a new input stream, will satisfy all the conditions imposed on $e(t)$ in the theorem proved in this section. The centring function of $s(t)$ is

$$P(t) = \int_0^t F(t-v)\,\mathrm{d}m(v) = \int_0^t (1 - G(t-v))(m + \varepsilon(v))\,\mathrm{d}v$$

$$= \int_0^t (m + \varepsilon_P(v))\,\mathrm{d}v, \quad \text{where } |\varepsilon_P(v)| < cv^{-\alpha}.$$

The normalizing multiplier is $A(T)$ and the limit process is written in (2.55). It is easy to verify that the remaining inequality (2.51) in condition (1) of the theorem is also satisfied. The processes $q(t)$ and $s(t)$ in the conditions of both sections 2.2 and 2.3 may be written in the form

$$q(t) = \int_0^t G(t-v)\,\mathrm{d}e(v) + \sqrt{T}\,\theta_T(t)$$

$$s(t) = \int_0^t F(t-v)\,\mathrm{d}e(v) - \sqrt{T}\,\theta_T(t)$$

where $\theta_T(t)$ does not depend on $e(t)$ and C-converges to the Gaussian process $\theta(t)$ defined in the statement of the theorem. This equality is valid in every case since the equality is *in distribution.*

These remarks will be used in the following section.

2.4.* The connection with branching processes with intensive immigration

We now consider, from a somewhat different viewpoint, the problem of the distribution of $q(t)$, which was discussed in the previous two sections. We represent $e(t)$ as a process which describes the birth of particles ($e(t)$ is equal to the number of particles born at time t). The jth particle lives for a time τ_j, then dies. Then, obviously, $q(t)$ may be interpreted (see the indicator representation (2.26) for $q(t)$) as the number of particles alive at time t and $s(t)$ as the number of dead particles.

Suppose now that particles at the end of their lifetime do not perish, but divide, creating a random number of particles of the same type. That is, the kth particle at a time τ_k after its birth divides into μ_k new particles, where the random variables μ_k are independent of $\{\tau_j\}$ and each other,

$$\mathbf{E}\mu_k = \mu < 1, \quad \mathbf{D}\mu_k = \sigma^2 < \infty.$$

We obtain, thus, a stream of particles $e_2(t)$ of the second generation. Each particle of the second generation again lives a random time, distributed as τ_1, and divides by the same law as the particles of the first generation independently of the fate of the other particles. The particles of the third and subsequent generations behave in precisely the same way.

We have therefore obtained a branching process with transformations depending on age and with immigration (see Sevast'yanov, 1971). Immigration is described by $e(t)$. If this process satisfies the conditions of the theorems of sections 2.2 and 2.3, then we will speak of *intensive* immigration.

Since $\mu = \mathbf{E}\mu_k < 1$, then the processes under discussion are *subcritical.* Here we consider a series array ($e(t)$ depends on an increasing parameter T), in which, however, the distribution $F(t)$ of the lifetime of a particle and the distribution of the number of descendants (the distribution of μ_k) remain fixed.

Thus, let an immigration process $e(t)$ be given, which satisfies the conditions of sections 2.2 and 2.3. Then, as remarked at the end of section 2.3, $s(t)$ may be represented in the form

$$s(t) = \int_0^t F(t - v)\, de(v) - \sqrt{T}\theta(t) = F * e(t) - \sqrt{T}\theta(t)$$

where $\theta(t)$ (for brevity the index T is omitted), in each fixed interval, C-converges to a Gaussian process with covariance function

$$R(t, t + u) = \int_0^t G(t + u - v)F(t - v)\, dm(v) = H_u * m(t) \tag{2.56}$$

where $H_u(t) = F(t)G(t + u)$.

Consider the increment $\Delta s(t) = \Delta E(t) - \sqrt{T}\Delta\theta(t)$ of $s(t)$ on $[t, t + \Delta]$. For small, but fixed, Δ this number increases without limit as $T \to \infty$ and is the number of particles in the first generation whose lifetime ends in this interval. The number of particles of the second generation appearing in this interval can be represented (in distribution) in the form

$$\mu\Delta s(t) - \sigma w(\Delta s(t)) \tag{2.57}$$

where w is a process C-converging to a standard Wiener process.

Supposing that for non-intersecting time intervals, and a fixed trajectory of $s(t)$, the numbers of particles born of the second generation are independent, then we can approximately (with the same limit distribution) write (2.57) as

$$\mu\Delta F * e(t) - \mu\sqrt{T}\Delta\theta(t) - \sigma\sqrt{T}\Delta w(F * m(t)).$$

This means that the total number of particles of the second generation appearing up to time t can (in distribution) be written

$$e_2(t) = \mu F * e_1(t) - \sqrt{T}[\mu\theta_1(t) + \sigma w_1(F * m_1(t))] \tag{2.58}$$

where $e_1 = e, \theta_1 = \theta, w_1 = w$, and $m_1 = m$ are the processes corresponding to the stream of particles of the first generation. We have provided them with the index 1 in order to distinguish them in future from the similar processes for the stream of the second, third, etc. generations. From the remarks at the end of section 2.3 it follows that $e_2(t)$, together with $e_1(t) = e(t)$, satisfies the conditions of the theorems in sections 2.2 and 2.3.

Formula (2.58) explains the rule by which the process of appearance of particles of the second generation is formed from the process $e_1(t)$ of particles of the first generation. But it is quite obvious that the number of particles of the kth generation will be similarly connected with the number of particles of the $(k-1)$st generation. Each time there will appear a new pair of processes θ_{k-1}, w_{k-1}, which do not depend on the previous ones.

Since the function $m_2(t) = \mu F * m_1(t)$ corresponds to $e_2(t)$, then for the particles of the third generation we obtain

$$\begin{aligned} e_3(t) &= \mu F * e_2(t) - \sqrt{T}[\mu\theta_2(t) + \sigma w_2(F * m_2(t))] \\ &= \mu^2 F^{(2)} * e_1(t) - \sqrt{T}[\mu^2 F * \theta_1(t) + \sigma\mu F * w_1(F * m_1)(t)] \\ &\quad - \sqrt{T}[\mu\theta_2(t) + \sigma w(\mu F^{(2)} * m_1(t))] \end{aligned}$$

where $F^{(k)}$, the kth convolution of F, is equal to the distribution of $\tau_1 + \cdots + \tau_k$. For the fourth generation

$$\begin{aligned} e_4(t) &= \mu^3 F^{(3)} * e(t) - \sqrt{T}[\mu^3 F^{(2)} * \theta_1(t) + \sigma\mu^2 F^{(2)} * w_1(F * m)(t)] \\ &\quad - \sqrt{T}[\mu^2 F * \theta_2(t) + \sigma\mu F * w_2(\mu F^{(2)} * m)(t)] \\ &\quad - \sqrt{T}[\mu\theta_3(t) + \sigma w_3(\mu^2 F^{(3)} * m(t))] \end{aligned}$$

etc. Carrying out a formal summation we obtain a representation for $\mathscr{E}(t) = \sum_{k=1}^{\infty} e_k(t)$, which describes the total appearance of particles of all generations. Namely, if we denote $\Phi(t) = \sum_{k=0}^{\infty} \mu^k F^{(k)}$, then

$$\begin{aligned} \mathscr{E}(t) &= \Phi * e(t) - \sqrt{T}\mu\Phi * (\theta_1 + \theta_2 + \cdots)(t) \\ &\quad - \sqrt{T}\sigma\Phi * (w_1(F * m) + w_2(\mu F^{(2)} * m) + \cdots)(t) \\ &= \Phi * e(t) - \sqrt{T}[\mu\Phi * \theta(t) + \sigma\Phi * w(F * \Phi * m)(t)] \end{aligned}$$

where, by the independence of θ_j and w_j, the limit processes θ and w are again respectively Gaussian and Wiener processes $(w_1(g(t)) + w_2(h(t))$ is distributed in the limit as $w_1(g(t) + h(t)))$. Here the correlation function $R(t, t+u)$ corresponding to $\theta(t)$ is equal to the sum of the correlation functions (2.55), where instead of m we successively substitute $m_1 = m$, $m_2 = \mu F * m$, $m_3 = \mu^2 F^{(2)} * m$, etc.:

$$R(t, t+u) = H_u * \Phi * m(t).$$

The function

$$(1-\mu)\Phi(t) = \sum_{k=0}^{\infty} \mu^k (1-\mu) F^{(k)}(t)$$

is the distribution function of the sum

$$S_v = \sum_{j=1}^{v} \tau_j \tag{2.59}$$

of a random number of τ_j, $j = 1, 2, \ldots$, where v does not depend on $\{\tau_j\}$ and has exponential distribution

$$\mathbf{P}(v = k) = (1-\mu)\mu^k, \quad k = 0, 1, \ldots \tag{2.60}$$

(we take $S_0 = 0$).

It follows from this that

$$\mu\Phi * \theta(t) + \sigma\Phi * w(F * \Phi * m)(t)$$

will also converge to a Gaussian process whose correlation function, on the basis of the preceding, is easy to write explicitly.

Thus, we have found a limit representation for the total number of particles which have appeared in the system up to time t.

We now find a similar representation for the basic characteristic of the processes under discussion—the number of live particles at time t.

As we already know, the number of particles of the first generation which do not live up to time t is

$$s_1(t) = s(t) = F * e_1(t) - \sqrt{T}\theta_1(t).$$

The number of particles of the second generation which do not live up to time t is equal to

$$\begin{aligned} s_2(t) &= F * e_2(t) - T\theta_2(t) \\ &= \mu F^{(2)} * e(t) - \sqrt{T}[\mu F * \theta_1(t) + \sigma F * w_1(F * m)(t)] \\ &\quad - \sqrt{T}\theta_2(t) \end{aligned}$$

etc. Summing as before we obtain a process $\mathscr{S}(t)$ which expresses the total number of particles of all generations which do not live up to time t:

$$\mathscr{S}(t) = F * \Phi * e(t) - \sqrt{T}[\Phi * \theta(t) + F * \Phi * w(F * \Phi * m)(t)].$$

Hence we find the desired expression for the number of particles $\mathcal{Q}(t) = \mathcal{E}(t) - \mathcal{S}(t)$ at time t:

$$\mathcal{Q}(t) = G * \Phi * e(t) + \sqrt{T}[(1-\mu)\Phi * \theta(t) - \sigma G * \Phi * w(F * \Phi * m)(t)].$$

Under the conditions of sections 2.2 and 2.3 this relation allows one to obtain the form of the limit processes for $\mathcal{Q}(t)$.

For example, under the conditions of theorem 1 of section 2.2, the process

$$\frac{\mathcal{Q}(t) - TG * \Phi * m(t)}{A(T)}$$

will C-converge to

$$BG * \Phi * \xi(t) + K[(1-\mu)\Phi * \theta(t) - G * \Phi * w(F * \Phi * m)(t)]$$

(to avoid introducing new notation here we have denoted the limit processes by the same letters θ and w).

Under the conditions of section 2.3, as $T \to \infty$, $t \to \infty$, we have

$$\begin{aligned} (1-\mu)\Phi * m(t) &= \int_0^t (m + \varepsilon_1(v))\,\mathrm{d}v \\ F * (1-\mu)\Phi * m(t) &= \int_0^t (m + \varepsilon_2(v))\,\mathrm{d}v \end{aligned} \tag{2.61}$$

where $\varepsilon_1(v)$ and $\varepsilon_2(v)$ satisfy the same conditions as the function $\varepsilon(v)$ in the theorem of section 2.3:

$$R(t, t+u) = H_u * \Phi * m(t) \to R(u) = \frac{m}{1-\mu}\int_0^\infty G(u+v)F(v)\,\mathrm{d}v = \frac{m}{1-\mu}R(u)$$

where $R(u)$ is defined in the theorem of section 2.3. The correlation function of $(1-\mu)\Phi * \theta(t)$, for large t, will converge to

$$R^*(u) = (1-\mu)^2\int_0^\infty\int_0^\infty R(u+v-z)\,\mathrm{d}\Phi(v)\,\mathrm{d}\Phi(z) = \mathbf{E}R(u+S_v-S'_v)$$

where S'_v is distributed as S_v and does not depend on S_v.

Since, by (2.61),

$$G * \Phi * m(t) = \frac{m\mathbf{E}\tau_j}{1-\mu} + O(t^{-\alpha})$$

then for $T \to \infty$ and $t_0^{2\alpha} \gg T$

$$\frac{\mathcal{Q}(t_0+t) - Tm\mathbf{E}\tau_j/(1-\mu)}{A(T)}$$

will C-converge on each finite interval of values of t to the stationary process

$$B\int_{-\infty}^{t}(G*\Phi)(t-v)\,\mathrm{d}\xi(v)$$
$$+K\left[\theta^*(t)-\sigma\sqrt{\frac{m}{1-\mu}}\int_{-\infty}^{t}G*\Phi(t-v)\,\mathrm{d}w(v)\right] \quad (2.62)$$

where $w(t)$ is a standard Wiener process and $\theta^*(t)$ is a standard Gaussian process with correlation function

$$R^*(u)=\frac{m}{1-\mu}\mathbf{E}R(u+S_\nu-S'_\nu).$$

R is defined in the theorem of section 2.3 and S_ν, S'_ν are independent and identically distributed (see (2.59), (2.60)). All three processes $\xi(t)$, $\theta^*(t)$, and $w(t)$ in (2.62), which arise by the randomness of the input stream, the lifetime, and the number of descendants, respectively, are independent.

We may also note here that $G*\Phi$ can be represented as a difference

$$G*\Phi(t)=\mathbf{P}(S_\nu>t)-\mathbf{P}(S_{\nu+1}>t).$$

The arguments of this section, although fairly clear, are not rigorous. This is because the missing foundation for the limit transitions (the problem is essentially technical) would significantly increase the size of the section. Therefore we mention only the main area which requires some effort to make the arguments precise—the foundation of the formal summation of the processes $e_k(t)$ and $s_k(t)$. However, the problem here is essentially simplified by the exponential decrease of the terms.

2.5.* On limit processes for loaded systems with refusals and with queues

We have seen (sections 2.2 and 2.3) that if the number of service channels $n=n_T$ is such that

$$\frac{n-TQ(t_0)}{A(T)}\to\infty \quad (2.63)$$

as $T\to\infty$, then the form of the service system (the presence of refusals or queues) does not influence the limit distribution of the number of busy lines $q(t)$. However, if

$$n=TQ(t_0)+DA(T) \quad (2.64)$$

where D is finite and does not depend on T, then the position is quite different.

Case (2.64) is naturally called a *transition* between the problems considered in sections 2.2 and 2.3 and the cases when the number of channels is considerably less than the critical value $TQ(t_0)$ (see section 2.7).

Here the type of service system becomes significant and its behaviour will be different depending on the behaviour of the requests which make the system busy.

For simplicity, we suppose here that $m(t) = t$, $B(T) = A(T) = \sqrt{T}$.

If the question is of an *essentially nonstationary regime* (t_0 fixed, $Q(t_0) < \mathbf{E}\tau_j$) and here $Q'(T_0) > 0$ exists, then the passage to case (2.64) is an insipid problem. In fact, under these conditions, for $t < t_0 - \delta_T$, where δ_T decreases just slower than $A(T)T^{-1}$, we may assume that condition (2.63) is satisfied. From here it is easy to obtain that the distribution of $(q(t_0) - TQ(t_0))/\sqrt{T}$ will, as before, (in the theorems of section 2.2) be described by the distribution of

$$\zeta(t_0) = \theta(t_0) + \int_0^{t_0} G(t_0 - u)\, \mathrm{d}\xi(u)$$

for systems with queues, and by the distribution of

$$\min(D, \zeta(t_0))$$

for systems with refusals.

The situation is much more complex with the description of *stationary* distributions of $q(t)$ in case (2.64). Suppose, for simplicity, that τ_j is bounded with probability 1 by some constant c, and consider values $t > c$

$$n = T\mathbf{E}\tau_j + D\sqrt{T}.$$

We will give here only some ideas and nonrigorous arguments with the aim of characterizing the form of the limit distribution of $q(t)$ in case (2.64) for systems with refusals and systems with queues.

We first consider systems with refusals and present one of the possible methods of constructing limit processes. For this we assume that $n = \infty$ and introduce 'randomized' refusals, which are of independent interest. Namely, we suppose that each request joining the system is refused with probability ϕ independently (for a fixed value of q at the time of arrival of the request) of the behaviour of the other requests. We will suppose that ϕ is continuously dependent on the normalized number of busy lines $z = (q - T\mathbf{E}\tau_j)/\sqrt{T}$. Then of the requests $\Delta e(u) = e(u + \Delta) - e(u) = T\Delta + T\Delta x(u)$ which arrive in $(u, u + \Delta)$, by the De Moivre–Laplace theorem approximately

$$\Delta e(u)G(t-u)[1-\phi(z(u))] + \sqrt{\Delta e(t)G(t-u)[1-\phi(z(u))][1-G(t-u)(1-\phi(z(u)))]}\,w \quad (2.65)$$

will remain in the system up to time t, where w is normally distributed with parameters $(0, 1)$. Since the variables w are independent for different Δu, then $\sqrt{\Delta}w$ can be represented as the increment $\Delta w(u)$ of some Wiener process. Summing over Δu we arrive at the following rough representation for $q(t)$:

$$T\int_0^t G(t-u)[1-\phi(z(u))]\,du+\sqrt{T}\int_0^t G(t-u)[1-\phi(z(u))]\,d\xi(u)$$
$$+\sqrt{T}\int_0^t \sqrt{G(t-u)[1-\phi(z(u))][1-G(t-u)(1-\phi(z(u)))]}\,dw(u).$$

If we now suppose that

$$\phi(z)=\frac{1}{\sqrt{T}}\,\phi_0(z)$$

where ϕ_0 does not depend on T, then for the limit $\zeta(t)$ of the normalized process $z(t)=(q(t)-T\mathrm{E}\tau_j)/\sqrt{\mathrm{T}}$ we obtain

$$\zeta(t)=-\int_0^t G(t-u)\phi_0(\zeta(u))\,du+\int_0^t G(t-u)\,d\xi(u)+\theta(t) \tag{2.66}$$

where $\theta(t)$ is the same Gaussian process as in sections 2.2 and 2.3.

Assume that there is a solution of this equation for any smooth nondecreasing function ϕ_0. Then the process $\zeta^*(t)$, for systems with refusals, may be constructed as the limit as $N\to\infty$ of solutions $\zeta_N(t)$ of (2.66) for the function $\phi_0^N(z)=0$ for $z\leqslant D$, $\phi_0^N(z)=N$ for $z>D+1/N$.

Thus the problem here lies in establishing the closeness of the distributions of $z(t)$, for systems with refusals, and $\zeta_N(t)$ for large T and N (which is intuitively clear), and in proving the existence of a solution to (2.66).

This whole discussion is significantly simplified if

$$G(t)=\mathrm{e}^{-\alpha t}.$$

In this case, for systems with refusals, the limit process $\zeta(t)$ is a Markov diffusion process with known coefficients and reflection at the boundary D (see chapter 3).

In the general case $\zeta(t)$ may be very complex.

No less difficult is the description of $\zeta(t)$ for *systems with queues* in case (2.64). Here, side by side with

$$e(t)=Tt+\sqrt{T}x(t)\quad x\underset{C}{\Rightarrow}\xi$$

$$q(t)=n+\sqrt{T}y(t)\qquad (y(t)=z(t)-D)$$

we introduce $H(t)$, the intensity of the busy channels, and $L(t)$, the intensity of the free channels. $H(t)$ is equal to the number of requests accepted into service up to time t (when $n=\infty$, $H(t)=e(t)$). For simplicity we suppose $H(0)=0$. $L(t)$ is defined similarly, that is, is the number of requests leaving the system before service up to time t. The processes H, L, y, and e are connected by

$$H(t)=\int_0^t \delta(y(u)\,de(u))+\int_0^t 1-\delta(y(u))\,dL(u) \tag{2.67}$$

where

$$\delta(y) = \begin{cases} 1 & \text{for } y < 0 \\ 0 & \text{for } y \geqslant 0. \end{cases}$$

The meaning of this equality is very simple: the intensity H in a small interval Δ coincides with the intensity e, when not all channels are busy, and coincides with the intensity of the free channels when there are queues.

We denote

$$\frac{H(t) - Tt}{\sqrt{T}} = h(t), \qquad \frac{n + L(t) - Tt}{\sqrt{T}} = l(t).$$

Regarding $l(t)$ we may restrict ourselves to the semiaxis (l_0, ∞), where $l_0 = \inf(u\colon q(u) \geqslant n)$. (For $t < l_0$ the values $l(t)$ are unboundedly increasing.)

Then (2.67) can be rewritten in terms of h and l as

$$h(t) = \int_0^t \delta(y(u))\,\mathrm{d}x(u) + \int_0^t [1 - \delta(y(u))]\,\mathrm{d}l(u). \tag{2.68}$$

In addition, since $q(t) = e(t) - L(t)$, then $y(t) = x(t) - l(t)$.

Finally we may obtain a third relation between y, x, l, and h (where y, l, and h are unknown) if we consider that $L(t)$ arises from them. As for systems with refusals consider an interval $(u, u + \Delta)$ in the segment $[0, t]$. From the number

$$\Delta H(u) = \Delta T + \sqrt{T}\Delta h(u)$$

of channels which are busy in this interval, by the De Moivre–Laplace theorem, approximately

$$F(t - u)\Delta H(u) + \sqrt{F(t - u)G(t - u)\Delta H(u)}w$$

become free up to time t, where w is normally distributed with parameters $(0, 1)$. As before this expression may be written

$$TF(t - u)\Delta + \sqrt{T}F(t - u)\Delta h(u) + \sqrt{TF(t - u)G(t - u)}\Delta w(u)$$

where w is a Wiener process. Summing over Δu, we arrive at the following approximate relation:

$$L(t) = Th - T\mathbf{E}\tau_j + \sqrt{T}\left[\int_0^t F(t - u)\,\mathrm{d}h(u) + \theta(t)\right]. \tag{2.69}$$

Here $\theta(t)$ is a Gaussian process, not depending on $x(t)$, whose correlation function, as is easily verified, is the same as the correlation function of the $\theta(t)$ in section 2.2. Since

$$n = T\mathbf{E}\tau_j + D\sqrt{T}$$

then (2.69) implies

$$l(t) = D + \int_0^t F(t-u)\,\mathrm{d}h(u) + \theta(t). \tag{2.70}$$

Assuming that the processes $y(t)$, $h(t)$, and $l(t)$ converge in an appropriate sense (so that the distributions of the integrals in (2.68) and (2.70) converge) to processes $\kappa(t)$, $\eta(t)$, and $\lambda(t)$ respectively, then for these limit processes we have

$$\begin{aligned} \kappa(t) &= \xi(t) - \lambda(t) \\ \lambda(t) &= D + \int_0^t F(t-u)\,\mathrm{d}\eta(u) + \theta(t) \\ \eta(t) &= \int_0^t \delta(\kappa(u))\,\mathrm{d}\xi(u) + \int_0^t [1 - \delta(\kappa(u))]\,\mathrm{d}\lambda(u). \end{aligned} \tag{2.71}$$

Clearly the question of the existence and properties of the solutions of these equations is very difficult.

An essential simplification comes only in the case

$$G(t) = \mathrm{e}^{-\alpha t}.$$

Then the process $\zeta = \kappa + D$ turns out to be a diffusion with constant diffusion and drift coefficient (see chapter 3)

$$a(x) = \max(-ax, -aD), \quad a > 0.$$

Since it is doubtful whether a simple solution to (2.66) and (2.71) can be hoped for then this section shows that the investigation of the asymptotic behaviour of systems with queues or with refusals in case (2.64) and for arbitrary F has some very essential difficulties.

If the number of channels n is such that

$$\frac{n - T\mathbf{E}\tau_j}{T} \to -\infty$$

then for systems with queues the latter will be unboundedly increasing with the growth of t. The study of the limit distribution of $q(t)$ in this case is not very difficult. For systems with refusals for $n < rT\mathbf{E}\tau_j$, $r < 1$, the number of free channels will have, under very broad assumptions, a proper limit distribution which can be found explicitly (see section 2.7). The above description of systems with queues and with refusals via diffusion processes is contained in chapter 3.

2.6.* Generalization of the basic results of sections 2.2 and 2.3 to the case of dependent waiting times

In this section we consider the problem of the asymptotic behaviour of the number of busy lines $q(t) = q_T(t)$ under the conditions of section 2.2 with, however, the essential difference that now the service times τ_j are dependent variables. Namely, we assume that the sequence $\{\tau_j\}$ does not depend on the input process $\{e(t)\}$ and is a *stationary* sequence satisfying the *uniform strong mixing* condition (u.s.m.)

$$\sup_{A\in\mathfrak{S}_1^k, B\in\mathfrak{S}_{k+s}^\infty} |\mathbf{P}(B/A) - \mathbf{P}(B)| = \phi(s) \to 0 \tag{2.72}$$

as $s \to \infty$ where $\mathfrak{S}_k^l$ is the σ-algebra generated by $\tau_k, \ldots, \tau_l$. Condition (2.72) may equivalently be written (see Ibragimov and Linnik, 1965)

$$\operatorname*{ess\,sup}_{\omega} |\mathbf{P}_{\mathfrak{S}_1^k}(B) - \mathbf{P}(B)| \leqslant \phi(s) \to 0$$

where $B \in \mathfrak{S}_{k+s}^\infty$, or

$$\operatorname{ess\,sup} |\mathbf{E}_{\mathfrak{S}_1^k}\xi - \mathbf{E}\xi| \leqslant \phi(s) \to 0 \tag{2.73}$$

where ξ is measurable with respect to $\mathfrak{S}_{k+s}^\infty$, $\mathbf{P}(|\xi| \leqslant 1) = 1$. As before we denote

$$G(t) = \mathbf{P}(\tau_j > t), \quad Q(t) = \int_0^t G(t-u)\,\mathrm{d}m(u).$$

In addition, put

$$G_k(u, v) = \mathbf{P}(\tau_1 > u, \tau_{k+1} > v), \quad -\infty < k < \infty$$

where $\ldots, \tau_{-2}, \tau_{-1}, \tau_0, \ldots$ is an extension of $\{\tau_j\}_{j=0}^\infty$ to a sequence $\{\tau_j\}_{j=-\infty}^\infty$ on the 'whole axis'.

THEOREM 1. *Suppose that $G(t)$ is continuous and that $\phi(s)$ in the u.s.m. condition satisfies*

$$\phi = \sum_{s=1}^\infty \phi(s) < \infty.$$

Further, let the sequence

$$x(u) = \frac{e(u) - Tm(u)}{B(T)}$$

where $m(u)$ is some monotone increasing function, C-converge on $[0, t_0]$ to a process $\{\xi(u), 0 \leqslant u \leqslant t_0\}$.

We denote, as before, $A(T) = \max(B(T), \sqrt{T})$ and suppose that $B(T)/A(T) \to B \geqslant 0$, $\sqrt{T}/A(T) \to K \geqslant 0$. Then, if the number of service channels $n = \infty$, the finite-dimensional distributions of

$$z(t) = \frac{q(t) - TQ(t)}{A(T)}, \quad 0 \leqslant t \leqslant t_0$$

weakly converge to the distributions of

$$\zeta(t) = K\theta(t) + B\int_0^t G(t-u)\,\mathrm{d}\xi(u) \tag{2.74}$$

where $\theta(t)$ is a centred Gaussian process, not depending on $\{\xi(t)\}$, with covariance function

$$\mathbf{E}\theta(t)\theta(t+u) = \int_0^t \sum_{k=-\infty}^{\infty} [G_k(t-v, t+u-v) - G(t-v)G(t+u-v)]\,\mathrm{d}m(v). \tag{2.75}$$

From previous formulae it follows that the variance of $\theta(t)$ is equal to

$$\mathbf{D}\theta(t) = \int_0^t \sum_{k=-\infty}^{\infty} [G_k(t-v, t-v) - G^2(t-v)]\,\mathrm{d}m(v). \tag{2.76}$$

Remark 1. The theorem shows that the nature of the limit distribution of $q(t)$ for dependent τ_j is as before—the same as for independent τ_j. Only the covariance function of $\theta(t)$ is changed.

The question of C-convergence of the distribution of $z(t)$ we put to one side, since its proof under the conditions of this theorem would be long. This proof, evidently, can be implemented following the arguments of section 2.3.

The extension of the theorem to a finite number of channels, satisfying $[n - TQ(t_0)]A^{-1}(T) \to \infty$, is as in section 2.2.

We will make some further remarks on this theorem after the proof.

Proof. As before, we will use the representation

$$q(t) = \sum_{k=1}^{e(t)} \chi_k(t)$$

$$\chi_k(t) = \chi_k(\tau_k + v_k > t) = \begin{cases} 1 & \text{if } \tau_k + v_k > t \\ 0 & \text{otherwise} \end{cases}$$

where v_k is the arrival time of the kth request. In the case of dependent τ_j, in order to avoid cumbersome formal operations with conditional expectations relative to the σ-algebra $\mathfrak{E}$ generated by $e(t)$, we will often fix (at some time) the trajectory $e(u)$, $0 \leqslant u \leqslant t$, and regard it as nonrandom.

Denote

$$e(t) = N, \quad b_k = t - v_k$$

$$x_k = \chi(\tau_k > b_k) - \mathbf{P}_{\mathfrak{E}}(\tau_k > b_k) = \chi(\tau_k > b_k) - G(b_k).$$

Then

$$\mathbf{E}_{\mathfrak{E}}q(t) = \sum_{k=1}^{N} G(b_k) = \int_0^t G(t-u)\,\mathrm{d}e(u)$$

and the centred number of busy lines

$$V = V(t) = q(t) - \mathbf{E}_{\mathfrak{E}} q(t) = q(t) - \int_0^t G(t-u)\,\mathrm{d}e(u) \tag{2.77}$$

will be written in the form

$$V = \sum_{k=1}^{N} x_k. \tag{2.78}$$

We divide this sum into m sums

$$V = \sum_{j=1}^{m} z_j^*, \quad z_j^* = \sum_{k=N_{j-1}+1}^{N_j} x_k$$

where $N_0 = 0$, $N_m = N$ and the values $n_j = N_j - N_{j-1}$ differ from each other by not more than 1, so that $\min_j n_j \leqslant N/m \leqslant \min_j n_j + 1$ (for simplicity in what follows, it is possible to take $n_1 = \cdots = n_m = N/m$).

We assume that the fixed trajectory $e(u)$, $u \leqslant t_0$, has the properties $\sup_{u \leqslant t_0} |x(u)| \leqslant T^{\frac{1}{4}}$, $\max_j |v_{j+1} - v_j| \to 0$ as $T \to \infty$. The set B of such trajectories (of course B, as also $\mathfrak{E}$, in a series array depends on T) will have probability $\mathbf{P}(B) \to 1$ as $T \to \infty$. This is easily derived from the C-convergence of the processes $x(t)$.

Further, it is easy to see that for $e(u) \in B$ the number of terms $N = e(t)$ in (2.78) increases without limit and has the same order of growth as T. To study the distribution of (2.78) we may use the following lemma due to Bernstein (1944).

LEMMA 1. *Let $z_1, \ldots, z_m$ be random variables,*

$$Z_m = \sum_{k=1}^{m} z_k, \quad \mathbf{E}z_k = 0, \quad \mathbf{E}Z_m^2 = B_m^2.$$

Denote

$$\operatorname*{ess\,sup}_{\omega} |\mathbf{E}_{\sigma_k} z_{k+1}| = \alpha_k$$

$$\operatorname*{ess\,sup}_{\omega} |\mathbf{E}_{\sigma_k} z_{k+1}^2 - M z_{k+1}^2| = \beta_k$$

$$\operatorname*{ess\,sup}_{\omega} \mathbf{E}_{\sigma_k} |z_{k+1}^3| = \gamma_k$$

where σ_k is the σ-algebra generated by $z_1, \ldots, z_k$.

Then, if as $m \to \infty$

$$\sum_{k=1}^{m} \alpha_k = o(B_m), \quad \sum_{k=1}^{m} \beta_k = o(B_m^2), \quad \sum_{k=1}^{m} \gamma_k = o(B_m^3)$$

then Z_m/B_m has normal $(0, 1)$ limit distribution.

To ensure the direct application of this lemma we use a method, also due to Bernstein, which consists of 'truncating' the sum (2.78). Namely, alongside the sum of the z_j^* we consider the sum z_j of size $n_j - L$

$$z_j = \sum_{k=N_{j-1}+L+1}^{N_j} x_k = z_j^* - \sum_{k=N_{j-1}+1}^{N_{j-1}+L} x_k \tag{2.79}$$

and denote

$$V' = \sum_{j=1}^{m} z_j.$$

We will show that L here may be chosen so that*

$$\mathbf{D}(V - V') = o(\mathbf{D}V) \tag{2.80}$$

and the z_j satisfy the conditions of the lemma. It will then follow that $V/\sqrt{\mathbf{D}V}$ is distributed the same as $V'/\sqrt{\mathbf{D}V}$, that is, asymptotically normal with parameters $(0, 1)$.

We first calculate the variance $\mathbf{D}(V)$. Since $\mathbf{E}x_j = 0$, then

$$\mathbf{D}V = \mathbf{E}V^2 = \mathbf{E}\left(\sum_{j=1}^{N} x_j\right)^2 = \sum_{j=1}^{N} \mathbf{E}x_j^2 + 2\sum_{i<j}^{N} \mathbf{E}x_i x_j$$

$$= \sum_{j=1}^{N} [G(b_j) - G^2(b_j)] + 2\sum_{i<j} r_{ij} \tag{2.81}$$

where

$$r_{ij} = \mathbf{P}(\tau_i > b_i, \tau_j > b_j) - \mathbf{P}(\tau_i > b_i)\mathbf{P}(\tau_j > b_j). \tag{2.82}$$

For fixed k consider

$$R_k = \sum_{i=1}^{N-k} r_{i,i+k} = \sum_{i=1}^{N-k} [\mathbf{P}(\tau_1 > b_i, \tau_{k+1} > b_i) - \mathbf{P}^2(\tau_1 > b_i)]$$

$$+ \sum_{i=1}^{N-k} \delta_i'(b_i, b_{i+k}) + \sum_{i=1}^{N-k} \delta_i''(b_i, b_{i+k}).$$

Here the functions δ_i' and δ_i'' are defined in a natural way as the corresponding differences. Since $G_k(t, u) = \mathbf{P}(\tau_1 > t, \tau_{k+1} > u)$ is continuous with respect to u and $e(u)$ belongs to B we have

$$\max_i \delta_i'(b_i, b_{i+k}) \to 0, \quad \max_i \delta_i''(b_i, b_{i+k}) \to 0$$

as $T \to \infty$.

Therefore

$$R_k = \int_0^t [G_k(t - v, t - v) - G^2(t - v)]\, de(v) + o(N). \tag{2.83}$$

* For simplicity we omit the index $\mathfrak{E}$ on conditional moments, regarding the trajectory $e(u)$, $u \leqslant t_0$, as fixed.

Since $e(u) \in B$ we obtain

$$R_k = T \int_0^t [G_k(t-v, t-v) - G^2(t-v)]\, \mathrm{d}m(v) + o(T).$$

We note, further, that $|r_{i,j}| \leqslant \phi(j-i)$ and, consequently,

$$R_k \leqslant (N-k)\phi(k)$$

$$\sum_{k=s}^{N-1} R_k \leqslant N \sum_{k=s}^{N-1} \phi(k) + \sum_{k=s}^{N-1} k\phi(k). \tag{2.84}$$

Since $\sum_{k=s}^{\infty} \phi(k) \to 0$ as $s \to \infty$ and $\sum_{k=1}^{N-1} k\phi(k) = o(N)$ as $N \to \infty$, then $(1/T)\sum_{k=s}^{N-1} R_k$ can be made arbitrarily small by the choice of s and N. This means that

$$\lim_{T\to\infty} \frac{1}{T} \sum_{i<j} r_{ij} = \lim_{T\to\infty} \frac{1}{T} \sum_{k=1}^{N-1} R_k$$

$$= \sum_{k=1}^{\infty} \int_0^t [G_k(t-v, t-v) - G^2(t-v)]\, \mathrm{d}m(v)$$

exists. Therefore we have found that

$$\mathbf{D}V = T \sum_{k=-\infty}^{\infty} \int_0^t [G_k(t-v, t-v) - G^2(t-v)]\, \mathrm{d}m(v) + o(T) \tag{2.85}$$

where $G_0(u, u) = G(u)$, $G_{-k}(u, u) = G_k(u, u)$.

We now move to the proof of (2.80) and the verification of the conditions of the lemma.

We have

$$\mathbf{D}(V - V') = \mathbf{E}\left(\sum_{j=0}^{m-1} \sum_{k=N_j+1}^{N_j+L} x_k\right)^2.$$

Let Λ denote the set of indices $(1, \ldots, L, N_1 + 1, \ldots, N_1 + L, N_2 + 1, \ldots)$ of the x_k in the sum $V - V'$. The number of indices in this set is equal to Lm. We have (see (2.81), (2.82))

$$\mathbf{D}(V - V') = \sum_{k\in\Lambda} \mathbf{E}x_k^2 + 2 \sum_{\substack{i<j \\ i\in\Lambda \\ j\in\Lambda}} r_{ij}.$$

We have just seen that the contribution given by the sum of the r_{ij}, even when the x_i *are summed in succession* (with no omissions), is estimated by the number of terms in the sum $\sum x_j$. Obviously this situation is largely preserved when the sum of the x_k is truncated and, consequently, the connections between the terms are weakened. Therefore, omitting calculations identical to those used in the estimate of $\mathbf{D}V$, we obtain

$$\mathbf{D}(V - V') \leqslant cLm = o(\mathbf{D}V)$$

if

$$Lm = o(T) \tag{2.86}$$

or, what is the same thing, if $L = o(n_1)$.

Put

$$n_1 = \sqrt{T}\alpha(T) \to \infty, \quad \alpha(T) \to 0 \tag{2.87}$$

where $\alpha(t)$ is to be chosen later. Under these conditions we will verify the conditions of the lemma. From the above it follows that $B_m = \mathbf{D}V' \sim \mathbf{D}V = O(T)$. For an estimate of $\mathbf{E}_{\sigma_k} z_{k+1}$ it is sufficient to discuss the behaviour of ($\sigma_k \subset \mathfrak{S}_{N_k}$, where $\mathfrak{S}_k = \mathfrak{S}^k_{-\infty}$)

$$\mathbf{E}_{\mathfrak{S}_l} \sum_{j=l+L}^{l+n_1} x_j = \sum_{j=l+L}^{l+n_1} \mathbf{E}x_j + \sum_{j=l+L}^{l+n_1} r(\omega, l, j)$$

where $\mathbf{E}x_j = 0$ and by u.s.m. $|r(\omega, l, j)| \leqslant \phi(j - l)$. Consequently, in the notation of the lemma,

$$\alpha_k \leqslant \sum_{j=l+L}^{\infty} \phi(j - l) = \psi(L) \to 0$$

as $L \to \infty$.

Let $L = L(T) = [T^{\frac{1}{4}}]$. Take

$$\alpha(T) = \sqrt{\psi(L)}.$$

Then

$$\sum_{k=1}^{m} \alpha_k \leqslant \psi(L)m = \frac{\psi(L)N}{n_1} = O\left(\frac{\psi(L)T}{\sqrt{T}\alpha(T)}\right) = O(\sqrt{T\psi(L)}) = o(\sqrt{T})$$

$$= o(B_m).$$

Further, we estimate the deviation of $\mathbf{E}_{\sigma_k} z^2_{k+1}$ from $\mathbf{E}z^2_{k+1}$ with the help of the expression ($\Lambda = (l + L, \ldots, l + n_1)$)

$$\mathbf{E}_{\mathfrak{S}_l}\left(\sum_{j\in\Lambda} x_j\right)^2 = \sum_{j\in\Lambda} \mathbf{E}_{\mathfrak{S}_l} x_j^2 + 2 \sum_{\substack{i<j \\ i\in\Lambda \\ j\in\Lambda}} \mathbf{E}_{\mathfrak{S}_l} x_i x_j.$$

Since

$$|\mathbf{E}_{\mathfrak{S}_l} x_j^2 - \mathbf{E}x_j^2| < \phi(j - l)$$

$$|\mathbf{E}_{\mathfrak{S}_l} x_i x_j - \mathbf{E}x_i x_j| < \phi(i - l) \quad \text{for } i < j$$

then

$$\left|\mathbf{E}_{\mathfrak{S}_l}\left(\sum_{j\in\Lambda} x_j\right)^2 - \mathbf{E}\left(\sum_{j\in\Lambda} x_j\right)^2\right| \leqslant \sum_{s=L}^{\infty} \phi(s) + \sum_{s=L}^{n} (n_1 - L - s)\phi(s).$$

Hence, in the notation of the lemma, it follows that

$$\sum_{k=1}^{m} \beta_k \leqslant m n_1 \psi(L) = o(T) = o(B_m^2).$$

Finally, we consider the behaviour of the conditional, relative to $\mathfrak{S}_l$, third moments of the variable

$$\left|\sum_{j\in\Lambda} x_j\right|^3 \leqslant \left(\sum_{s\in\Lambda} |x_s|\right)\left(\sum_{i\in\Lambda} x_i^2 + 2\sum_{i\in\Lambda} x_i \sum_{\substack{j>i\\ i\in\Lambda}} x_j\right)$$

$$= \sum_{\substack{s\in\Lambda\\ i\in\Lambda}} |x_s| x_i^2 + 2\sum_{s=l+L}^{l+n_1} |x_s| \sum_{i=s}^{l+n_1} x_i \sum_{j=i+1}^{l+n_1} x_j$$

$$+ 2\sum_{i=l+L}^{l+n_1} x_i\left(\sum_{j=i+1}^{l+n_1} x_j\right)\left(\sum_{l=i+1}^{l+n_1} |x_l|\right). \tag{2.88}$$

By the u.s.m. condition the conditional expectation $\mathbf{E}_{\mathfrak{S}_l}$ of the last two terms in (2.88) differs from the absolute expectation by no more than

$$n_1^2 \sum_{s=l+L}^{l+n_1} \phi(s-k) + n_1^2 \sum_{i=l+L}^{l+n_1} \phi(i-l) \leqslant 2n_1^2 \sum_{s=L}^{\infty} \phi(s) = 2n_1^2 \psi(L).$$

The conditional expectation of the first sum in (2.88) is estimated by n_1^2. It remains for us to consider the absolute expectations

$$\left|\mathbf{E} \sum_{s=l+L}^{l+n_1} |x_s| \sum_{i=s}^{l+n_1} x_i \sum_{j=i+1}^{l+n_1} x_j\right| \leqslant \sum_{s=l+L}^{l+n_1} \sum_{i=s}^{l+n_1} \sum_{j=i+1}^{l+n_1} \phi(j-i) \leqslant \phi n_1^2$$

and

$$\left|\mathbf{E} \sum_{i=l+L}^{l+n_1} x_i\left(\sum_{j=i+1}^{l+n_1} x_j\right)\left(\sum_{s=i+1}^{l+n_1} |x_s|\right)\right|.$$

In the latter there are $2(l+n_1-i-t)+1$ products $x_j|x_s|$, for which $\min[(j-i),(l-i)] = t$. Therefore it is not less than

$$2\sum_{i=l+L}^{l+n_1} \sum_{t=1}^{l+n_1-i} (l+n_1-i-t)\phi(t) \leqslant 2n_1^2\phi.$$

Thus we have established that

$$\mathbf{E}_{\mathfrak{S}_l}\left|\sum_{j\in\Lambda} x_j\right|^3 \leqslant c n_1^2$$

and, consequently, in the notation of the lemma,

$$\sum_{k=1} \gamma_k \leqslant c n_1^2 m = c n_1 N = O(T^{3/2}\alpha(T)) = o(T^{3/2}) = o(B_m^3).$$

Therefore, it has been established that $V' = \sum_{j=1}^{m} z_j$ (see (2.79)) satisfies the conditions of the lemma and, consequently, is asymptotically normal with parameters $(0, B_m)$. As already noted, it then follows that V from (2.78) is asymptotically normal with the same parameters. In other words, the conditional, relative to $\mathfrak{C}$, distribution of

$$\frac{V}{A(T)} = \frac{q(t) - TQ(t)}{A(T)} - \frac{B(T)}{A(T)} \int_0^t G(t-u)\,dx(u)$$

on B, $P(B) \to 1$, is asymptotically normal with parameters $(0, K^2\mathbf{D}\theta(t))$, where $\mathbf{D}\theta(t)$ is defined in (2.76) (see (2.85)). Hence it follows that the absolute distribution of $z(t)$ coincides with the distribution of $\zeta(t)$ in (2.74), where $\theta(t)$ is a process with normal one-dimensional distributions and variance defined in (2.76).

It remains to prove that not only the one-dimensional but also arbitrary finite-dimensional distributions of the limit process $\theta(t)$ are also normal and have covariance function (2.75). Since this, in principle, repeats the latter proof we will state it only briefly.

We first calculate

$$\mathbf{E}V(t)V(t+u)$$

where $V(t)$ is defined in (2.77). Using a representation of the form (2.78) for $V(t+u)$ we obtain

$$\mathbf{E}V(t)V(t+u) = \mathbf{E}\left(\sum_{i=1}^{e(t)} x_i\right)\left(\sum_{j=1}^{e(t+u)} y_j\right) \tag{2.89}$$

where

$$y_j = \chi(\tau_j > b_j + u) - G(b_j + u), \quad b_j = t - v_j.$$

Since

$$\mathbf{E}x_i y_{i+k} = \mathbf{P}(\tau_i > b_i, \tau_{i+k} > b_{i+k} + u) - \mathbf{P}(\tau_i > b_i)\mathbf{P}(\tau_{i+k} > b_{i+k} + u)$$

then, by the continuity of $G(v)$, as in (2.83), we find

$$\sum^h \mathbf{E}x_i y_{i+k} = \int_0^t [G_k(t-v, t-v+u) - G(t-v)G(t-v+u)]\,de(v) + o(T).$$

Finally, summing over k and using estimates similar to (2.84), we obtain

$$\mathbf{E}V(t)V(t+u)$$
$$= T\int_0^t \sum_{k=-\infty}^{\infty} [G_k(t-v, t-v+u) - G(t-v)G(t-v+u)]\,dm(v) + o(T).$$

For the proof of normality of the finite-dimensional distributions of $\theta(t)$ we first consider the two-dimensional distributions of $V(t)$. It is sufficient to establish the asymptotic normality of any linear combination

$$V_{a_1 a_2} = a_1 V(t) + a_2 V(t+u).$$

But $V_{a_1a_2}$ may be written as a sum

$$V_{a_1a_2} = \sum_{i=1}^{e(t+u)} y_i^*$$

where

$$y_i^* = \begin{cases} a_1x_i + a_2y_i & \text{for } 1 \leqslant i \leqslant e(t) \\ a_2y_i & \text{for } e(t) \leqslant i \leqslant e(t+u). \end{cases}$$

For fixed $\mathfrak{E}$ the y_i^* are functions only of τ_i. This means that the dependence between the y_i^* is the same as the dependence between the x_i. This in turn allows us to prove the asymptotic normality of $V_{a_1a_2}$ by repeating, with the obvious changes, the proof of normality for $V(t)$.

Since the same arguments apply to the sum

$$\sum_{j=1}^{k} a_jV(t+u_j)$$

for any fixed k, then we may regard the theorem as completely proved. □

COROLLARY. *Let $m(v) = v$ and let $\xi(u)$ be a process with strictly stationary increments. Then, if for some $c_\tau = \text{const.}$*

$$\mathbf{P}(\tau_j \leqslant c_\tau) = 1$$

then for $t > c_\tau$ we obtain the stationary limit distribution of

$$z(t) = \frac{q(t) - T\mathbf{E}\tau}{A(T)}$$

coinciding with the distribution of

$$\zeta(t) = K\theta(t) + B\int_{-\infty}^{t} G(t-v)\,\mathrm{d}\xi(v) \tag{2.90}$$

where $\theta(t)$ is a centred stationary Gaussian process not depending on $\{\xi(u)\}$ and having covariance function

$$\mathbf{E}\theta(t)\theta(t+u) = \int_0^\infty \sum_{k=-\infty}^{\infty} [G_k(v, v+u) - G(v)G(v+u)]\,\mathrm{d}v. \tag{2.91}$$

It is easy to see that such a representation of the stationary limit process (see (2.90), (2.91)) is preserved in the general case when τ_j is not bounded.

Remark 1. The requirement of continuity of $G(u)$, in the conditions of the theorem, is not essential and was included to simplify the arguments. The reader may easily verify that the proof is the same if $G(u)$ is discontinuous and its set of points of discontinuity can be included in a finite number of intervals of arbitrarily small sum of lengths. Then, for example, in the first part of the proof, when studying $\sum_{i=1}^{e(t)} x_i$, it is sufficient to simply omit those x_i for which b_i is in one of the selected intervals. As we have already seen the limit distribution of $\sum_{i=1}^{e(t)} x_i$ is not changed by this.

Remark 2. The theorem was proved under the assumption that τ_j satisfies the u.s.m. condition. However, there is reason to suppose that the theorem still holds under the weaker strong mixing (s.m.) condition where, instead of (2.72), it is required that

$$\sup_{A\in\mathfrak{S}_1^k, B\in\mathfrak{S}_{k+s}^\infty} |\mathbf{P}(AB) - \mathbf{P}(A)\mathbf{P}(B)| = \phi(s) \to 0 \tag{2.92}$$

as $s \to \infty$. (As before we will suppose that $\sum \phi(s) < \infty$.) Indeed, the calculation which produced the estimates of $\mathbf{D}V$ and $\mathbf{D}(V - V')$ stays without changes. The same kind of calculation shows that the variables z_j, defined in (2.79), will be asymptotically uncorrelated. On the other hand, for smooth (for example, linear) $G(u)$ and the chosen values n_j, the terms x_k in

$$z_j = \sum_{k=N_{j-1}+L+1}^{N_j} x_k$$

may be replaced by stationary terms $x_k^* = \chi(\tau_k > b_{k_0}) - G(b_{k_0})$, where k_0 may be put equal, for example, to $k_0 = N_{j-1} + L + 1$. Then, using known theorems on asymptotic normality of a sum of stationarily connected variables (see, for example, Ibragimov and Linnik, 1965), we may establish that z_j will be asymptotically normal. The facts that z_j is asymptotically uncorrelated and normal and z_j and z_k are asymptotically independent for large $|k - j|$, evidently, imply asymptotic normality of $V' = \sum z_j$ and, consequently, V.

2.7. The distribution of the number of free channels for overloaded systems

To conclude this chapter we consider the distribution of the number of free channels in an 'overloaded' system with refusals. These are systems in which the number of service channels n satisfies

$$\frac{n}{T\mathbf{E}\tau_j} \to r < 1.$$

It turns out that under very broad assumptions there is a proper limit distribution of the nonnormalized number of free channels $Z(t)$ up to time t, and that this distribution can be completely characterized.

We will assume that the following conditions hold.

(I) *The processes* $e(u) = e_T(u)$ *have stationary increments,* $e(u)/T \xrightarrow[C]{} u$, u *in* $[0, t_0]$, *and the processes* $e(v + u/T) - e(v)$, *for each fixed* $v < t_0$, *have finite-dimensional distributions which converge weakly to the distributions of some proper process* $E(u)$, *with stationary increments,* $\mathbf{E}(E(1) - E(0)) = 1$.

This condition will obviously be satisfied if $e(u)$ is obtained from a process with stationary increments by a time contraction.

(II) *As sequence* $\{\tau_j\}$, *controlling the service of requests, we consider at first a sequence of independent identically distributed variables with exponential distribution*

$$\mathbf{P}(\tau_1 > x) = G(x) = e^{-\alpha x}.$$

THEOREM 1. *Let* $\{\Pi(u)\}$ *be a Poisson process with parameter* $r < 1$, *not depending on* $\{E(u)\}$. *Then, for any* $v > 0$ *and* $Z(0) = o(T)$ *the finite-dimensional distributions of the number of free channels* $Z(v + u/T)$ *converge as* $T \to \infty$ *to the finite-dimensional distributions of*

$$Y(u) = \sup_{s \leqslant u} (X(u) - X(s)), \quad u \geqslant 0$$

where $X(u) = \Pi(u) - E(u)$.

This result concerning the limit distribution of the number of free channels is in fact valid under much broader conditions, when instead of condition (II) we suppose only that $\{\tau_j\}$ is *stationary* and satisfies the u.s.m. condition (see the last section), or that *the service times in each channel are stationarily connected.* We continue this discussion at the end of the section.

Proof. The number of channels freed in the time interval $(v, v + u)$, under the condition $Z(v) = h$, behaves locally as a Poisson process with parameter $(n - k)\alpha$. More precisely, if $s(v + u) - s(v)$, $u \geqslant 0$, denotes the 'freeing process of channels' (the number of channels freed in $(v, v + u)$ without regard for requests joining before v), then this may be written as a sum of a decreasing number of independent Poisson processes $\eta_j(u)$ with parameter α:

$$s(v + u) - s(v) = \sum_{j=1}^{q_v^*(u)} \eta_j(u)$$

where

$$q_v^*(0) = q(v) = n - Z(v) = n - k$$

$$q_v^*(u) = q_v^*(0) - [s(v + u) - s(v)].$$

We have as $\Delta \to 0$, $n\Delta \to 0$

$$\mathbf{P}(s(v+u)-s(v)=j/Z(v)=k)=\begin{cases} 1-(n-k)\alpha\Delta+o(\Delta n) & \text{for } j = 0 \\ (n-k)\alpha\Delta+o(n\Delta) & \text{for } j = 1 \\ o(n\Delta) & \text{for } j \geqslant 2. \end{cases}$$

This process will be of most interest for $k = 0$. Therefore on the same probability space it is possible to construct a process Z^* such that

$$Z^*(0) = Z(0), \quad Z^*(v) \geqslant Z(v) \tag{2.93}$$

where

$$Z^*(v + dv) = \max [0, Z^*(v) - de(v) + d\Pi^*(v)]$$

and $\Pi^*(v)$ is a Poisson process with parameter $n\alpha$, not depending on $\{e(v)\}$. In a similar way to Borovkov (1976a, chapter 1) it is easily established that the solution of this equation is

$$Z^*(v) = \sup_{0 \leqslant s \leqslant v} [y(v), y(v) - y(s)]$$

where

$$y(v) = Z^*(0) + [\Pi^*(v) - \Pi^*(0)] - [e(v) - e(0)].$$

Since for $Z^*(0) = o(T)$ we have $y(v)/T \underset{C}{\rightarrow} (r-1)v < 0$, then on an ω-set of probability close to 1, $Z^*(v)$ coincides (see Borovkov, 1976a) with

$$\sup_{s \leqslant v} ([\Pi^*(v) - \Pi^*(s)] - [e(v) - e(s)]). \tag{2.94}$$

We now expand time by T (replacing v and s in (2.94) by $v + u/T$ and $v + w/T$ respectively). Then $\Pi^*(v + u/T) - \Pi^*(v + w/T)$ will converge to $\Pi(u) - \Pi(w)$, where $\Pi(u)$ is a Poisson process with parameter $r < 1$ (here we always have in mind the convergence of the finite-dimensional distributions). The process $e(v) - e(s)$ will converge to $E(u) - E(w)$. Since the convergence of Π^* occurs together with the convergence of expectations ($n\alpha/T \rightarrow r$), then by the stability theorem (see section 4.1) the finite-dimensional distributions of the process (2.94) (with v and s replaced by $v + u/T$ and $v + w/T$) will converge to the distributions of

$$\sup_{w \leqslant u} ([\Pi(u) - \Pi(w)] - [E(u) - E(w)]). \tag{2.95}$$

In a similar way it is possible to construct a process $Z^{**}(v)$ which approximates $Z(u)$ from below. From the results of the last section it follows that the event

$$B_T = \left\{ \sup_{0 \leqslant v \leqslant t_0} Z(v) < \varepsilon_T T \right\}$$

for some $\varepsilon_T = o(1)$, has probability $\mathbf{P}(B_T) \rightarrow 1$ (this fact can be derived from the estimate $Z(v) \leqslant Z^*(v)$). But on B_T, for $v + u/T \leqslant t_0$

$$Z\left(v + \frac{u}{T}\right) \geqslant Z^{**}\left(v + \frac{u}{T}\right) \tag{2.96}$$

where Z^{**} is constructed in the same way as Z^* but via a Poisson process Π^{**} with parameter $(n - \varepsilon_T T)\alpha$, $\varepsilon_T = o(1)$. Repeating the above arguments relative to Z^{**} we obtain that the finite-dimensional distributions of $Z^{**}(v + u/T)$ converge, as $T \rightarrow \infty$, to the distributions of (2.95). By (2.93) and (2.96) this completes the proof of the theorem. □

We noted at the beginning of the section that this theorem still holds in the more general case when the τ_j have arbitrary distribution and the independence of the $\{\tau_j\}$ is replaced by stationary connectedness with the u.s.m. condition. Moreover, this result holds in another very general situation when the service in each channel is controlled by its own stationary sequence $\{\tau_j^{(l)}\}_{j=1}^{\infty}$ (l being the number of the channel). The possibility of such an essential generalization follows from the theorem of Grigolionis that under broad conditions the sum of truncated point processes converges to a Poisson process (see, for example, Gnedenko and Kovalenko, 1966).

The proof of these facts, within the confines of this book, would require a disproportionate amount of space. Therefore we limit ourselves to giving a reference to Borisov and Borovkov (1980).

In conclusion we consider the important special case when $E(u)$ is a process with independent increments. Here the stationary distribution of the number of free channels $Z(u)$ coincides with the distribution of the maximum on $[0, \infty)$ of the trajectory of the generalized Poisson process $\Pi(u) - E(u)$, which is continuous from above. If

$$\mathbf{E}\, e^{i\lambda E(u)} = \exp\left\{u \sum_{k=0}^{\infty} (e^{i\lambda k} - 1)p_k\right\}, \quad \sum_{k=0}^{\infty} p_k = 1$$

then

$$\mathbf{E}\, e^{i\lambda(\Pi(u) - E(u))} = \exp\left\{(e^{i\lambda} - 1)ru + u \sum_{k=0}^{\infty} (e^{-i\lambda k} - 1)p_k\right\}$$

and the limit distribution of $Z(t)$ will be the same as the distribution of

$$Y = \max_{k \geqslant 0} \sum_{j=0}^{k} \xi_j$$

where the ξ_j are independent, $\mathbf{P}(\xi_j = 1) = r/(r+1)$, $P(\xi_j = -j) = p_j/(r+1), j = 0, 1, \ldots$. It is known (see Borovkov, 1976a), that in this case

$$\mathbf{P}(Y \geqslant x) = d^{-x} \quad (x \text{ an integer})$$

where d is the largest root of the equation $p(d) = 1$, $p(y) = \sum_k y^k \mathbf{P}(\xi_1 = k)$.

If $E(u)$ is also a Poisson process ($p_1 = 1$), then $d = r^{-1}$, $\mathbf{P}(Z(u) \geqslant x) = r^x, r < 1$.

Once again we note that the above simple formulae will be valid for a broad class of stationary controls of the service process.

CHAPTER 3

The Description of Service Systems by Diffusion Processes

The simplest situation in which approximation by a diffusion process appears in an absolutely natural way is in the study of a prestationary queue of a loaded single-channel system. In fact, it is well known that for such systems, under the zero initial condition $w_0 = 0$, the distribution of the waiting time w_n of the nth request is the same as $\max_{k \leqslant n} X_k$, where $X_k = \sum_{j=1}^{k} \xi_j$, ξ_j the independent differences $\xi_j = \tau_j^s - \tau_j^e$ between the intervals τ_j^e, which separate the time of appearance of requests and the service times τ_j^s. (For details see, for example, Borovkov, 1976a.) The condition of critical loading means that $\mathbf{E}\tau_j^s = \mathbf{E}\tau_j^e$ or, what is the same, $\mathbf{E}\xi_j = 0$. But, by the Donsker–Prokhorov invariance principle (section 1.2), the distribution of $\max_{k \leqslant n} X_k/\sigma\sqrt{n}$, as $n \to \infty$, will approach the distribution of $\sup_{t \leqslant 1} w(t)$, where $\sigma^2 = \mathbf{D}\xi_j$ and $w(t)$ is a standard Wiener process. Consequently the distribution of $w_n/\sigma\sqrt{n}$ will be approximately the same.

The situation is rather more complex when $\mathbf{E}\xi_j$ is different from 0 but still small. Here it is already necessary to consider two simultaneously varying parameters: n and $\mathbf{E}\xi_j$. The passage to the limit here may be carried out in various ways. If $\mathbf{E}\xi_j < 0$, then we may first let n tend to infinity (i.e. consider stationary distributions), and then let $\mathbf{E}\xi_j$ tend to 0. It is even possible to consider simultaneous variation of n and $\mathbf{E}\xi_j$. Of most importance here is the direction $\mathbf{E}\xi_{1,n} = \delta/\sqrt{n}$, $\delta = \text{const.}$, $n \to \infty$. Under these conditions (we now consider a series array where the distribution of $\xi_j = \xi_{j,n}$ depends on n) the distribution of $w_n/\sigma\sqrt{n}$ or, equivalently, the distribution of

$$\max_{k \leqslant n} \frac{X_k}{\sigma\sqrt{n}} = \max_{k \leqslant n} \left[\frac{X_k - k\mathbf{E}\xi_{1,n}}{\sigma\sqrt{n}} + \frac{k\delta}{n\sigma} \right]$$

by the invariance principle will approach the distribution of $\max_{t \leqslant 1} [w(t) + t\delta/\sigma]$.

A detailed analysis, in a somewhat different form, has been carried out by Prokhorov (1956).

After this work appeared it was fairly quickly realized that it is possible to find an approximation by diffusion processes for the whole sequence $w_1, \ldots, w_n$ as processes, considering, for example, the polygonal arcs constructed relative to the points $(k/n, w_k/\sigma\sqrt{n})$, $k = 0, 1, \ldots, n$. A similar picture holds even for multichannel systems (see Barrer, 1957; Blomqvist, 1973; Borovkov, 1964; Cohen, 1973; Gaver, 1968; Geza, 1974; Gnedenko and Kovalenko, 1966; Iglehart, 1965, 1973; Iglehart and Whitt, 1970; Kyprianou, 1971; Loulou, 1973; Samandarov, 1963; Szcotha, 1976; Whitt, 1974). However, in all of these papers an essential role is played both by the independence of the elements of the controlling sequence and by a knowledge of a concrete and fairly simple service algorithm. Progress in this area for more complex systems by the methods in the above mentioned papers is either very difficult or impossible.

In this third chapter we consider a more general approach to the problem of diffusion approximation which will include, alongside all the already known results, a whole series of new possibilities.

We have already seen in the previous chapter that it is possible to obtain very interesting and useful theorems on the asymptotic behaviour of service systems, without fixing the concrete nature of the input stream of requests. It was sufficient to suppose that as a random process the stream satisfied some very broad conditions, which are easy to verify and are satisfied in most cases of interest.

It turns out that the detailed picture is generally the same if, travelling further down this route, we also omit from the discussion the concrete nature of the service algorithm. In this situation the utility of and preference for the general definition of service processes, discussed at the beginning of the second chapter, will become very noticeable.

At the same time the absence of a visible concrete service system does create some difficulties and, in particular, forces the stipulation of certain peculiarities in the interrelations between the components $e(t)$, $r(t)$, and $s(t)$ of the service process $\mathbf{S}(t)$ (see the introduction to chapter 2). For example, it is necessary to formalize such notions as 'independence of input', 'independence of output', etc., which before (usually, implicitly) followed from the concrete construction of the given service system.

These notions, together with the notion of stochastic control, are introduced and discussed in section 3.1.

In section 3.2 we describe conditions which, within the confines of this chapter (i.e. for diffusion approximation), the service process must satisfy and we also discuss various possibilities which arise in the problem of diffusion approximation.

Section 3.3 contains general theorems on the convergence of processes of 'occupation' to diffusion processes, which, in sections 3.4–3.6, are then made concrete for various types of service process. In particular, we consider multichannel systems with intensive input, stochastic control of refusals, loaded systems. Section 3.7, in which we consider a number of examples, completes the chapter.

Before passing to the basic account, in which we study the asymptotic behaviour of the 'length of queue' or 'occupation' $q(t) = e(t) - r(t) - s(t)$, we consider two examples of service systems which will hereafter frequently be used as illustrations.

Example 1. Systems $\langle G_I, G_I, G_I, G_I \rangle$ (in the notation of Borovkov, 1976a). These are single-channel systems with queues, controlled by a sequence of independent variables

$$\{\tau_j^e, \nu_j^e, \tau_j^s, \nu_j^s\}_{j=1}^{\infty}.$$

On entry to this system, through intervals of time $\tau_1^e, \tau_2^e, \ldots; \mathbf{P}(\tau_j^e < x) = F_e(x)$, groups of requests of sizes $\nu_1^e, \nu_2^e, \ldots; \mathbf{P}(\nu_j^e < x) = H_e(x)$ are received. These requests are served also in groups of sizes $\nu_1^s, \nu_2^s, \ldots; \mathbf{P}(\nu_j^s < x) = H_s(x)$ (or in smaller groups if there are not enough requests in the queue). A time τ_j^s; $\mathbf{P}(\tau_j^s < x) = F_s(x)$, is spent on the service of the jth group. If there is no queue, then the system begins operation by the entry of at least one request. If there is a queue, then the serving mechanism operates uninterruptedly, serving ν_j^s requests in time τ_j^s, until the queue disappears. For this system the refusal process is $r(t) \equiv 0$; the input process is

$$e(t) = \sum_{j=1}^{\eta_e(t)-1} \nu_j^e, \quad \eta_e(t) = \min\{k: \tau_1^e + \cdots + \tau_k^e > t\}.$$

The evolution of the service process $\mathbf{S}(t)$ here is conveniently characterized by constructing a Markov process

$$X(t) = \{\mathbf{S}(t), \gamma^e(t), \gamma^s(t)\}$$

where

$$\gamma^e(t) = \inf\{s \geqslant 0: e(t-s+0) - e(t-s-0) > 0\} \tag{3.1}$$

is the time up to the last jump before t of $e(t)$ (or, what is the same, is the size of 'defect' or 'undershoot' up to t in a random walk with jumps $\tau_1^e, \tau_2^e, \ldots$). The process $\gamma^s(t)$ is defined by analogy with (3.1), as the distance from t to the nearest $s < t$ at which the service of the next request began.

Another way of supplementing $\mathbf{S}(t)$ to make a Markov process consists of adding to $\mathbf{S}(t)$ the processes $\chi^e(t)$ and $\chi^s(t)$, where $\chi^e(t)$ is the value of the 'excess' or 'overshoot' of the level t in a random walk with jumps $\tau_1^e, \tau_2^e, \ldots$:

$$\chi^e(t) = \inf\{s > 0: e(t+s+0) - e(t+s-0) > 0\}.$$

$\chi^s(t)$ is defined similarly with respect to $s(t)$.

We mention also the following two modifications of the system described in example 1.

Example 1a. Systems with autonomous service. The only difference between these and the systems of example 1 is that service only begins at times $0, \tau_1^s, \tau_1^s + \tau_2^s, \tau_1^s + \tau_2^s + \tau_3^s, \ldots$ independently of the presence of a queue.

Example 1b. Systems with bounded queues. These are the systems described in example 1 in which, however, requests which find a queue of size N (N the maximal admissible length of queue) on arrival obtain refusal and leave the discussion.

We now consider

Example 2. Systems $\langle G_I, G_I, E/m, 1 \rangle$ (see Borovkov, 1976a) are systems with queues with the same input stream of requests as in example 1 and with m independent service channels, in each of which requests are served singly ($v_j^s \equiv 1$), and with service time exponentially distributed

$$F_s(x) = \mathbf{P}(\tau_j^s < x) = 1 - e^{-\alpha x}.$$

Here the evolution of the process may be described with the help of the Markov processes

$$X(t) = \{\mathbf{S}(t), \gamma^e(t)\} \quad \text{or} \quad Y(t) = \{\mathbf{S}(t), \chi^e(t)\}$$

where $\gamma^e(t)$ and $\chi^e(t)$ are defined relative to the input process in the same way as in example 1.

In the following section we introduce some useful notions which will be used below.

3.1. The notions of independence of input and output streams and stochastic control

It will be simpler in what follows to suppose that the process

$$\mathbf{S}(t) = (e(t), r(t), s(t))$$

is defined on the whole of the semiaxis $t \geqslant 0$ ($U = \infty$). We also recall that the trajectories of $\mathbf{S}(t)$ are assumed to be continuous from the right.

Suppose that we are given families of nested σ-algebras

$$\mathfrak{M}(u) \quad (\mathfrak{M}(u_1) \subset \mathfrak{M}(u_2) \text{ for } u_1 \leqslant u_2)$$

$$\mathfrak{E}(u) \subset \mathfrak{M}(u) \quad (\mathfrak{E}(u_1) \subset \mathfrak{E}(u_2) \text{ for } u_1 \leqslant u_2)$$

$$\mathfrak{S}(u) \subset \mathfrak{M}(u) \quad (\mathfrak{S}(u_1) \subset \mathfrak{S}(u_2) \text{ for } u_1 \leqslant u_2)$$

$$\mathfrak{R}(u) \subset \mathfrak{M}(u) \quad (\mathfrak{R}(u_1) \subset \mathfrak{R}(u_2) \text{ for } u_1 \leqslant u_2)$$

having the property that the values of $\mathbf{S}(t)$ are measurable with respect to $\mathfrak{M}(u)$ for $t \in [0, u]$, and the values of $e(t)$, $s(t)$, and $r(t)$ are measurable, respectively, with respect to $\mathfrak{E}(u)$, $\mathfrak{S}(u)$, and $\mathfrak{R}(u)$ for $t \in [0, u]$.

Frequently it is convenient to take $\mathfrak{M}(u)$, $\mathfrak{E}(u)$, $\mathfrak{S}(u)$, and $\mathfrak{R}(u)$ as the σ-algebras generated by events related to their respective processes on $[0, u]$, for example, by events of the form $\{e(v) < x\}$, $v \leqslant u$ for $\mathfrak{E}(u)$.

In this case continuity from the right of $\mathbf{S}(t)$ *will mean continuity from the right of the corresponding streams of σ-algebras* (cf. section 1.3):

$$\mathfrak{M}(t) = \mathfrak{M}(t + 0) = \bigcap_{k=1}^{\infty} \mathfrak{M}\left(t + \frac{1}{k}\right).$$

In fact, consider, for example, $e(t)$. Its trajectories do not decrease, are piecewise-constant and are continuous from the right. Consequently, for given t and n there is $\varepsilon = \varepsilon(t, n)$ such that

$$\mathbf{P}(e(t + \varepsilon) \neq e(t)) < n^{-2}.$$

Now take any $A \in \mathfrak{E}(t+0) \subset \mathfrak{E}(t+\varepsilon)$. Since $\mathfrak{E}(t+\varepsilon)$ is generated by the algebra of 'cylindrical' sets of the form

$$\tilde{A} = \{(e(t_1), \ldots, e(t_N)) \in B_N\}, \quad B_N \subset R^N; \quad 0 \leqslant t_j \leqslant t + \varepsilon$$

then there is a set $\tilde{A}_n$ of this form such that the probability of the symmetric difference of A and $\tilde{A}_n$ satisfies

$$\mathbf{P}(A - \tilde{A}_n) + \mathbf{P}(\tilde{A}_n - A) < n^{-2}$$

(see, for example, Borovkov, 1976c). Put $t_j^* = \min(t, t_j)$ and consider the set A_n^*, equal to $\tilde{A}_n$ with t_j replaced by t_j^*. Then $\tilde{A}_n\{e(t+\varepsilon) = e(t)\} = A_n^*\{e(t+\varepsilon) = e(t)\}$ and, consequently,

$$\mathbf{P}(\tilde{A}_n - A_n^*) + \mathbf{P}(A_n^* - \tilde{A}_n) < 2n^{-2}$$
$$\mathbf{P}(A - A_n^*) + \mathbf{P}(A_n^* - A) < 3n^{-2}.$$

But $A_n^* \in \mathfrak{E}(t)$ and any of the sets

$$A_* = \liminf A_n^* = \bigcup_{N=1}^{\infty} \bigcap_{n=N}^{\infty} A_n \in \mathfrak{E}(t)$$

or

$$A^* = \limsup A_n^* = \bigcap_{N=1}^{\infty} \bigcup_{n=N}^{\infty} A_n^* \in \mathfrak{E}(t)$$

differ from A only on a set of measure 0. For example,

$$\mathbf{P}(A^* - A) = \lim_{N \to \infty} \mathbf{P}\left(\bigcup_{n=N}^{\infty} A_n^* - A\right) \leqslant \lim_{N \to \infty} \sum_{n=N}^{\infty} \mathbf{P}(A_n^* - A)$$
$$\leqslant \lim_{N \to \infty} \sum_{n=N}^{\infty} 3n^{-2} = 0$$

$$\mathbf{P}(A - A^*) \leqslant \lim_{N \to \infty} \mathbf{P}\left(A - \bigcup_{n=N}^{\infty} A_n^*\right) \leqslant \lim_{N \to \infty} \sum_{n=N}^{\infty} \mathbf{P}(A - A_n^*)$$
$$\leqslant \lim_{N \to \infty} \sum_{n=N}^{\infty} 3n^{-2} = 0.$$

Thus for any A from $\mathfrak{E}(t+0)$ there is a set A^* from $\mathfrak{E}(t)$ which differs from A on a set of measure 0. Since we are assuming the completeness of these σ-algebras then the required relation $\mathfrak{E}(t+0) = \mathfrak{E}(t)$ is proved.*

Later we will always assume, where necessary, that the σ-algebras $\mathfrak{M}(t)$, $\mathfrak{E}(t)$, $\mathfrak{S}(t)$, and $\mathfrak{R}(t)$, together with the trajectories of the processes under discussion, are *continuous from the right*.

* We have given a proof due to Yurinskii. Without the assumption of completeness of $\mathfrak{M}(t)$ the equality $\mathfrak{M}(t+0) = \mathfrak{M}(t)$ will, in general, be false, just as, however, without the assumption that the set of values of $\mathbf{S}(t)$ is discrete.

DEFINITION 1. *We will say that the input process (or stream)* $\{e(t)\}$ *of the system is independent, if for any* $t > 0$ *the* σ*-algebras* $\mathfrak{E}(\infty)$ *and* $\mathfrak{M}(t)$ *are conditionally independent* with respect to* $\mathfrak{E}(t)$. *That is, if for any* $M \in \mathfrak{M}(t)$

$$\mathbf{P}_{\mathfrak{E}(\infty)}(M) = \mathbf{P}_{\mathfrak{E}(t)}(M) \tag{3.2}$$

or, equivalently, for any $E \in \mathfrak{E}(\infty)$

$$\mathbf{P}_{\mathfrak{M}(t)}(E) = \mathbf{P}_{\mathfrak{E}(t)}(E).$$

This equality may also be written in the form: if $E \in \mathfrak{E}(\infty)$, $M \in \mathfrak{M}(t)$, then

$$\mathbf{P}_{\mathfrak{E}(t)}(EM) = \mathbf{P}_{\mathfrak{E}(t)}(E)\mathbf{P}_{\mathfrak{E}(t)}(M).$$

This definition can be formulated equivalently in terms of random variables. If ξ_1 is measurable with respect to $\mathfrak{E}(\infty)$ and ξ_2 with respect to $\mathfrak{M}(t)$, then

$$\mathbf{E}_{\mathfrak{E}(t)}\xi_1\xi_2 = \mathbf{E}_{\mathfrak{E}(t)}\xi_1\mathbf{E}_{\mathfrak{E}(t)}\xi_2.$$

We note that this definition, like many subsequent definitions connected with service systems, is essentially related to the chosen family of σ-algebras so that, say, the same system may or may not have independent input according to the σ-algebras introduced. To avoid this problem, henceforth a service process will mean $\mathbf{S}(t)$ and the assignment of the necessary families of σ-algebras $\mathfrak{M}(t)$, $\mathfrak{E}(t)$, $\mathfrak{S}(t)$, and $\mathfrak{R}(t)$ which, as already noted, we assume to be continuous from the right.

Remark 1. Let $\mathfrak{M}'$ be any σ-algebra such that

$$\mathfrak{E}(t) \subset \mathfrak{M} \subset \mathfrak{M}(t).$$

Then, for a system with independent input, $\mathfrak{M}(t)$ and $\mathfrak{E}(\infty)$ are conditionally independent relative to $\mathfrak{M}'$. This follows since for $E \in \mathfrak{E}(\infty)$

$$\mathbf{P}_{\mathfrak{M}(t)}(E) = \mathbf{P}_{\mathfrak{E}(t)}(E) = \mathbf{P}_{\mathfrak{M}'}(E).$$

As in (3.2), it is sometimes convenient to state this independence as the conditional independence of $\mathfrak{M}(t)$ and the σ-algebras $\mathfrak{E}(t, \infty)$ generated by the random variables $e(t + u) - e(t)$, $u \geqslant 0$.

Conditional independence means that the future flow of $e(t)$ on $[t, \infty)$ does not depend on the past of $\mathbf{S}(t)$ on $[0, t]$ provided that the past history of $e(u)$ on $[0, t]$ is fixed. In other words, it is only important to know the *proper* past history of $e(u)$ on $[0, t]$.

It is obvious that the systems in examples 1 and 2 have independent inputs. The systems studied in chapter 2 also have independent inputs, as do the majority of systems of practical interest.

* *σ-algebras $\mathfrak{B}_1$ and $\mathfrak{B}_2$ are conditionally independent relative to $\mathfrak{B}$ if* $\mathbf{P}_{(\mathfrak{B},\mathfrak{B}_1)}(A) = \mathbf{P}_{\mathfrak{B}}(A)$ *for any* $A \in \mathfrak{B}_2$. The indices 1 and 2 here may be exchanged (for details see Loeve, 1962). $(\mathfrak{B}, \mathfrak{B}_1)$ denotes the σ-algebra generated by $\mathfrak{B}$ and $\mathfrak{B}_1$.

Now let τ be a stopping time of the input process: $\{\tau \leqslant t\} \in \mathfrak{E}(t)$. We introduce the σ-algebras of 'random past and present' $\mathfrak{E}(\tau)$ and $\mathfrak{M}(\tau)$. As is usual, *we will say that* $A \in \mathfrak{M}(\infty)$ *and*

$$A \cap \{\tau \leqslant t\} \in \mathfrak{M}(\tau).$$

We will say that $B \in \mathfrak{E}(\tau)$ *if and only if* $B \in \mathfrak{E}(\infty)$,

$$B \cap \{\tau \leqslant t\} \in \mathfrak{E}(t).$$

Since the latter holds for $B = \{\tau \leqslant t\}$, then τ is measurable with respect to $\mathfrak{E}(\tau)$.

DEFINITION 2. *The input $e(t)$ of a system is called strictly independent if the σ-algebras $\mathfrak{E}(\infty)$ and $\mathfrak{M}(\tau)$ are conditionally independent with respect to $\mathfrak{E}(\tau)$ for any stopping time τ of the input process.*

As before the property of strict independence may be written as follows. Let $\mathfrak{M}'$ be any σ-algebra such that

$$\mathfrak{E}(\tau) \subset \mathfrak{M}' \subset \mathfrak{M}(\tau).$$

Then for any $E \in \mathfrak{E}(\infty)$, $M \in \mathfrak{M}(\tau)$

$$\mathbf{P}_{\mathfrak{M}(\tau)}(E) = \mathbf{P}_{\mathfrak{M}'}(E) \tag{3.3}$$

$$\mathbf{P}_{(\mathfrak{M}', \mathfrak{E}(\infty))}(M) = \mathbf{P}_{\mathfrak{M}'}(M) \tag{3.4}$$

($(\mathfrak{M}', \mathfrak{E}(\infty))$ denotes the σ-algebra $\sigma(\mathfrak{M}', \mathfrak{E}(\infty))$ generated by $\mathfrak{M}'$ and $\mathfrak{E}(\infty)$), or, what is the same,

$$\mathbf{P}_{\mathfrak{M}'}(EM) = \mathbf{P}_{\mathfrak{M}'}(E)\mathbf{P}_{\mathfrak{M}'}(M).$$

Equivalent statements can be given in terms of random variables.

We note that since $e(t)$ is continuous from the right then $e(\tau + t)$, for all t, are random variables. In fact, in this case, $e(\tau + t)$ may be represented as the limit as $n \to \infty$ of the sequence of random variables

$$e_n(\tau + t) = \sum_{k=0}^{\infty} e\left(\frac{k}{2^n} + t\right) I\left(\frac{k}{2^n} \leqslant \tau < \frac{k+1}{2^n}\right). \tag{3.5}$$

Using this remark we can define the σ-algebra of 'random future', $\mathfrak{E}(\tau, \infty)$, as generated by the family of random variables τ and $e(\tau + t) - e(\tau)$ for $t \geqslant 0$. Then

$$\mathfrak{E}(\infty) = (\mathfrak{E}(\tau), \mathfrak{E}(\tau, \infty))$$

and the strict independence of the input can be also defined as the conditional independence of $\mathfrak{E}(\tau, \infty)$ and $\mathfrak{M}(\tau)$ with respect to $\mathfrak{E}(\tau)$. Such a version of the definition, it seems to us, better reflects the essence of these ideas. ('The random future' of $e(t)$ does not depend on the 'random past' of $\mathbf{S}(t)$ on $[0, \tau]$ provided the past history of $e(t)$ on $[0, \tau]$ is fixed.)*

A natural question arises: does independence of input imply strict independence? An affirmative answer is contained in the following:

THEOREM. *If the flows of σ-algebras $\mathfrak{E}(u)$ and $\mathfrak{M}(u)$ are continuous from the right and for any $E \in \mathfrak{E}(\infty)$*

$$\mathbf{P}_{\mathfrak{M}(u)}(E) = \mathbf{P}_{\mathfrak{E}(u)}(E)$$

then for every stopping time τ (relative to $\{E(u)\}$)

$$\mathbf{P}_{\mathfrak{M}(\tau)}(E) = \mathbf{P}_{\mathfrak{E}(\tau)}(E).$$

Proof. Fix an $E \in \mathfrak{E}(\infty)$ and consider the martingales

$$\xi(u) = \mathbf{P}_{\mathfrak{E}(u)}(E) = \mathbf{E}_{\mathfrak{E}(u)} I(E)$$

$$\zeta(u) = \mathbf{P}_{\mathfrak{M}(u)}(E) = \mathbf{E}_{\mathfrak{M}(u)} I(E)$$

which, according to the premises, will be stochastically equivalent

$$\mathbf{P}(\xi(u) = \zeta(u)) = 1$$

and continuous from the right. For such processes the values $\xi(\tau)$ and $\zeta(\tau)$ at random moments of time will equal

$$\begin{aligned} \xi(\tau) &= \lim_{n\to\infty} \sum_{k=0}^{\infty} \xi\left(\frac{k}{2^n}\right) I\left(\frac{k}{2^n} \leqslant \tau < \frac{k+1}{2^n}\right) \\ \zeta(\tau) &= \lim_{n\to\infty} \sum_{k=0}^{\infty} \zeta\left(\frac{k}{2^n}\right) I\left(\frac{k}{2^n} \leqslant \tau < \frac{k+1}{2^n}\right). \end{aligned} \tag{3.6}$$

It is known (see Lipster, 1974, p. 69, theorem 3.6) that these values coincide

* To prove $\mathfrak{E}(\infty) = (\mathfrak{E}(\tau), \mathfrak{E}(\tau, \infty))$ we note that for any t and x

$$\{e(t)<x\} = \{\tau>t\}\{e(t)<x\} \cup \{\tau\leqslant t\}\left[\bigcup_j \{e(\tau)=j\}\{e(\tau+(t-\tau))-e(\tau)<x-j\}\right].$$

Here

$$\{\tau>t\}\{e(t)<x\} \in \mathfrak{E}(\tau), \quad \{\tau\leqslant t\}\{e(\tau)=j\} \in \mathfrak{E}(\tau)$$

$$I(\tau\leqslant t)[e(\tau+(t-\tau))-e(\tau)] = \lim_{n\to\infty} \sum_{0}^{\infty} I\left(\frac{k}{2^n}\leqslant t-\tau<\frac{k+1}{2^n}\right)\left(e\left(\tau+\frac{k}{2^n}\right)-e(\tau)\right).$$

The latter equality, valid by the continuity from the right of $e(u)$, shows that the random variable $I(\tau\leqslant t)[e(\tau+(t-\tau))-e(\tau)]$ is measurable with respect to $\mathfrak{E}(\tau, \infty)$. This means that

$$\{e(t)<x\} \in (\mathfrak{E}(\tau), \mathfrak{E}(\tau, \infty)), \quad \mathfrak{E}(\infty) \subset (\mathfrak{E}(\tau), \mathfrak{E}(\tau, \infty)).$$

Since the reverse inclusion is obvious then

$$\mathfrak{E}(\infty) = (\mathfrak{E}(\tau), \mathfrak{E}(\tau, \infty)).$$

respectively with

$$\mathbf{E}_{\mathfrak{E}(\tau)}\xi(\infty) = \mathbf{P}_{\mathfrak{E}(\tau)}(E) \quad \text{and} \quad \mathbf{E}_{\mathfrak{M}(\tau)}\zeta(\infty) = \mathbf{P}_{\mathfrak{M}(\tau)}(E).$$

Consequently it is sufficient to prove that $\xi(\tau) = \zeta(\tau)$ a.s. But this follows immediately from the coincidence of the limits in (3.6). □

We now pass on to the definition of *independent output* from a system. It is somewhat different, since even in the simplest examples the σ-algebras $\mathfrak{M}(t)$ and $\mathfrak{S}(t, \infty)$ are not conditionally independent relative to $\mathfrak{S}(t)$. In example 1 with sparse input stream $e(u)$ the idle time on $[t, t+v]$ will be large if $q(t)$ is zero at t and small if $q(t)$ is large. Consequently, for a fixed past history of $s(v)$, $v \in [0, t]$, the value $s(t+u) - s(t)$ will depend essentially on the past history of $e(v)$ (that is, on the values $e(v)$ for $0 \leqslant v \leqslant t$).

However, if during some period of time $(t, t+u)$ the system is loaded and there is no idle time, then the number of requests served in this period will no longer depend on $\mathfrak{M}(t)$ and $\mathfrak{E}(t+u)$ for fixed $\mathfrak{S}(t)$ (again we have in mind the systems of example 1 and nearby systems). It is just this situation which is taken at the basis of the definition given below.

Let $\mathfrak{E}^*(t) = (\mathfrak{E}(t), \mathfrak{R}(t))$, so that the 'reduced' input stream

$$e^*(u) = e(u) - r(u)$$

on $[0, t]$ is measurable relative to $\mathfrak{E}^*(t)$.

We denoted by Γ_t an event of the form

$$\Gamma_t = \{q(t) \in K\} \tag{3.7}$$

where K is a set of integers situated above some given fixed number $d \geqslant 0$, and

$$q(t) = e(t) - r(t) - s(t)$$

is the value of the 'queue' at t. We consider a set K and two events

$$V_{e^*} \in \mathfrak{E}^*(t+u) \quad \text{and} \quad V_s \in \mathfrak{S}(t+u)$$

(depending on K), such that for the product $\Gamma_t V_{e^*} V_s$ we have

$$\Gamma_t V_{e^*} V_s \subset \left\{ \inf_{t \leqslant v \leqslant t+u} q(v) \geqslant d \right\}. \tag{3.8}$$

DEFINITION 3. *The 'output' of the system $\{\mathbf{S}(t), t \in T\}$ is called independent in the domain $q \geqslant d$, if for any $u \geqslant 0$, $t > 0$, $S \in \mathfrak{S}(t+u)$ and for any K, V_{e*}, V_s satisfying* (3.8) *we have*

$$I(\Gamma_t V_{e^*})\mathbf{P}_{(\mathfrak{M}(t), \mathfrak{E}^*(t+u))} V_s S = I(\Gamma_t V_{e^*})\mathbf{P}_{\mathfrak{S}(t)} V_s S. \tag{3.9}$$

This relation may be called *conditional independence in the domain $\{q \geqslant d\}$ of the σ-algebras $\mathfrak{S}(t+u)$ and $\mathfrak{M}(t)\mathfrak{E}^*(t+u)$ given $\mathfrak{S}(t)$*. This can equivalently be written as the equality

$$\mathbf{P}_{\mathfrak{S}(t)}\Gamma_t V_{e^*} V_s BS = \mathbf{P}_{\mathfrak{S}(t)} V_s S \mathbf{P}_{\mathfrak{S}(t)}\Gamma_t V_{e^*} B \tag{3.10}$$

for any $S \in \mathfrak{S}(t+u)$, $B \in (\mathfrak{M}(t)\mathfrak{E}^*(t+u))$. To establish (3.10) it is sufficient to write the left-hand side in the form

$$\mathbf{E}_{\mathfrak{S}(t)} I(\Gamma_t V_{e^*} B)\mathbf{P}_{(\mathfrak{M}(t)\mathfrak{E}^*(t+u))} V_s S$$

and use (3.9). Conversely, if (3.10) is true, then dividing both sides of (3.10) by $\mathbf{P}_{\mathfrak{S}(t)}(B)$ and passing to conditional probabilities we obtain, by the arbitrariness of $B \in \mathfrak{M}(t)\mathfrak{E}^*(t+u)$

$$\mathbf{P}_{(\mathfrak{M}(t)\mathfrak{E}^*(t+u))}\Gamma_t V_{e^*} V_s S = \mathbf{P}_{(\mathfrak{M}(t)\mathfrak{E}^*(t+u))}\Gamma_t V_{e^*} \mathbf{P}_{\mathfrak{S}(t)} V_s S \tag{3.11}$$

which coincides with (3.9).

Roughly speaking independence of output means that if the 'past' of $s(v)$ on $[0, t]$ is known, then its 'future' in $[t, t+u]$ does not depend on events from $\mathfrak{M}(t)\mathfrak{E}^*(t+u)$, provided that $q(v)$ in $(t, t+u)$ does not touch the domain $q < d$.

Remark 2. In definition 3, as in previous definitions, formulations equivalent to (3.9) may be given in terms of random variables. In addition, an analogue to remark 1 is also valid here with the only difference that as intermediate σ-algebra it is possible to take any σ-algebra $\mathfrak{M}'$, $\mathfrak{S}(t) \subset \mathfrak{M}' \subset \mathfrak{M}(t)\mathfrak{E}^*(t+u))$.

DEFINITION 3A. *The output of a system is called independent in the domain $D \geqslant q \geqslant d$, if the condition of definition 3 is satisfied for arbitrary K and $V_{e^*} \in \mathfrak{E}^*(t+u)$ and $V_s \in \mathfrak{S}(t+u)$ such that for $\Gamma_t V_{e^*} V_s$ we have $\Gamma_t V_{e^*} V_s \subset \{d \leqslant q(v) \leqslant D\}$ for all $v \in [t, t+u]$.*

Here independence is preserved up to the first output of the queue q from the interval $[d, D]$.

As an illustration consider example 1, in which, for simplicity, we put $v_j^s \equiv 1$. After the explanations given in definition 3, it must be clear that in this example the system has independent output in the domain $q \geqslant 1$. We seek, however, a more formal verification. The σ-algebra $\mathfrak{S}(u)$ here is the σ-algebra generated by the random variables $\eta_e(u)$, $\gamma^e(u)$; $\tau_1^e, \ldots, \tau_{\eta_e(u)}^e$; $v_1^e, \ldots, v_{\eta_e(u)}^e$. The random variables $s(u), \gamma^s(u), \tau_1^s, \ldots, \tau_{s(u)}^s$ will be measurable with respect to $\mathfrak{S}(u)$. It is clear that in this example $\mathfrak{M}(u) = (\mathfrak{E}(u)\mathfrak{S}(u))$, $\mathfrak{E}^*(u) = \mathfrak{E}(u)$.

Take an 'elementary' event from $(\mathfrak{M}(u)\mathfrak{E}(t+u))$. This is a 'trajectory' $\omega = (\omega_1, \omega_2)$, where ω_1 is formed by a trajectory of $e(v)$ for $0 \leqslant v \leqslant t+u$ and ω_2 is formed by a trajectory of $s(v)$ for $0 \leqslant v \leqslant t$ and the variables $\tau_1^s, \ldots, \tau_{s(u)}^s$. Let Γ_t, V_e, and V_s be any events defined in (3.7) and (3.8) for $d = 1$. Then the conditional probability of $V_s S$ under the condition $\omega \in \Gamma_t V_e$ (on the intersection of $V_s S$ and ω service on $[u, u+t]$ is not interrupted), by the independence of $\{\tau_j^e, v_j^e\}$ and $\{\tau_j^s\}$, will be determined by ω_2 (i.e. it remains unchanged for different ω_1 satisfying $\omega \in \Gamma_t V_e$). It is clear that this implies (3.11).

Precisely the same is the case for the system of example 1a.

In a similar way we show that the system in example 2 will have independent output only in the domain $q \geqslant n$. If $n = \infty$, then the system will not have independent output in any domain.

The system of example 1b will have independent output in the domain $1 \leqslant q \leqslant N - 1$.

To conclude this section we consider the so-called *stochastic dependence of the refusal process* $r(t)$ *on the length of queue.*

Definitions 2 and 3 single out the class of service systems which are important in what follows and are very widespread in applications, in which either the input or output is independent. Such systems are more accessible to study.

At the same time, in problems of various kinds on systems with control, input streams are propagated whose intensity depends on the length of queue. We pick out an important class of realizations of such control with the help of a refusal process in which each request arriving in the system obtains refusal with probability depending on the length of the queue at the time of its arrival. It is clear here that if the input stream $e(t)$ is independent, then the 'reduced' input stream $e^* = e - r$ will no longer have this property.

In what follows we will adhere to the following natural terminology. The epochs of jumps of $e(t)$ will be called *arrival epochs of requests* and the value of the jump will be the *number of requests arriving at the given epoch.* In other words, to each unit increment of $e(t)$ we associate a certain object, named a 'request', and consider $\{e(u)\}$ as a process describing the arrival of requests in time, so that the arrival time of the kth request is $\min\{t: e(t) \geqslant k\}$. We use similar terminology when the jumps of $s(t)$ are considered.

Let τ be a Markov epoch relative to $\{\mathfrak{E}(t)\}$. We denote by $\mathfrak{M}(\tau -)$ the σ-algebra generated by the events

$$A \cap \{\tau > t\}, \quad A \in \mathfrak{M}(t).$$

This is the σ-algebra of events *strictly preceding the epoch* τ. The σ-algebras $\mathfrak{M}(\tau)$ and $\mathfrak{M}(\tau -)$, in general, do not coincide, $\mathfrak{M}(\tau -) \subset \mathfrak{M}(\tau)$ (see Dellacherie, 1972); the random variable $\mathbf{S}(\tau - 0)$ is measurable with respect to $\mathfrak{M}(\tau - 0)$. This follows from the representation

$$\mathbf{S}(\tau - 0) = \lim_{n \to \infty} \sum_k \mathbf{S}\left(\frac{k-1}{n}\right) I\left(\frac{k-1}{n} < \tau \leqslant \frac{k}{n}\right)$$

where, obviously, the random variables

$$\mathbf{S}\left(\frac{k-1}{n}\right) I\left(\frac{k-1}{n} < \tau \leqslant \frac{k}{n}\right) = \mathbf{S}\left(\frac{k-1}{n}\right) I\left(\tau > \frac{k-1}{n}\right) - \mathbf{S}\left(\frac{k-1}{n}\right) I\left(\tau > \frac{k}{n}\right)$$

are measurable with respect to $\mathfrak{M}(\tau -)$.

DEFINITION 4. *Let v be a random number of request such that its arrival time $\mu(v)$ is a stopping time relative to $\{\mathfrak{E}(t)\}$. We denote by $\mathfrak{M}^*(\mu)$ the σ-algebra generated by the events from $\mathfrak{M}(\mu -)$, the random variables $e(\mu) - e(\mu - 0)$ and $(\Delta_1, \ldots, \Delta_{v-1})$, where $\Delta_k = 0$ or 1 according to whether or not the* kth *request is refused. We will say that the refusal process $r(t)$ depends stochastically on $q(t)$, if*

$$\mathbf{P}_{\mathfrak{M}^*(\mu)} \text{ (the v-th request is refused)} = 1 - p(q(\mu - 0)). \tag{3.12}$$

The function $p(x)$, $0 \leqslant p(x) \leqslant 1$, is supposed given. We will call the difference $1 - p(x)$ the probability of refusal.

Why is the conditional probability in (3.12) taken with respect to $\mathfrak{M}^*(\mu)$ and not with respect to $\mathfrak{M}(\mu)$? The fact is that the event $\{$*v-th request is refused*$\}$ belongs to $\mathfrak{M}(\mu)$ ($\mathbf{S}$ is continuous from the right), whereas the σ-algebra $\mathfrak{M}^*(\mu)$ 'delineates' precisely the whole past history prior to the arrival of the vth request. The essence of definition 4 is that the probability of refusal under stochastic control does not depend on the past history, if the length of the queue at the arrival epoch of the request is prescribed.

If the jumps of $e(t)$ are larger than 1, then there is also the possibility of the control when the right-hand side of (3.12) is equal to

$$1 - p(q(\mu - 0) + \Delta_{l+1} + \cdots + \Delta_{v-1}) \tag{3.13}$$

where $l + 1, \ldots, l + e(\mu + 0) - e(\mu - 0)$ are the numbers of the requests arriving with the vth request. However, from the point of view of asymptotic methods, these modes of control, (3.12) and (3.13), are indistinguishable.

3.2. Preliminary remarks on approximation by diffusion processes

In the last section we considered some interrelations between the components of a service process, the latter itself being constructed arbitrarily. We now consider conditions on $e(t)$, $r(t)$, and $s(t)$ which allow us to approximately describe the behaviour of $q(t)$.

The asymptotic analysis of explicit formulae for systems with refusals and a large number of service channels, the known limit theorems for loaded systems, and a whole series of other facts and considerations show that the most prevalent type of process by which it is possible to approximate the limit processes of service* is the diffusion process. The remainder of this chapter is devoted to the search for the broadest and most general conditions on service systems, under which $q(t)$, suitably normalized, will converge to some diffusion process or other.

We will study the behaviour of $\{q(t)\}$ on large time intervals $[0, vT]$ as $T \to \infty$.

* We have in mind systems which are somehow close to critical. Here we refer to systems with queues and load close to 1, systems with intensive input stream and a large number of service channels (both with queues and refusals), and many others. Examples of such systems are considered in sections 3.4–3.7.

Here we will consider a series array, that is, a sequence of processes $\mathbf{S}_T(t)$ depending on a parameter T (we will often omit the index T) and defined on $[0, \infty)$.

The basis of the subsequent discussions of this chapter are the theorems on convergence to a diffusion, presented in sections 1.5–1.10. Convergence of processes here will mean C-convergence of their distributions. (Recall that a sequence of processes $y_T(u)$ C-converges to $w(u)$ on $[0, v]$ if, for any measurable functional f, continuous in the uniform metric at points of C, we have $\mathbf{P}(f(y_T) < y) \Rightarrow \mathbf{P}(f(w) < y)$.)

The convergence conditions in question are mainly concerned with the first two moments of the increments of $e(u)$, $s(u)$, and $r(u)$ on large intervals of time t. The increments of $x(u)$ on $(u, u + t)$ will be denoted

$$x_{u,t} = x(t + u) - x(u)$$

where x stands for e, s, r, etc., so that, for example,

$$e_{u,t} = e(t + u) - e(t).$$

To explain here (before discussing the examples in sections 3.4–3.7) the naturality of the statement of the problem, used later, consider the following situation. Suppose that in some ω-set of high probability the conditional expectation $\mathbf{M}_{\mathfrak{M}(u)}e_{u,t}$ increases asymptotically linearly with t. Then under very broad conditions on $s(u)$ the queue $q(t)$ after time T can attain values of the order of T. Suppose further that the control system (i.e. the processes $e(u)$, $s(u)$, and $r(u)$) depends on the length of queue, that is to say, in units of the full scale of possible values of $q(t)$ (i.e. in a scale comparable with the size of T). Namely, we suppose that functions $a_e(q)$, $a_r(q)$, and $a_s(q)$ are given such that, with probability close to 1,

$$\mathbf{E}_{\mathfrak{M}(u)}x_{u,t} = ta_x\left(\frac{q(u)}{T}\right) + o\left(\frac{t}{\sqrt{T}}\right) \tag{3.14}$$

as $t \to \infty$, $t = o(T)$. Here x takes the values e, r, and s.

If a_e, a_s, and a_r are continuous, then this means that the intensity of the input and output streams will be essentially changed only by a change in the queue length of order T. Such dependence of the control flow on $q(t)$ is called '*rough*'.

It is natural to take as fundamental the case when *the function* $A(q) = a_e(q) - a_r(q) - a_s(q)$ *is non-increasing* (the larger the queue the more likely are refusals and the more intense is the service). If $A(0) \geqslant 0$, then, after some time, $q(u)/T$ is close to a *point of equilibrium* Q, which is a solution of

$$A(Q) = 0. \tag{3.15}$$

In the general case we suppose that there is at least one solution of (3.15) and we distinguish two cases of 'rough' dependence:

(1) $Q > 0$.
(2) *$A(0)$ is small and depends on T in such a way that* $A(0)\sqrt{T} \to A_0$. We agree to classify this as the case '$Q = 0$'.

First we consider the version $Q > 0$ and denote

$$Q(u) = \frac{q(u)}{T}.$$

Alongside (3.14), for values of $Q(u)$ close to Q, we require that the second moments of

$$x_{u,t} - ta_x(Q(u))$$

($x = e, r, s$) also behave asymptotically regularly (grow linearly with t):

$$\mathbf{E}_{\mathfrak{M}(u)}[x_{u,t} - ta_x(Q(u))][y_{u,t} - ta_y(Q(u))] = tb_{xy}(Q(u)) + o(t), \quad x, y = e, r, s. \quad (3.16)$$

Then the process

$$z(u) = \frac{q(uT) - QT}{\sqrt{T}}$$

under certain other, less essential, conditions will C-converge to a diffusion process $w(t)$ on $[0, v]$ with drift coefficient $xA'(Q)$ and diffusion constant $B(Q)$, where $B = b_{ee} + b_{ss} + b_{rr} - 2b_{es} - 2b_{er} - 2b_{rs}$.

This process can be explicitly described using a standard Wiener process (see section 3.3). The corresponding precise result is given in theorem 4 of section 3.3, and examples of systems satisfying the conditions of this theorem are considered in section 3.4. Related to these, for example, are the systems with intensive input stream and with exponential service time in n service channels, where $n - QT \gg \sqrt{T}$.

In the second case, $Q = 0$, under similar conditions we obtain convergence to a diffusion with reflection from the zero boundary. This process also can be explicitly described using a Wiener process. The precise result is given in theorem 5 of section 3.3. As examples of systems satisfying the conditions of this theorem we can take the systems with a heavy load (see section 3.6). Here, as we will see below, the practical calculation of the coefficients a_x, b_{xy} in conditions (3.14) and (3.16) (as also in case $Q > 0$) presents no great difficulties.

It is also possible to mention examples of systems (see sections 3.5–3.7) whose limit processes are diffusion processes with reflections from two boundaries.

In all of these examples we deal with dependence of the control systems on the length of the queue '*in units of the full scale*' of possible values of $q(t)$ on the time interval $[0, T]$ (i.e. of the scale whose length is comparable to T). This dependence is accomplished by the assignment of the function a_x, b_{xy}, $x, y = e, r, s$, so that, for example, $a_e(q(u)/T)$ describes the principal part of the increment of the input process $e(t)$ on the interval $(u, u+t)$. Here we only need to know the behaviour of this function in a neighbourhood of an equilibrium point Q at which

$$A(Q) = a_e(Q) - a_r(q) - a_s(q) = 0$$

and the parameters of the limit process $w(t)$ are completely determined by the values $A'(Q)$ and $B(Q)$.

However, even amongst systems with rough dependence of control on queue length, there are systems for which the derivative A' has a break in a neighbourhood of Q (related to this, for example, are the multichannel systems with queues, in which the number of channels n is such that $(n - QT)/\sqrt{T}$ is bounded (see section 3.4, and also section 2.5)). Regarding the dependence of the control on $q(t)$, it need not have the form (3.14), (3.16).

Therefore it is natural to consider the following rather more general statement of the problem.

Let the initial value $q(0)/T$ be in a neighbourhood of some Q. We assume that in a neighbourhood of this point the control of the system depends on the length of the queue $q(t)$, as in (3.14), (3.16) but with a_x and b_{xy} changed into

$$\begin{aligned} a_x(Q(u)) &= a_x^0 + \frac{\alpha_x(\zeta(u))}{\sqrt{T}} + o\left(\frac{1}{\sqrt{T}}\right) \\ b_{xy}(Q(u)) &= \beta_{xy}(\zeta(u)) + o(1) \end{aligned} \tag{3.17}$$

where $\zeta(u) = (q(u) - QT)/\sqrt{T}$, $a_e^0 - a_r^0 - a_s^0 = 0$, and α_x, β_{xy} are arbitrary functions.

It is clear that in conditions (3.14) and (3.16), with continuous a'_x and b_{xy}, we are dealing with a particular case of (3.17) since, for rough dependence of control on $q(u)$

$$\begin{aligned} a_x(Q(u)) &= a_x(Q) + (Q(u) - Q)a'_x(Q) + o(|Q(u) - Q|) \\ &= a_x(Q) + \frac{a'_x(Q)\zeta(u)}{T} + o\left(\frac{1}{\sqrt{T}}\right) \end{aligned} \tag{3.18}$$

$$b_{xy}(Q(u)) = b_{xy}(Q) + o(1)$$

which corresponds to the values $\alpha_x(\zeta) = \zeta a'_x(Q)$, $\beta_{xy}(\zeta) = b_{xy}(Q)$.

Clearly formulae (3.17) define the dependence of control on $q(u)$ only in units of $\sqrt{T}$ in a neighbourhood of TQ. This turns out to be sufficient for the definition and description of the limit process.

In the following section we give a precise statement of the problem and prove the theorems on convergence to a diffusion process.

3.3. General theorems on convergence of the normalized 'occupation' $q(t)$ to a diffusion process

We first consider conditions for convergence to an unbounded diffusion.

As before we denote

$$Q(u) = \frac{q(u)}{T}, \quad \zeta(u) = \frac{q(u) - TQ}{\sqrt{T}} \tag{3.19}$$

and by $\theta = \theta(T)$ we mean a sequence

$$\theta \to \infty, \quad \theta = o(T)$$

as $T \to \infty$.

$N > 0$ and $\varepsilon > 0$ denote arbitrary (respectively large and small) numbers. $\Gamma_u^N \in \mathfrak{M}(u)$ will be defined by

$$\Gamma_u^N = \{|q(u) - QT| < N\sqrt{T}\} = \{|\zeta(u)| < N\}. \tag{3.20}$$

THEOREM 1. *Let there be a sequence θ and sets $\Omega_{u,\theta} \in \mathfrak{M}(u), 0 \leqslant u \leqslant vT$, such that**

$$\mathbf{P}\left(\bigcap_{u \leqslant vT} \Omega_{u,\theta}\right) \to 1 \tag{3.21}$$

and for each $u \leqslant vT - \theta$ on $\Omega_{u,\theta} \cap \Gamma_u^N$ we have

(I) $\mathbf{E}_{\mathfrak{M}(u)} x_{u,\theta} = \theta . \left(a_x^0 + \frac{\alpha_x(\zeta(u))}{\sqrt{T}} + \frac{r_x(\omega)}{\sqrt{T}}\right)$

(II) $\mathbf{E}_{\mathfrak{M}(u)} x_{u,\theta}^0 y_{u,\theta}^0 = \theta . (\beta_{xy}(\zeta(u)) + r_{xy}(\omega))$

(III) $\mathbf{E}_{\mathfrak{M}(u)}[(x_{u,\theta}^0)^2; |x_{u,\theta}^0| > \delta\sqrt{T}] < \varepsilon\theta$

for any $\delta > 0$. Here

$$x_{u,\theta}^0 = x_{u,\theta} - \theta . \left(a_x^0 + \frac{\alpha_x(\zeta(u))}{\sqrt{T}}\right)$$

x and y take the values e, r, and s and for any fixed $\varepsilon > 0$

$$|r_x(\omega)| < \varepsilon \quad |r_{xy}(\omega)| < \varepsilon.$$

In this connection we assume that if (I)–(III) *are satisfied for a sequence θ, then they remain valid also for any sequence $\theta_1 = c\theta$, $\frac{1}{2} \leqslant c \leqslant 2$.*

In addition, suppose we have

(IV) $a_e^0 - a_s^0 - a_r^0 = 0$

* Here we suppose that $\bigcap_{u \leqslant s} \Omega_{u,\theta}$ is an event. A precise statement of the condition requires the existence of events included in this product and having probability tending to 1.

and $\alpha = \alpha_e - \alpha_r - \alpha_s$ *and* $\beta = \beta_{ee} + \beta_{rr} + \beta_{ss} - 2\beta_{es} - 2\beta_{er} - 2\beta_{sr}$ *satisfy condition* (IV) *of theorem 1 of section 1.5 (on the existence of a diffusion process* $w(t)$ *with drift coefficient* α *and diffusion coefficient* β*).*

(V) *The probability that the right continuous processes e, r, and s have at least one jump in* $[0, vT]$ *of size greater than* $\delta\sqrt{T}$ *converges to* 0 *as* $T \to \infty$ *for any* $\delta > 0$.

Then, if

$$q_T(0) = QT + z_0\sqrt{T} \tag{3.22}$$

and the distribution of z_0 *converges to* p_0 *as* $T \to \infty$, *then under conditions* (I)–(V) *the process*

$$z(t) = \frac{q_T(tT) - QT}{\sqrt{T}} = \zeta(tT), \quad t \in [0, v] \tag{3.23}$$

C-converges on $[0, v]$ *to the diffusion process* $w(t)$ *(referred to in condition* (IV)*) with initial distribution* p_0.

Remark 1. In order to study $q(t)$ it may be convenient to consider the 'reduced' service process

$$\mathbf{S}^*(t) = (e^*(t) \equiv e(t) - r(t), 0, s(t))$$

with the conditions for convergence imposed on the components of the reduced process. The form of these conditions is easy to obtain from theorem 1; here β will be $\beta = \beta_{ee^*} - 2\beta_{e^*s} + \beta_{ss}$.

Remark 2. The condition $a_e^0 - a_r^0 - a_s^0 = 0$ in theorem 1 can obviously be replaced by the following condition: in Γ_u^N

$$a_e^0 - a_r^0 - a_s^0 = o\left(\frac{1}{\sqrt{T}}\right).$$

But if

$$a_e^0 - a_r^0 - a_s^0 = \frac{A^0}{\sqrt{T}} + o\left(\frac{1}{\sqrt{T}}\right)$$

then, as is easy to verify by looking through the proof of theorem 1 (see also the proof of theorem 5), the limit process for $z(t)$ will also be a diffusion process, but with coefficients $(\alpha + A^0, \beta)$.

The proof of theorem 1 is based completely on the use of theorems 1 and 2 of section 1.5. We recommend that the reader should now look again at the statements of these theorems and the conditions in them. The latter, in both form and content, are very close to the conditions of theorem 1.

Thus, for the proof of theorem 1, it is sufficient to show that $q(t)$ and its increments

$$q_{u,\theta} = e_{u,\theta} - s_{u,\theta} - r_{u,\theta}$$

satisfy the conditions of theorems 1 and 2 of section 1.5.

Under the conditions of theorem 1 we will have on $\Omega_{u,\theta} \cap \Gamma_u^N$

$$\begin{aligned}\mathbf{E}_{\mathfrak{M}(u)} q_{u,\theta} &= \mathbf{E}_{\mathfrak{M}(u)}(e_{u,\theta} - r_{u,\theta} - s_{u,\theta}) \\ &= \frac{\theta\alpha(\zeta(u))}{\sqrt{T}} + \frac{\theta}{\sqrt{T}}(r_e(\omega) - r_r(\omega) - r_s(\omega)) \\ &= \frac{\theta}{\sqrt{T}}(\alpha(\zeta(u)) + r_1(\omega))\end{aligned}$$

$$\begin{aligned}\mathbf{E}_{\mathfrak{M}(u)} q_{u,\theta} &= \mathbf{E}_{\mathfrak{M}(u)}\left(e_{u,\theta} - s_{u,\theta} - r_{u,\theta} - \frac{\theta\alpha(\zeta(u))}{\sqrt{T}}\right)^2 \\ &\quad - \frac{\theta^2\alpha^2(\zeta(u))}{T} + 2\frac{\theta^2\alpha(\zeta(u))}{T}(\alpha(\zeta(u)) + r_1(\omega)).\end{aligned} \tag{3.24}$$

Since $|\alpha(z)| < c(1 + |z|)$ (see condition (IV)), then in $|\zeta(u)| < N$ the last two terms in (3.24) do not exceed

$$c_1 \frac{\theta^2 N^2}{T} = o(\theta).$$

Since the first term in (3.24) is equal to

$$\mathbf{E}_{\mathfrak{M}(u)}(e_{u,\theta}^0 - s_{u,\theta}^0 - r_{u,\theta}^0)^2$$

then (II) of theorem 1 implies (PII) of theorems 1 and 2 of section 1.5, as also (PI), will be satisfied. Here the coefficients, depending on $\zeta(u)$, in these conditions are the functions α and β defined in condition (IV) of theorem 1.

For the proof of condition (PIII) of theorems 1 and 2 of section 1.5, we note first that for any X and Y, $X^2 \leqslant 2(X - Y)^2 + 2Y^2$ and, consequently, in Γ_u^N

$$\begin{aligned}&\mathbf{E}_{\mathfrak{M}(u)}(q_{u,\theta}^2; |q_{u,\theta}| > \delta\sqrt{T}) \\ &\quad\leqslant 2\mathbf{E}_{\mathfrak{M}(u)}\left[\left(q_{u,\theta} - \frac{\theta\alpha(\zeta(u))}{\sqrt{T}}\right)^2 + \frac{c^2\theta^2(1 + N^2)}{T}; \right. \\ &\qquad\qquad \left. \left|q_{u,\theta} - \frac{\theta\alpha(\zeta(u))}{\sqrt{T}}\right| > \frac{\delta}{2}\sqrt{T}\right]\end{aligned}$$

for sufficiently large T. If we denote

$$D_x = \left\{|x_{u,\theta}^0| > \frac{\delta}{6}\sqrt{T}\right\}$$

then

$$\left\{\left|q_{u,\theta} - \frac{\theta\alpha(\zeta(u))}{\sqrt{T}}\right| > \frac{\delta}{2}\sqrt{T}\right\} \subset D_e + D_r + D_s.$$

In addition, from condition (III) of theorem 1 and the Chebyshev inequality, it follows that on $\Omega_{u,\theta} \cap \Gamma_u^N$

$$\mathbf{P}_{\mathfrak{M}(u)} D_x \leqslant \varepsilon \frac{\theta}{T}. \tag{3.25}$$

Thus, on $\Omega_{u,\theta} \cap \Gamma_u^N$,

$$\mathbf{E}_{\mathfrak{M}(u)}(q_{u,\theta}^2; |q_{u,\theta}| > \delta\sqrt{T})$$

$$\leqslant 6 \sum_{x,y=e,r,s} \mathbf{E}_{\mathfrak{M}(u)}((x_{u,\theta}^0)^2; D_y) + 6 \frac{c^2\theta^2(1+N^2)}{T} \cdot \frac{\varepsilon\theta}{T}. \tag{3.26}$$

The second term here is obviously $o(\theta)$.

To estimate the terms in the sum on the right-hand side of (3.26) we use the following simple lemma.

LEMMA 1. *If* $\mathbf{E}_{\mathfrak{M}(u)}(\xi^2; |\xi| > \delta\sqrt{T}) < \varepsilon\theta$ *on some* ω*-set* Ω' *and the event* D *is such that* $\mathbf{P}_{\mathfrak{M}(u)}(D) \leqslant \varepsilon\theta/T$, *then on* Ω'

$$\mathbf{E}_{\mathfrak{M}(u)}(\xi^2; D) \leqslant \varepsilon\theta(1+\delta^2).$$

Proof.

$$\mathbf{E}_{\mathfrak{M}(u)}(\xi^2; D) = \mathbf{E}(\xi^2; D \cap \{|\xi| < \delta\sqrt{T}\}) + \mathbf{E}(\xi^2; D \cap \{|\xi| \leqslant \delta\sqrt{T}\})$$

$$\leqslant \varepsilon\theta + \frac{\delta^2 T\varepsilon\theta}{T} = \varepsilon\theta + \delta^2\varepsilon\theta. \quad \square$$

From this lemma, inequality (3.25), and condition (III) of theorem 1, it follows that on $\Omega_{u,\theta} \cap \Gamma_u^N$

$$\mathbf{E}_{\mathfrak{M}(u)}[(x_{u,\theta}^0)^2; D_y] \leqslant \varepsilon\theta\left[1 + \left(\frac{\delta}{6}\right)^2\right].$$

Together with (3.26) this proves (PIII) for $q_{u,\theta}$.

Conditions (PIV) and (PV) are satisfied in an obvious way. Theorem 1 is proved. $\square$

In the same way, from the theorems of sections 1.9 and 1.10, we may obtain the result on convergence to a diffusion with reflection. Any level may serve as reflecting boundary. However, to simplify the statement we discuss the case when the limit process is nonnegative and is reflected from the zero boundary (this relates to the example of rough dependence of control on $q(u)$ in the case $Q = 0$).

In the following two theorems we take $Q = 0$, so that

$$\Gamma_u^N = \{|q(u)| < N\sqrt{T}\} = \{|\zeta(u)| < N\}.$$

THEOREM 2. *Let conditions* (I)–(III) *of theorem* 1 *be satisfied, but only in the domain*

$$\{\zeta(u) \geqslant \varepsilon^*\} = \{q_T(u) \geqslant \varepsilon^* \sqrt{T}\}$$

for any $\varepsilon^* > 0$.

In addition, let condition (V) *of theorem* 1 *be satisfied and*

(IV) $a_e^0 - a_s^0 - a_r^0 = 0$, *but* α *and* β (*see condition* (IV) *of theorem* 1) *satisfy* (PRIV) *of section* 1.9 (*on the existence of a diffusion process* $w_R(t)$ *with coefficients* α *and* β *and reflection from the boundary* $\zeta = 0$).

(VI) *For any sufficiently small* $\varepsilon^{**} > 0$ *there exist* $p > 0$, $c > 0$ *such that for* $\theta_1 \sim (\varepsilon^{**})^2 Tc$ *in* $\Omega_{u,\theta} \cap \{\zeta(u) < \varepsilon^{**}\}$ *we have*

$$\mathbf{P}_{\mathfrak{M}(u)}(q_{u,\theta_1} > \varepsilon^{**}\sqrt{T}) > p.$$

Then, if the distribution of $\zeta(0) = q_T(0)/\sqrt{T}$ *converges to the distribution* p_0, *then under conditions* (I)–(VI) *the process*

$$z(t) = \frac{q(tT)}{T} = \zeta(tT)$$

C-converges on $[0, v]$ *to a diffusion process* w_R *with reflection from the boundary* $\zeta = 0$, *with coefficients of drift and diffusion respectively* α *and* β *and with initial distribution* p_0.

In a similar way we have a theorem on convergence to a diffusion with two reflecting boundaries. As such boundaries we take the levels $\zeta = 0$ and $\zeta = R > 0$.

THEOREM 3. *Let conditions* (I)–(III) *of theorem* 1 *be satisfied, but only in the domain*

$$\{R - \varepsilon^* \geqslant \zeta(u) \geqslant \varepsilon^*\} = \{(R - \varepsilon^*)\sqrt{T} \geqslant q_T(u) \geqslant \varepsilon^* \sqrt{T}\}$$

for any $\varepsilon^* > 0$.

In addition, let condition (V) *of theorem* 1 *hold and also*

(IV) $a_e^0 - a_r^0 - a_s^0 = 0$, *and* α *and* β (*see condition* (IV) *of theorem* 1) *satisfy* (PRRIV) *of section* 1.10 (*on the existence of a diffusion process* $w_{RR}(t)$ *with coefficients* α *and* β *and with reflections from the boundaries* 0 *and* R).

(VI) *The processes* $q_T(t)$ *and* $R\sqrt{T} - q_T(t)$ *satisfy condition* (VI) *of theorem* 2.

Then, if the distribution of $\zeta(0) = q_T(0)/\sqrt{T}$ *converges to the distribution* p_0, *then under conditions* (I)–(VI),

$$z(t) = \frac{q(tT)}{\sqrt{T}} = \zeta(tT)$$

C-converges on $[0, v]$ *to a diffusion process* w_{RR} *with reflection from the levels* 0 *and* R *and with initial distribution* p_0.

The proofs of theorems 2 and 3 are completely based on the theorems of sections 1.9 and 1.10. Since the conditions (IV)–(VI) are taken from these latter theorems to theorems 2 and 3 without change, then it is sufficient to verify that conditions (I)–(III) in theorems 2 and 3 imply conditions (I)–(III) in the theorems of sections 1.9 and 1.10. But this is exactly the same verification as in the proof of theorem 1. Thus, we may regard theorems 2 and 3 as proved. □

We pass on to the analogue of theorems 1–3 in the important special case of rough dependence of the control system on the queue length. This case was described in section 3.2; its diffusion processes, which appear in theorems 1–3, can be found explicitly.

First consider convergence to an unbounded diffusion when the 'point of equilibrium' Q (defined as the solution of $A(Q) = a_e(Q) - a_r(Q) - a_s(Q) = 0$ (see section 3.2)) is positive. As before, let

$$Q(u) = \frac{q(u)}{T}, \quad \Gamma_u^N = \left\{|Q(u) - Q| < \frac{N}{\sqrt{T}}\right\}.$$

THEOREM 4. *Let there be a sequence θ and sets $\Omega_{u,\theta} \in \mathfrak{M}(u)$ such that*

$$\mathbf{P}\left(\bigcap_{u \leqslant vT} \Omega_{u,\theta}\right) \to 1$$

and on $\Omega_{u,\theta} \cap \Gamma_u^N$ we have

$$\text{(I)}\ \mathbf{E}_{\mathfrak{M}(u)} x_{u,\theta} = \theta a_x(Q(u)) + \frac{\theta}{\sqrt{T}} r_x(\omega)$$

$$\text{(II)}\ \mathbf{E}_{\mathfrak{M}(u)} x_{u,\theta}^0 y_{u,\theta}^0 = \theta b_{xy}(Q(u)) + \theta r_{xy}(\omega)$$

$$\text{(III)}\ \mathbf{E}_{\mathfrak{M}(u)}[(x_{u,\theta}^0)^2;\ |x_{u,\theta}^0| > \delta\sqrt{T}] < \varepsilon\theta$$

for any $\delta > 0$. Here

$$x_{u,\theta}^0 = x_{u,\theta} - \theta a_x(Q(u))$$

x and y take the values e, r, and s; r_x, r_{xy}, and ε have the same sense as in theorem 1.

(IV) *The function $A = a_e - a_s - a_r$ is continuously differentiable at Q and*

$$B = b_{ee} + b_{rr} + b_{ss} - 2b_{er} - 2b_{es} - 2b_{rs}$$

is continuous at Q, $B(Q) > 0$.

Then, if $z(0) = (q_T(0) - QT)/T$, $t \in [0, v]$, has limit distribution p_0, then the process

$$z(t) = \frac{q_T(t) - QT}{T}, \quad t \in [0, v]$$

C-converges to a diffusion process $w(t)$ with drift coefficient $A'(Q)x$, diffusion coefficient $B(Q)$ ($B = b_{ee} + b_{rr} + b_{ss} - 2b_{er} - 2b_{es} - 2b_{rs}$), and initial distribution p_0.

The proof of this theorem is virtually a repeat of the proof of theorem 1.* Minor changes occur only in the calculation of the moments of $q_{u,\theta}$:

$$\mathbf{E}_{\mathfrak{M}(u)} q_{u,\theta} = \theta A(Q(u)) + \frac{\theta}{\sqrt{T}}(r_e - r_s - r_r)$$

$$= \frac{\theta\zeta(u)}{\sqrt{T}} A'(Q) + \theta(Q(u) - Q)(A'(\tilde{Q}) - A'(Q))$$

$$+ \frac{\theta}{\sqrt{T}}(r_e - r_s - r_r). \tag{3.27}$$

Since A' is continuous, then $\tilde{Q} \in [Q(u), Q]$ and, consequently, in the domain $|Q(u) - Q| \leqslant N\sqrt{T}$, the last two terms in (3.27) will be uniformly small relative to $\theta/\sqrt{T}$.

Similarly we calculate $\mathbf{E}_{\mathfrak{M}(u)} q^2_{u,\theta}$ (cf. (3.24)) and estimate $\mathbf{E}_{\mathfrak{M}(u)}(q^2_{u,\theta};\ |q_{u,\theta}| > \delta\sqrt{T})$. Thus (PI)–(PIII) of theorem 2 of section 1.5 will be satisfied. Since, in our case, the coefficients of drift and diffusion are, respectively, $\alpha(x) = xA'(Q)$ and $\beta(x) = B(Q)$, then (PIV) will also be satisfied (at least by the sufficient conditions stated in section 1.5). In addition (PIV) is easily derived from the form of the transition function of the limit process $w(t)$ given below.

Thus, the conditions of theorem 2 of section 1.5 are satisfied and theorem 4 is proved. □

The limit process $w(t)$ may be written in the following explicit form (see (1.26))

$$w(t) = w(0)\,\mathrm{e}^{at} + \sigma\,\mathrm{e}^{at}\int_0^t \mathrm{e}^{-av}\,\mathrm{d}w_0(v)$$
$$w(t) = w(0)\,\mathrm{e}^{at} + \sigma\,\mathrm{e}^{at} w_0\left(\frac{1 - \mathrm{e}^{-2at}}{2a}\right) \tag{3.28}$$

where $w_0(t)$ is a standard Wiener process, $a = -A'(Q)$, $\sigma^2 = B(Q)$.

This is easy to verify directly by calculating the infinitesimal operator of the processes (3.28). The transition function of this process has the form (see (1.26))

$$\mathbf{P}(z, t, \mathrm{d}y) = \mathbf{P}(w^{(z)}(t) \in \mathrm{d}y) = \frac{1}{\sqrt{2\pi\sigma_t}} \exp\left\{-\frac{1}{2\sigma_t^2}(y - z\,\mathrm{e}^{at})^2\right\}\mathrm{d}y$$
$$\sigma_t^2 = \frac{\sigma^2}{2a}(\mathrm{e}^{2at} - 1) \tag{3.29}$$

* However, theorem 4 is not a special case of theorem 1, since in theorem 4, for example, the smoothness of $a_x(Q)$, which is necessary for condition (I) of theorem 1, is not required.

where, as before, $w^{(z)}$ denotes the process w with initial value z. The right-hand side of (3.29) is a fundamental solution of the heat equation with coefficients ax and σ^2.

As t $\to \infty$ and for $a < 0$ the process (3.28) approaches the stationary process

$$w_c(t) = \sigma \int_{-\infty}^{t} e^{a(t-v)}\, dw_0(v) \quad \left(w_c(t) = \sigma\, e^{at} w_0\left(-\frac{e^{-2at}}{2a}\right)\right)$$

whose one-dimensional distributions are, obviously, normal with parameters $(0, -\sigma/2a)$.

The explicit form, (3.28), of the limit process $w(t)$ in theorem 4 allows us to answer many questions concerning the behaviour of the trajectory of $q(t)$. For example, in reliability studies there often arises the question of the distribution of $\sup_{t \leqslant T} q(t)$. By theorem 1

$$\mathbf{P}\left(\sup_{t \leqslant vT} q(t) \leqslant QT + c\sqrt{T}\right) \xrightarrow[T \to \infty]{} \mathbf{P}\left(\sup_{t \leqslant v} w(t) < c\right).$$

We may also give the asymptotic (large v) form of the upper and lower functions for the process $z(t) = (q(tT) - QT)/\sqrt{T}$, $z(0) = \text{const}$. In fact, by theorem 4, for smooth $g_1(u)$ and $g_2(u)$, such that $z(0) \in (g_1(0), g_2(0))$, we have

$$\begin{aligned}
\mathbf{P}(g_2(u) < z(u) < g_1(u), 0 \leqslant u \leqslant v) &\xrightarrow[T \to \infty]{} G(g_1, g_2, v) \\
&\equiv \mathbf{P}(g_2(u) < w(u) < g_1(u), 0 \leqslant u \leqslant v) \\
&= \mathbf{P}\left((g_2(u)\, e^{-au} - z(0)) < \sigma w_0\left(\frac{1 - e^{-2au}}{2a}\right)\right. \\
&\quad \left. < (g_1(u)\, e^{-au} - z(0)), 0 \leqslant u \leqslant v\right).
\end{aligned}$$

Putting $u = -(1/2a) \log(1 - 2at)$ we can write this in the form

$$G(g_1, g_2, v) = \mathbf{P}\left(f_2(t) < w_0(t) < f_1(t), 0 \leqslant t \leqslant \frac{1 - e^{-2av}}{2a}\right)$$

where

$$f_j(t) = \frac{1}{\sigma}\left[g_j\left(-\frac{1}{2a}\log(1 - 2at)\right)\sqrt{1 - 2at} - z(0)\right], \quad j = 1, 2.$$

Hence, clearly, for $a < 0$ it is possible to speak of the upper and lower functions for $w(u)$ as $u \to \infty$. Using the law of the iterated logarithm for w_0 it is not difficult to guess the nature of these functions. Equating

$$f_j(t) = (-1)^j \sqrt{2t \log\log t}$$

we find that as such functions we may use

$$g_j(t) = (-1)^j \sigma \sqrt{\frac{\log t}{-a}}.$$

Remark 1. Condition (I) of theorem 4 can be replaced by the following condition: in $\Omega_{u,\theta} \cap \Gamma_u^N$

$$\mathbf{E}_{\mathfrak{M}(u)} x_{u,\theta} = \theta a_x(Q(u)) + a_x^*(Q(u), T) + \frac{\theta}{\sqrt{T}} r_x(\omega)$$

where $a_e^* - a_s^* - a_r^* = A^*(Q, T)$, for sufficiently large T, satisfies $|A^*(Q, T)| < \theta\varepsilon/\sqrt{T}$ in Γ_u^N.

In exactly the same way as theorem 4, and theorems 2 and 3, we obtain the following two results on convergence to a diffusion with reflection in the case of rough dependence of the control on $q(u)$. Here we take

$$Q = 0, \quad \Gamma_u^N = \left\{|Q(u)| < \frac{N}{\sqrt{T}}\right\} = \{|\zeta(u)| < N\}.$$

In addition, it is convenient to assume that the functions a_x and b_{xy}, in conditions (I)–(III) of theorem 4, also depend on T (the same modification could be introduced in theorems 1–4, but there is not so necessary).

THEOREM 5. *Let conditions* (I)–(III) *of theorem* 4 *hold, but only in the domain*

$$\{\zeta(u) \geqslant \varepsilon^*\}$$

for any $\varepsilon^* > 0$.

In addition, let condition (V) *of theorem* 1, *and condition* (VI) *of theorem* 2, *hold and also*

(IV′) *The coefficients* a_x, b_{xy} *are such that*

$$\lim_{T\to\infty} A(0)\sqrt{T} = A_0, \quad \lim_{T\to\infty} A'(0) = A'_0, \quad \lim_{T\to\infty} B(0) = B_0 > 0$$

and $A'(x)$ *and* $B(x)$ *are equicontinuous from the right at* $x = 0$.

Then, if the distribution of $\zeta(0) = q_T(0)/\sqrt{T}$ *converges to* p_0, *the process* $z(t) = q(tT)/\sqrt{T}$ *C-converges to a diffusion process* $w_R(t)$ *with drift coefficient* $A_0 + xA'_0$, *diffusion coefficient* B_0, *reflection from* $z = 0$, *and with initial distribution** p_0.

In section 1.3, with $A'_0 \neq 0$, for the unbounded diffusion $w^{(z)}(t)$ with these

* In subsequent theorems, for brevity, we will not mention the initial distribution but will take it as understood.

coefficients, we obtained the representation

$$w^{(z)}(t) = -\frac{A_0}{a} + \sigma\, e^{at} w_0^{(z_1)}\left(\frac{1 - e^{-2at}}{2a}\right)$$

where

$$z_1 = \frac{z}{\sigma} + \frac{A_0}{a\sigma}, \quad a = A_0', \quad \sigma^2 = B_0$$

$w_0^{(z)}$ is a standard Wiener process with initial value z. This allows us to write the required process $w_R^{(z)}(t)$ ($w_R(t)$ with the initial condition $z \geqslant 0$) in the form

$$w_R^{(z)}(t) = -\frac{A_0}{a} + \sigma\, e^{at} w_{0R}^{(z_1)}\left(\frac{1 - e^{-2at}}{2a}\right) \tag{3.30}$$

where $w_{0R}^{(z_1)}$ is a standard Wiener process with initial value $z_1 = z/\sigma + A_0/a\sigma$ and with upwards reflection from the boundary

$$g(u) = \frac{A_0}{a\sigma}\sqrt{1 - 2au}. \tag{3.31}$$

For $a = A_0' = 0$, the process $w_R(t)$ in theorem 5 will be a Wiener process with coefficients (A_0, σ^2) and reflection from the zero boundary, so that

$$w_{0R}(t) = \frac{1}{\sigma}\, w_R(t) \tag{3.32}$$

will have coefficients $(A_0/\sigma, 1)$ and also reflection from the zero boundary.

From these expressions it is clear that when either $a = A_0' = 0$ or $A_0 = 0$ then as $w_{0R}(u)$ we obtain a Wiener process with reflection from the zero boundary (in case (3.32) this process has drift).

The transition function $\mathbf{P}(w_{0R}(t) \in E \mid w_{0R}(0) = x) = \mathbf{P}(w_{0R}^{(x)}(t) \in E)$ of such a process $w_{0R}(u)$ can be found, for example, by a passage to the limit in the problem of a random walk with jumps ± 1 and reflecting barrier at zero (see Feller, 1950, p. 388, the probability of jumps being suitably chosen). We then obtain

$$\begin{aligned}\mathbf{P}(x, t, dy) &= \mathbf{P}(w_{0R}^{(x)}(t) \in dy) \\ &= \left\{ 2h^- e^{2h^- y} + \frac{2}{\pi} \exp\left[-\frac{h^2 t}{2} + h(y - x)\right] \right. \\ &\quad \times \int_0^\infty e^{-(t^2 z^2/2)}(z \cos zx + h \sin zx) \\ &\quad \left. \times (z \cos zy + h \sin zy) \frac{dz}{h^2 + z^2} \right\} dy \end{aligned} \tag{3.33}$$

where $h^- = \min(0, h)$, h is the drift coefficient of $w_{0R}(u)$ ($h = 0$ in case (3.30), $h = A_0/\sigma$ in case (3.32)).

From these equations it is easy to obtain the form of the stationary distribution when $h < 0$ (that is, when $a = A'_0 = 0$, $A_0 < 0$). Letting t tend to infinity we find

$$\begin{aligned} \mathbf{P}(w_{0R}^{(z)}(\infty) \in \mathrm{d}y) &= -2h\,\mathrm{e}^{2hy}\,\mathrm{d}y \\ \mathbf{P}(w_R(\infty) \in \mathrm{d}y) &= -2\frac{A_0}{\sigma^2}\,\mathrm{e}^{(2A_0 y/\sigma^2)}\,\mathrm{d}y. \end{aligned} \tag{3.34}$$

If the drift coefficient $h = A_0 = 0$, $a \neq 0$, then we may give expressions for the transition function of $w_R(t)$ which are simpler than those based on (3.33). It is sufficient, for example, to recall that in this case (see section 1.3) the distribution of $w_R^{(z)}(t)$ coincides with the distribution of $|w^{(z)}(t)|$, where $w^{(z)}$ is an unbounded diffusion process with coefficients (ax, σ^2). So, by (3.28),

$$w_R^{(z)}(t) = \left| z\,\mathrm{e}^{at} + \sigma\,\mathrm{e}^{at} w_0\left(\frac{1-\mathrm{e}^{-2at}}{2a}\right)\right| = \left|\sigma\,\mathrm{e}^{at} w_0^{(z/\sigma)}\left(\frac{1-\mathrm{e}^{-2at}}{2a}\right)\right|.$$

Hence, and from (3.29), for the transition function of $w_R(t)$ we obtain

$$\mathbf{P}(w_R^{(z)}(t)\,\mathrm{d}y)$$

$$= \frac{\mathrm{d}y}{\sqrt{2\pi}\sigma_t}\left[\exp\left\{-\frac{1}{2\sigma_t^2}(y - z\,\mathrm{e}^{at})^2\right\} + \exp\left\{-\frac{1}{2\sigma_t^2}(y + z\,\mathrm{e}^{at})^2\right\}\right]$$

where

$$\sigma_t^2 = \frac{\sigma^2}{2a}(\mathrm{e}^{2at} - 1).$$

For $a < 0$ there is a *stationary* process $w_R(t)$ with coefficients (ax, σ^2), representable in the form

$$\mathrm{e}^{at}\left|w_0\left(-\frac{\mathrm{e}^{-2at}}{2a}\right)\right|.$$

We now give a result on convergence to a diffusion with two reflecting boundaries, 0 and R, in the case of rough dependence of the control on $q(u)$.

THEOREM 6. *Let* $0 \leqslant q(t) \leqslant R\sqrt{T}$ *and let conditions* (I)–(III) *of theorem* 4 *be satisfied, but only in the domain*

$$\{R - \varepsilon^* \geqslant \zeta(u) \geqslant \varepsilon^*\}$$

for any $\varepsilon^* > 0$.

In addition let condition (IV) *of theorem* 5, *and condition* (V) *of theorem* 1, *hold and also*

(VI) $q(t)$ *and* $R\sqrt{T} - q(t)$ *satisfy condition* (VI) *of theorem* 2.

Then $z(t) = q(tT)/\sqrt{T}$ *C-converges to a diffusion process* $w_{RR}(t)$ *with reflection from the levels* 0 *and* R *and with drift and diffusion coefficients equal to* $A_0 + A'_0 x$ *and* B_0 *respectively.*

As under the conditions of theorem 5, the process $w_{RR}^{(z)}$ with initial value z, $0 \leqslant z \leqslant R$, for $a = A_0' = 0$, admits a representation (3.30) where $w_{0R}^{(z_1)}(u)$ is replaced by $w_{0RR}^{(z_1)}(u)$, which is a standard Wiener process with initial value $z_1 = z/\sigma + A_0/a\sigma$ and reflections from the boundaries

$$g_1(u) = \frac{A_0}{a\sigma}\sqrt{1-2au}, \quad g_2(u) = \frac{1}{\sigma}\left(\frac{A_0}{a} + R\right)\sqrt{1-2au}.$$

For $a = 0$, as for w_R, we can find an explicit form for the transition function of w_{RR} using a passage to the limit in the problem of random walk with jumps ± 1 and two reflecting boundaries (see Feller, 1950, p. 388, in the formula below $\alpha = A_0/\sigma^2$)

$$\mathbf{P}(w_{RR}^{(x)}(t) \in \mathrm{d}y) = \left[\frac{2\alpha\, \mathrm{e}^{2\alpha y}}{\mathrm{e}^{2\alpha R} - 1} + \frac{2}{R}\exp\left\{-\frac{\alpha^2 t^2 \sigma^2}{2} + \alpha(y - x)\right\}\right.$$

$$\times \sum_{r=1}^{\infty} \mathrm{e}^{-(t\sigma^2/2)(\pi r/R)^2} \frac{1}{\alpha^2 + (\pi r/R)^2}$$

$$\left.\times \left(\frac{\pi r}{R}\cos\frac{\pi r x}{R} + \alpha \sin\frac{\pi r x}{R}\right)\left(\frac{\pi r}{R}\cos\frac{\pi r y}{R} + \alpha\sin\frac{\pi r y}{R}\right)\right]\mathrm{d}y.$$

Hence it is clear, in particular, that as $t \to \infty$ the distribution of $w_{RR}(t)$ converges to a stationary distribution, which coincides with the truncated exponential distribution

$$\lim_{t\to\infty} \mathbf{P}(w_{RR}^{(x)}(t)\,\mathrm{d}y) = \frac{2\alpha\,\mathrm{e}^{2\alpha y}}{\mathrm{e}^{2\alpha R} - 1}\,\mathrm{d}y, \quad y \in [0, R]$$

with exponent $2\alpha = 2A_0/\sigma^2$ (cf. (3.34)).

We note that the results of theorems 1–6 have a 'collective' nature. This means in calculating the limit distributions in these theorems that we can use the '*invariance principle*' in the following sense. Suppose that alongside the system under investigation the conditions of some theorem are also satisfied by another, very simple, system (say with Poisson input and output streams), which admits explicit formulae for the distribution of $q(t)$. Then these explicit formulae can be used to obtain the limit distribution of the original system.

Before passing on to the applications of theorems 1–6 in more concrete problems, we give also an auxiliary result which essentially repeats lemma 2 of section 1.4, and which will help us to verify conditions (I) and (II) of the above theorems in concrete examples.

The essence of this result is: on the basis of the inequality

$$\mathbf{E}_{\mathfrak{M}(u)}[x_{u,t}^2; |x_{u,t}| > \delta\sqrt{T}] < \varepsilon\theta \tag{3.35}$$

satisfied on $\Omega_{u,\theta} \cap \Gamma_u^N$ for $\theta \leqslant t \leqslant 2\theta$ and following, for $x = q, e^0, s^0, r^0$, from (IV) of theorems 1–6, we establish the necessary estimate for the probability

$$\mathbf{P}_{\mathfrak{M}(u)}\left(\sup_{s\leqslant\theta}|x_{u,s}|>\delta\sqrt{T}\right).$$

Here $\Gamma_u^N=\{|x(u)|<N\sqrt{T}\}$.

We put, for brevity,

$$\Omega(u)=\Omega_{u,\theta}\cap h(u,\delta)$$

where

$$h(u,\delta)=\bigcap_{v\leqslant u}\{|x(v+0)-x(v-0)|\leqslant\delta\sqrt{T}\}$$

and, without loss of generality, we may take the sets $\Omega_{u,\theta}$ to be nested: $\Omega_{u,\theta}=\bigcap_{v\leqslant u}\Omega_{v,\theta}$.

For convenience, instead of $\Omega_{u,\theta}$, from the beginning we may take $\Omega(u)$, since $\mathbf{P}(\Omega(T))\to 1$.

LEMMA 2. *If* (3.35) *is satisfied, then on* $\Omega(u)\cap\Gamma_u^{N-1}$ *for any* $\varepsilon>0,\frac{1}{2}>\delta>0$ *and sufficiently large* T

$$\mathbf{P}_{\mathfrak{M}(u)}\left(\sup_{s\leqslant\theta}|x_{u,s}|>\delta\sqrt{T};\Omega(u+\theta)\right)\leqslant\frac{\varepsilon\theta}{T}\tag{3.36}$$

$$\mathbf{E}_{\mathfrak{M}(u)}\left(x_{u,\theta}^2;\sup_{s\leqslant\theta}|x_{u,s}|>\delta\sqrt{T};\Omega(u+\theta)\right)\leqslant\varepsilon\theta.\tag{3.37}$$

The proof of this lemma is based on the following inequality

$$\mathbf{P}_{\mathfrak{M}(u)}\left(\sup_{s\leqslant\theta}|x_{u,s}|\geqslant 2\delta\sqrt{\mathrm{T}};\Omega(u+\theta)\right)\leqslant 2\mathbf{P}_{\mathfrak{M}(u)}(|x_{u,2\theta}|\geqslant\delta\sqrt{T};\Omega(u))\tag{3.38}$$

which essentially follows from lemma 2 of section 1.4. In fact, it follows from (3.35) that on Γ_u^N

$$I(\Omega(u))\mathbf{P}_{\mathfrak{M}(u)}(|x_{u,t}|>\delta\sqrt{T})\leqslant\frac{\varepsilon\theta}{\delta^2T}<\frac{1}{2}\tag{3.39}$$

for $\theta\leqslant t\leqslant 2\theta$ and sufficiently large T. This allows us to repeat the calculations of the lemma of chapter 1, replacing there the absolute expectation by the conditional expectation.

For this we denote by τ the first passage time of $\{x_{u,t},t\geqslant 0\}$ through the level $2\delta\sqrt{T}$. Then, by lemma 1 of section 1.4,

$$\begin{aligned}
&I(\Gamma_u^{N-1})\mathbf{P}_{\mathfrak{M}(u)}(x_{u,2\theta}>\delta\sqrt{\mathrm{T}};\Omega(u))\\
&\qquad\geqslant I(\Gamma_u^{N-1})\mathbf{E}_{\mathfrak{M}(u)}I(\tau\leqslant\theta)I(\Omega(u+\tau))\mathbf{E}_{\mathfrak{M}(u+\tau)}I(x_{u+\tau,2\theta-\tau}\geqslant-\delta\sqrt{T})\\
&\qquad\geqslant I(\Gamma_u^{N-1})\mathbf{E}_{\mathfrak{M}(u)}I(\tau\leqslant\theta)I(\Omega(u+\theta))\tfrac{1}{2}\\
&\qquad\geqslant\tfrac{1}{2}I(\Gamma_u^{N-1})\mathbf{P}_{\mathfrak{M}(u)}\left(\sup_{0\leqslant t\leqslant\theta}x_{u,t}>2\delta\sqrt{T};\Omega(u+\theta)\right).
\end{aligned}$$

We have been able to use lemma 1 of section 1.4 here, since $2\delta < 1$, $\Gamma_u^{N-1}\Omega(u+\tau) \subset \Gamma_{u+\tau}^N\Omega(u+\tau)$, and, consequently, it follows from (3.39) that on $\Gamma_u^{N-1}\Omega(u+\tau)$

$$\mathbf{P}_{\mathfrak{M}(u+1)}(x_{u+\tau,2\theta-\tau} \geqslant -\delta\sqrt{T}) \geqslant \tfrac{1}{2}.$$

In a similar way, estimating

$$\mathbf{P}_{\mathfrak{M}(u)}(x_{u,2\theta} < -\delta\sqrt{T}; \Omega(u))$$

we obtain (3.38).

To obtain lemma 2 from (3.38) it is necessary to note that (3.37) follows from (3.36) and lemma 1 of this section. The inequality (3.36) follows from (3.38) and (3.39). The lemma is proved. □

We make one more remark. In the conditions of theorems 1–6 we have required that for sufficiently large T

$$\mathbf{P}\left(\bigcup_{u \leqslant T} \overline{\Omega(u)}\right) = \mathbf{P}(\overline{\Omega(T)}) < \varepsilon. \tag{3.40}$$

Since in all subsequent problems the service processes have the same *homogeneity* in time, then, without loss of generality, and also for the sake of simplicity, we can assume that (3.40) is satisfied homogeneously in the following sense: the probability that condition $\Omega(u)$ will be first broken in the time interval $[t, t+\theta]$ of length θ, for all t, does not exceed $\varepsilon\theta/T$. More precisely, for all u

$$I(\Omega(u))\mathbf{P}_{\mathfrak{M}(u)}(\overline{\Omega(u+\theta)}) < \frac{\varepsilon\theta}{T}. \tag{3.41}$$

We note that instead of (3.41) we could have used the more general condition

$$P\left(\bigcup_{u \leqslant T} \{\Omega(u)\overline{\Omega}(u+\theta)\}\right) < \varepsilon$$

equivalent to (3.40). However, it would complicate the calculation here and attaining this generality would have questionable value (both from the point of view of real problems, and from the point of view of theorems 1–6 where conditions (I)–(III) have a homogeneous character).

If (3.41) holds, then lemma 2 implies

COROLLARY 1. *On* $\gamma(u) = \Omega(u) \cap \Gamma_u^{N-1}$

$$\begin{aligned} &\mathbf{P}_{\mathfrak{M}(u)}\left(\sup_{s \leqslant \theta} |x_{u,s}| > \delta\sqrt{T}\right) < \frac{\varepsilon\theta}{T} \\ &\mathbf{E}_{\mathfrak{M}(u)}\left(x_{u,\theta}^2; \sup_{s \leqslant \theta} |x_{u,s}| > \delta\sqrt{T}\right) < \varepsilon\theta. \end{aligned} \tag{3.42}$$

In accordance with the conditions of theorems 1–6 this corollary means that (3.42) will be satisfied for $x = e^0, r^0, s^0, q$ if conditions (III), (V) of those theorems are satisfied.

From conditions (I), (II) of theorems 1–6 and the homogeneity condition (3.41) also follows

COROLLARY 2. *If*

$$|I(\gamma(u))\mathbf{E}_{\mathfrak{M}(u)}x_{u,\theta}| \leqslant \varepsilon\sqrt{\theta} \quad \left(\frac{\theta}{T} = o(\sqrt{\theta})\right)$$

$$I(\gamma(u))|\mathbf{E}_{\mathfrak{M}(u)}x_{u,\theta}^2 - b\theta| \leqslant \varepsilon\theta$$

and (3.41) *holds, then for any* v, $\theta \geqslant v \geqslant 0$ *and* $\omega \in \gamma(u)$

$$|\mathbf{E}_{\mathfrak{M}(u)}x_{u+v,\theta}^2 - b\theta| \leqslant \varepsilon\theta$$

$$|\mathbf{E}_{\mathfrak{M}(u)}x_{u,v}^2 - bv| \leqslant \varepsilon\theta.$$

In fact, we estimate first of all $\mathbf{E}_{\mathfrak{M}(u)}x_{u+v,\theta}^2$. For brevity we denote $\gamma(u) = I(\gamma(u))$. Then $I(\bar{\gamma}(u)) = 1 - \gamma(u)$,

$$\gamma(u)\mathbf{E}_{\mathfrak{M}(u)}x_{u+v,\theta}^2$$

$$= \mathbf{E}_{\mathfrak{M}(u)}\gamma(u)\gamma(u+v)\mathbf{E}_{\mathfrak{M}(u+v)}x_{u+v,\theta}^2 + \mathbf{E}_{\mathfrak{M}(u)}\gamma(u)(1-\gamma(u+v))x_{u+v,\theta}.$$

The first term here is of the form $b\theta + r\theta$, $|r| < \varepsilon$; the second, by lemma 1 and (3.41), may be estimated by $\varepsilon\theta$.

To prove the second inequality of the corollary we note that

$$x_{u,\theta+v}^2 = x_{u,v}^2 + x_{u+v,\theta}^2 + 2x_{u,v}x_{u+v,\theta}. \tag{3.43}$$

Since

$$|\gamma(u)\gamma(u+v)\mathbf{E}_{\mathfrak{M}(u+v)}x_{u+v,\theta}| < \varepsilon\sqrt{\theta}$$

and, by the latter,

$$|\gamma(u)(1-\gamma(u+v))\mathbf{E}_{\mathfrak{M}(u+v)}x_{u+v,\theta}| \leqslant \sqrt{\gamma(u)(1-\gamma(u+v)\mathbf{E}_{\mathfrak{M}(u+v)}x_{u+v,\theta}^2} \leqslant \varepsilon\sqrt{\theta}$$

then, denoting

$$y^2 = \gamma(u)\mathbf{E}_{\mathfrak{M}(u)}x_{u,v}^2$$

we obtain

$$|\gamma(u)\mathbf{E}_{\mathfrak{M}(u)}x_{u,v}x_{u+v,\theta}| \leqslant \varepsilon\sqrt{\theta}y$$

and, further, from (3.43) (after multiplication by $\gamma(u)$)

$$b(\theta + v) = y^2 + b\theta + r_1\theta + r_2\sqrt{\theta}y, \quad \text{where } |r_i| < \varepsilon.$$

It is obvious that the solution of this equation has the form

$$y^2 = bv + r_3\theta, \quad |r_3| < \varepsilon. \quad \square$$

If, for the proof of lemma 2, we lean not on (3.35) but on an inequality of the form

$$I(\Omega(u))\mathbf{E}_{\mathfrak{M}(u)}x_{u,t}^2 < c\theta \tag{3.44}$$

satisfied for $0 \leqslant t \leqslant 2\theta$ and $\omega \in \Gamma_u^N$, then in a completely analogous way it is easy to establish

LEMMA 3. *If* (3.44) *holds, then on* $\Omega(u) \cap \Gamma_u^{N-1}$ *for sufficiently large* N_1

$$\mathbf{E}_{\mathfrak{M}(u)}\left(\sup_{s\leqslant\theta} x_{u,s}^2, \Omega(u+\theta)\right) \leqslant N_1\theta.$$

For the proof of this lemma we need to note that (3.44) implies, for $\omega \in \Gamma_u^N$, that

$$I(\Omega(u))\mathbf{P}_{\mathfrak{M}(u)}(|x_{u,t}| > z\sqrt{\theta}) \leqslant \frac{c\theta}{z\theta} < \frac{1}{2}$$

for sufficiently large z ($z \geqslant N_2$). This replaces (3.39). After this, as before, we show that on $\Omega(u) \cap \Gamma_u^{N-1}$ for $z \geqslant N_2$

$$\mathbf{P}_{\mathfrak{M}(u)}(\sup |x_{u,t}| > 2z\sqrt{\theta}; \Omega(u+\theta)) \leqslant 2\mathbf{P}_{\mathfrak{M}(u)}(|x_{u,2\theta}| > z\sqrt{\theta}).$$

This proves the lemma. □

3.4. Multichannel systems with intensive input streams

As a first illustration of theorems 1–6 of the preceding section we consider the system described in section 2.2, under the assumption that the service time in each channel is exponential.

Related to such systems are, for example, the systems of example 2 (systems $\langle G_I, G_I, E/n, 1\rangle$ in the notation of Borovkov, 1976a), when

$$\mathbf{E}\tau_j^e = 1/T, \quad \mathbf{P}(\tau_j^s > t) = \mathrm{e}^{-\lambda t}, \quad n = \infty.$$

Concerning more general systems we assume that the input stream, after a change of time scale (expansion) by T times will satisfy the conditions of type (I)–(III) and (V) of theorem 1 of section 3.3.

More precisely, the following holds.

THEOREM 1. *Let the input process* $e(t)$ *satisfy: there exist* θ *and a system of sets* $\Omega_{u,\theta}^e \in \mathfrak{E}(u)$ *such that*

$$\mathbf{P}\left(\bigcap_{u\leqslant T} \Omega_{u,\theta}^e\right) \to 1 \tag{3.45}$$

and for $\omega \in \Omega_{u,\theta}^e$

(I) $\mathbf{E}e_{u,\theta} = \theta a_e + \dfrac{\theta}{\sqrt{T}} r_e(\omega), \quad |r_e| < \varepsilon$

(II) $\mathbf{E}(e_{u,\theta} - \theta a_e)^2 = \theta b_{ee} + \theta r_{ee}(\omega), \quad |r_{ee}| < \varepsilon$

(III) $\mathbf{E}((e_{u,\theta} - \theta a_e)^2; |e_{u,\theta} - \theta a_e| > \delta\sqrt{T}) < \varepsilon\theta$

for any $\delta > 0$, $\varepsilon > 0$.

(IV) *The service system has a total of n channels in each of which the service times are independent and exponentially distributed*

$$\mathbf{P}(\tau_j^s > t) = \mathrm{e}^{-(\lambda/T)t}.$$

(V) *Side by side with* (3.45) *there is a condition of homogeneity in the decrease of the probability that at least one of* $\bar{\Omega}^e_{u,\theta}$ *or* $\bar{h}^e(u,\delta) = \bigcup_{v \leqslant u}\{|e(v+0) - e(v-0)| > \delta\sqrt{T}\}$ *occurs* (cf. (3.41)). *Namely, if we denote* $\Omega^e(u) = \Omega^e_{u,\theta} h^e(u,\delta)$ *(we will also assume the* $\Omega^e_{u,\theta}$ *to be nested:* $\Omega^e_{u,\theta} = \bigcap_{v \leqslant u} \Omega^e_{v,\theta}$*), then*

$$\mathbf{P}_{\mathfrak{M}(u)}(\Omega^e(u)\bar{\Omega}^e(u+\theta)) < \frac{\varepsilon\theta}{T}.$$

Then, if

$$\frac{n - Ta_e/\lambda}{\sqrt{T}} \to \infty \tag{3.46}$$

the normalized number of busy lines

$$z(t) = \frac{q(tT) - a_e T/\lambda}{\sqrt{T}}$$

C-converges to an unbounded diffusion with coefficients $(-\lambda x, b_{ee} + a_e)$.

This process and its explicit representation have been discussed in detail in the remarks after theorem 4 of section 3.3.

As shown in section 1.11, example 2, the systems considered at the beginning of this chapter (i.e. the systems $\langle G_I, G_I, E/n, 1\rangle$) will satisfy the conditions of this theorem if

$$\begin{gathered} \mathbf{E}\tau^e \geqslant c > 0, \quad \mathbf{D}\tau^e \geqslant c > 0 \\ \mathbf{E}[(\tau^e)^2, \tau^e \geqslant N] \to 0, \quad \mathbf{E}((v^e)^2;\ |v^e| \geqslant N) \to 0 \end{gathered} \tag{3.47}$$

uniformly with respect to T as $N \to \infty$ (the distribution of τ^e, v^e, τ^s may depend on T). In addition, we must have

$$\frac{\mathbf{E}v^e}{\mathbf{E}\tau^e} = a_e + o\left(\frac{1}{\sqrt{T}}\right), \quad \frac{\mathbf{D}v^e}{\mathbf{D}\tau^e} + \frac{\mathbf{E}^2 v^2 \mathbf{D}\tau^e}{(\mathbf{E}\tau^e)^3} \to \sigma_e^2 > 0. \tag{3.48}$$

As $\Omega^e_{u,\theta}$ (see the statement of theorem 2 of section 1.11 and theorem 1 of section 3.3) we may take $\Omega^e_{u,\theta} = \{\chi^e(u) < s\}$ for a suitable choice of s, and as $\mathfrak{M}(u)$ take the σ-algebra generated by $Y(t)$ for $t \leqslant u$ (see example 2 at the beginning of this chapter).

Conditions (3.47) and (3.48) are obviously always satisfied if τ^e and v^e do not depend on T.

In section 1.11 there are other examples of input streams which satisfy condition 1 of this section.

Since the normalized input process $(e(tT) - a_e tT)/\sqrt{b_{ee}T}$ C-converges to a standard Wiener process, then this theorem could also have been obtained from theorem 1 of section 2.3. However, the limit process for $z(t)$ was there described via an integral representation which is equivalent, as shown in section 3.3, to the description of this process as a diffusion.

Comparison of theorem 1 with the result of section 2.3 leads to an important remark. *If the distribution of τ_j^s (or τ_j^s/T) is not exponential, then the limit process for $z(t)$ will not be a Markov (and, consequently not a Markov diffusion) process.*

Proof of theorem 1. Our aim is to show that the system described in theorem 1 leads to a system with rough dependence of control on the number of busy lines $q(t)$ and to verify for it the conditions of theorem 4 of the preceding section.

For simplicity we first let the number of channels $n = \infty$.

Consider the increment $s_{u,\theta}$ and represent it in the form

$$s_{u,\theta} = \sum_{i=1}^{q(u)} \delta_i + \sum_{i=1}^{e_{u,\theta}} \delta_i^* \tag{3.49}$$

where $\delta_i = 1$ or 0 depending on whether or not the service of the ith request of the queue $q(u)$ is completed at $u + \theta$. The random variables δ_i^* are defined in a similar way. If $n + \theta_j$ is the arrival epoch of the jth request from the interval $(u, u + \theta)$, then obviously

$$\begin{aligned} \mathbf{P}_{\mathfrak{M}(u)\mathfrak{E}(u+\theta))}(\delta_j^* = 0) &= \mathrm{e}^{-\lambda(\theta-\theta_j)/T} \\ \mathbf{P}_{\mathfrak{M}(u)}(\delta_j = 0) &= \mathrm{e}^{-\lambda\theta/T}. \end{aligned} \tag{3.50}$$

We have $(Q(u) = q(u)/T)$

$$\begin{aligned} \mathbf{E}_{\mathfrak{M}(u)} s_{u,\theta} &= (1 - \mathrm{e}^{-\lambda\theta/T})q(u) + \mathbf{E}_{\mathfrak{M}(u)} \int_0^\theta (1 - \mathrm{e}^{-\lambda(\theta-v)/T}) \mathrm{d}_v e_{u,v} \\ &= \lambda\theta Q(u) - Q(u)T\left(\mathrm{e}^{-\lambda\theta/T} - 1 + \frac{\lambda\theta}{T}\right) + \int_0^\theta (1 - \mathrm{e}^{-\lambda(\theta-v)/T}) a_e \,\mathrm{d}v \\ &\quad + \frac{\lambda}{T}\int_0^\theta \mathrm{e}^{-\lambda(\theta-v)/T} \mathbf{E}_{\mathfrak{M}(u)}(e_{u,v} - a_e v)\,\mathrm{d}v. \end{aligned} \tag{3.51}$$

Here the second and third terms in the last sum will transform to

$$T\left(\frac{a_e}{\lambda} - Q(u)\right)\left(\mathrm{e}^{-\lambda\theta/T} - 1 + \frac{\lambda\theta}{T}\right).$$

Therefore, putting $Q = a_e/\lambda$, we obtain that in $|Q(u) - Q| < N/\sqrt{T}$ these terms, in absolute value, do not exceed

$$T\frac{N}{\sqrt{T}}\frac{\lambda^2\theta^2}{T}=\frac{\theta}{\sqrt{T}}\cdot\frac{\lambda^2 N\theta}{T}=o\left(\frac{\theta}{\sqrt{T}}\right).$$

In the last term of (3.51), by corollary 2 of the preceding section, we obtain under the integral sign for $\omega\in\Omega^e_{u,\theta}$

$$|\mathbf{E}_{\mathfrak{M}(u)}e^0_{u,v}|\leqslant\sqrt{\mathbf{E}_{\mathfrak{M}(u)}(e^0_{u,v})^2}\leqslant 2\sqrt{b_{ee}\theta}$$

and

$$\left|\frac{\lambda}{T}\int_0^\theta \mathrm{e}^{-\lambda(\theta-v)/T}\mathbf{E}_{\mathfrak{M}(u)}(e_{u,v}-a_e v)\,\mathrm{d}v\right|\leqslant\frac{2\lambda\theta^{3/2}\sqrt{b_{ee}}}{T}=o\left(\frac{\theta}{\sqrt{T}}\right).$$

Thus, on $\Omega^e_{u,\theta}\cap\Gamma^N_u$

$$\mathbf{E}_{\mathfrak{M}(u)}s_{u,\theta}=\lambda\theta Q(u)+\frac{\theta}{\sqrt{T}}r_s(\omega),\quad r_s(\omega)<\varepsilon.$$

We will see that relative to the behaviour of the first-order moments ($\mathbf{M}_{\mathfrak{M}(u)}e_{u,\theta}=a_e\theta+o(\theta/\sqrt{T})$) our system leads to a system with rough dependence of control of queue length. The 'point of equilibrium' Q here is the solution $Q=a_e/\lambda$ of $a_e-\lambda Q=0$.

Consider the second moments. For $e(u)$ they are well known. Further, by (3.50),

$$\begin{aligned}
&\mathbf{E}_{\mathfrak{M}(u)}(s_{u,\theta}-\mathbf{E}_{\mathfrak{M}(u)}s_{u,\theta})^2\\
&\quad=\mathbf{E}_{\mathfrak{M}(u)}\mathbf{E}_{(\mathfrak{M}(u)\mathfrak{E}(u+\theta))}(s_{u,\theta}-\mathbf{E}_{\mathfrak{M}(u)\mathfrak{E}(u+\theta))}s_{u,\theta})^2\\
&\qquad+\mathbf{E}_{\mathfrak{M}(u)}(\mathbf{E}_{(\mathfrak{M}(u)\mathfrak{E}(u+\theta))}s_{u,\theta}-\mathbf{E}_{\mathfrak{M}(u)}s_{u,\theta})^2\\
&\quad=\mathrm{e}^{-\lambda\theta/T}(1-\mathrm{e}^{-\lambda\theta/T})q(u)+\mathbf{E}_{\mathfrak{M}(u)}\int_0^\theta \mathrm{e}^{-\lambda(\theta-v)/T}(1-\mathrm{e}^{-\lambda(\theta-v)/T})\,\mathrm{d}_v e_{u,v}\\
&\qquad+\mathbf{E}_{\mathfrak{M}(u)}\left[\frac{\lambda}{T}\int_0^\theta \mathrm{e}^{-\lambda(\theta-v)/T}(e_{u,v}-\mathbf{E}_{\mathfrak{M}(u)}e_{u,v})\,\mathrm{d}v\right]^2.
\end{aligned}\tag{3.52}$$

Here in Γ^N_u the first term is equal to $\lambda\theta Q(u)+r_1\theta$, $|r_1|<\varepsilon$. In $\Omega^e_{u,\theta}$ the second term does not exceed

$$\frac{\lambda\theta}{T}\mathbf{E}_{\mathfrak{M}(u)}e_{u,\theta}=r_2\theta,\quad |r_2|<\varepsilon.$$

Finally for the last term in (3.52), for $\omega\in\Omega^e_{u,\theta}$, we have

$$\begin{aligned}
&\mathbf{E}_{\mathfrak{M}(u)}(\mathbf{E}_{(\mathfrak{M}(u)\mathfrak{E}(u+\theta))}s_{u,\theta}-\mathbf{E}_{\mathfrak{M}(u)}s_{u,\theta})^2\\
&\quad\leqslant\frac{\lambda^2\theta}{T^2}\int_0^\theta \mathbf{E}_{\mathfrak{M}(u)}(e_{u,v}-\mathbf{E}_{\mathfrak{M}(u)}e_{u,v})^2\,\mathrm{d}v\leqslant\frac{\lambda^2\theta}{T^2}b_{ee}\theta^2=o(\theta).
\end{aligned}\tag{3.53}$$

To estimate the integrand here we have used corollary 2 of the preceding section.

We obtain as a result that

$$\mathbf{E}_{\mathfrak{M}(u)}(s_{u,\theta}-\mathbf{E}_{\mathfrak{M}(u)}s_{u,\theta})^2$$
$$=\mathbf{E}_{\mathfrak{M}(u)}(s_{u,\theta}-\lambda\theta Q(u))+r_1(\omega)\theta+\lambda\theta Q(u)+r_2(\omega)\theta, \quad |r_i(\Omega)|<\varepsilon.$$

It remains for us to consider the mixed moment

$$\mathbf{E}_{\mathfrak{M}(u)}(s_{u,\theta}-\theta\lambda Q(u))(e_{u,\theta}-a_e\theta)$$
$$=\mathbf{E}_{\mathfrak{M}(u)}(e_{u,\theta}-a_e\theta)\mathbf{E}_{(\mathfrak{M}(u)\mathfrak{E}(u+\theta))}(s_{u,\theta}-\theta\lambda Q(u)). \quad (3.54)$$

As in (3.51) we obtain that the last multiplier here is

$$\frac{\lambda}{T}\int_0^\theta \mathrm{e}^{-\lambda(\theta-v)/T}(e_{u,v}-a_e v)\,\mathrm{d}v+\frac{\theta r}{\sqrt{T}}, \quad |r|<\varepsilon.$$

Therefore, in absolute value, (3.54) does not exceed

$$\frac{\theta\varepsilon}{\sqrt{T}}\sqrt{b_{ee}+\theta r_{ee}(\omega)}+\frac{\lambda}{T}\mathbf{E}_{\mathfrak{M}(u)}|e_{u,v}-a_e\theta|\int_0^\theta |e_{u,v}-a_e v|\,\mathrm{d}v$$
$$\leqslant o(\theta)+\frac{\lambda}{T}\left[\mathbf{E}_{\mathfrak{M}(u)}(e_{u,\theta}-a_e\theta)^2\mathbf{E}_{\mathfrak{M}(u)}\left(\int_0^\theta |e_{u,v}-a_e v|\,\mathrm{d}v\right)^2\right]^{\frac{1}{2}}$$
$$=o(\theta)+\frac{\lambda\sqrt{b_{ee}\theta}}{T}\left[\theta\int_0^\theta \mathbf{E}_{\mathfrak{M}(u)}(e_{u,v}-a_e v)^2\,\mathrm{d}v\right]^{\frac{1}{2}}.$$

Using corollary 2 of the previous section we obtain

$$o(\theta)+\frac{\lambda\sqrt{b_{ee}\theta}}{T}(b_{ee}\theta^3)^{\frac{1}{2}}=o(\theta).$$

Thus, conditions (I), (II) of theorem 4 of section 3.3 are satisfied. In this connection $a_e(z)=a_e=\text{const.}$, $a_s(z)=\lambda z$, $b_{ee}(z)=b_{ee}=\text{const.}$, $b_{ss}=\lambda z$, $b_{es}=0$. (The refusal process $r(u)$ is identically zero.) We verify condition (III) of theorem 4 of section 3.3. We have

$$s_{u,\theta}-\mathbf{E}_{\mathfrak{M}(u)}s_{u,\theta}=(s_{u,\theta}-\mathbf{E}_{(\mathfrak{M}(u)\mathfrak{E}(u+\theta))}s_{u,\theta})$$
$$+(\mathbf{E}_{(\mathfrak{M}(u)\mathfrak{E}(u+\theta))}s_{u,\theta}-\mathbf{E}_{\mathfrak{M}(u)}s_{u,\theta}).$$

By (3.53) it is sufficient to verify (III) for the first difference only

$$s_{u,\theta}-\mathbf{E}_{(\mathfrak{M}(u)\mathfrak{E}(u+\theta))}s_{u,\theta}=\sum_{j=1}^{q(u)}(\delta_j-p)+\sum_{j=1}^{e_{u,\theta}}(\delta_j^*-p_j^*) \quad (3.55)$$

(see (3.49), (3.50); the numbers p and p_j^*, equal to the probability of a single outcome in a Bernoulli scheme, are defined by (3.50)). Condition (III) for (3.55) will obviously be satisfied if we show that the fourth moment of (3.55) is bounded above by $c\theta^2$. Further, if the ξ_j are independent, $\mathbf{P}(\xi_j=1)=p_j=1-\mathbf{P}(\xi_j=0)=1-q_j$, then

$$\mathbf{E}\left[\sum_{j=1}^{n}(\xi_j-p_j)\right]^4=\sum_{j=1}^{n}p_jq_j(p_j^3+q_j^3)+6\sum_{i<j}p_iq_ip_jq_j$$
$$\leqslant\sum_{j=1}^{n}p_j+3\left(\sum_{j=1}^{n}p_j\right)^2\leqslant 4\left(\sum p_j\right)^2$$

if $\sum p_j\geqslant 1$. In accordance with (3.55)

$$\sum p_j=pq(u)+\sum_{j=1}^{e_{u,\theta}}p_j^*=\mathbf{E}_{(\mathfrak{M}(u)\mathfrak{E}(u+\theta))}s_{u,\theta}.$$

Therefore in $\Omega_{u,\theta}^e\cap\Gamma_u^N$

$$\mathbf{E}_{\mathfrak{M}(u)}(s_{u,\theta}-\mathbf{E}_{(\mathfrak{M}(u)\mathfrak{E}(u+\theta))}s_{u,\theta})^4\leqslant 4\mathbf{E}_{\mathfrak{M}(u)}[\mathbf{E}_{(\mathfrak{M}(u)\mathfrak{E}(u+\theta))}s_{u,\theta}]^2\leqslant 4\mathbf{E}_{\mathfrak{M}(u)}s_{u,\theta}^2$$
$$\leqslant 8(\lambda\theta Q(u))^2+8\mathbf{E}_{\mathfrak{M}(u)}(s_{u,\theta}-\lambda\theta Q(u))^2\leqslant c\theta^2.$$

Condition (III) is proved.

Since conditions (IV) and (V) of theorem 4 of section 3.3 are satisfied in an obvious way (for $\Omega_{u,\theta}=\Omega_{u,\theta}^e$), then the theorem is proved.

The validity of the theorem for finite n satisfying (3.46) is obvious. □

Remark. In a similar way we could consider a 'dependent' input stream e, when the coefficients a_e and b_{ee} depend on the number of busy lines $q(u)$.

THEOREM 2. *Let conditions* (I)–(V) *of theorem* 1 *be satisfied. Then, if the number of service channels n is equal to*

$$n=\frac{Ta_e}{\lambda}+D\sqrt{T},\quad |D|<\infty$$

and requests which on arrival find all n channels busy are refused (*leave the system*), *the normalized number of busy lines*

$$z(t)=\frac{q(tT)-a_eT/\lambda}{\sqrt{T}}$$

C-converges to a diffusion process with coefficients $(-\lambda x, b_{ee}+a_e)$ *and downwards reflection from the boundary* $z=D$.

Diffusion processes of this form have been considered in detail in the remarks to theorem 5 of the previous section.

Examples of input processes $e(t)$ which satisfy the conditions of theorem 2 are given above.

Proof. We will show that

$$\bar{q}(u) = \max\left(q(u), n - 2\bar{N}\sqrt{T}\right) \tag{3.56}$$

satisfies the conditions of theorem 5 (with reflecting barrier at D rather than 0). The number $\bar{N} \to \infty$ will be chosen later.

We have seen in the previous theorem that for $n = \infty$ the increments $q_{u,\theta}$ satisfy $(\zeta(u) = (q(u) - (a_e/\lambda)T)/\sqrt{T}, \omega \in \Omega_{u,\theta}\Gamma_u^N)$

$$\begin{aligned} \mathbf{E}_{\mathfrak{M}(u)} q_{u,\theta} &= -\lambda\zeta(u)\frac{\theta}{\sqrt{T}} + r_1\frac{\theta}{\sqrt{T}} \\ \mathbf{E}_{\mathfrak{M}(u)} q_{u,\theta}^2 &= \theta(b_{ee} + a_e) + r_2\theta \\ \mathbf{E}_{\mathfrak{M}(u)}(q_{u,\theta}^2; |q_{u,\theta}| > \delta\sqrt{T}) &< \varepsilon\theta \end{aligned} \tag{3.57}$$

$|r_i| < \varepsilon$. In addition, from corollary 1, lemma 2 of section 3.3, it also follows that

$$\mathbf{E}_{\mathfrak{M}(u)}\left(\sup_{s \leqslant \theta} |q_{u,s}| > \delta\sqrt{T}\right) < \frac{\varepsilon\theta}{T}. \tag{3.58}$$

We will show first of all that all these relations still hold for the increments $\bar{q}_{u,\theta}$ of $\bar{q}(u)$, but only in the domain

$$\Omega_{u,\theta} \cap \{-N < \zeta(u) < D - \delta\} \tag{3.59}$$

for any $\delta > 0$. In this domain, obviously,

$$\left\{\sup_{s \leqslant \theta} |\bar{q}_{u,s}| < \delta\sqrt{T}\right\} = \left\{\sup_{s \leqslant \theta} |q_{u,s}| < \delta\sqrt{T}\right\}$$

(in these calculations $q_{u,s}$ means the increment of $q(u)$ when $n = \infty$). Therefore, by (3.58),

$$\begin{aligned} &\mathbf{E}_{\mathfrak{M}(u)}(\bar{q}_{u,\theta}^2; |\bar{q}_{u,\theta}| > \delta\sqrt{T}) \\ &\qquad \leqslant \mathbf{E}_{\mathfrak{M}(u)}\left(\bar{q}_{u,\theta}^2; \sup_{s \leqslant \theta} |\bar{q}_{u,s}| \geqslant \delta\sqrt{T}\right) \leqslant (2\bar{N}\sqrt{T})^2\frac{\varepsilon\theta}{T} \leqslant \varepsilon_1\theta \end{aligned} \tag{3.60}$$

where $\varepsilon_1 = \varepsilon^{1/3}$, $2\bar{N} = \varepsilon^{-1/3}$ (since in our discussions ε is an arbitrarily small number, then we may suppose that $\varepsilon \to 0$, $\bar{N} \to \infty$ as $T \to \infty$). Further, in the same domain (3.59)

$$\begin{aligned} \mathbf{E}_{\mathfrak{M}(u)}\bar{q}_{u,\theta}^2 &= \mathbf{E}_{\mathfrak{M}(u)}\left(\bar{q}_{u,\theta}^2; \sup_{s \leqslant \theta} |\bar{q}_{u,s}| \geqslant \delta\sqrt{T}\right) \\ &\quad + \mathbf{E}_{\mathfrak{M}(u)}\left(q_{u,\theta}^2; \sup_{s \leqslant \theta} |q_{u,s}| < \delta\sqrt{T}\right). \end{aligned}$$

From (3.58), (3.60), and (3.57) it follows that this is

$$\theta(b_{ee} + a_e) + \bar{r}_2\theta, \quad |r_2| < \varepsilon_1.$$

$\mathbf{E}_{\mathfrak{M}(u)}\bar{q}_{u,\theta}$ is estimated in exactly the same way. Thus (3.57) and (3.58) for $\bar{q}_{u,\theta}$ are proved.

We now verify the reflection condition. For sufficiently small ε^* we must estimate the behaviour of

$$\mathbf{P}_{\mathfrak{M}(u)}(q_{u,r_1} < -\varepsilon^*\sqrt{T})$$

in $\Omega_{u,\theta} \cap \{\zeta(u) > D - \varepsilon^*\}$ for $\theta_1 = (\varepsilon^*)^2 T$.

For those requests which have been refused in the interval $(u, u + \theta)$ we introduce special supplementary service channels with the same properties as the main n channels.

It is obvious that if we now form a new queue $q^*(u)$ taking the new channels together with $q(u)$, then we obtain

$$q^*_{u,\theta_1} \geqslant q_{u,\theta_1}.$$

Consequently

$$\begin{aligned}\mathbf{P}_{\mathfrak{M}(u)}(q_{u,\theta_1} < -\varepsilon^*\sqrt{T}) &= \mathbf{P}_{\mathfrak{M}(u)}(q_{u,\theta_1} < -\sqrt{\theta_1}) \\ &\geqslant \mathbf{P}_{\mathfrak{M}(u)}(q^*_{u,\theta_1} < -\sqrt{\theta_1}).\end{aligned}$$

But the increment q^*_{u,θ_1}, obviously, behaves just the same as for the system with $n = \infty$. Therefore, if we put $\tilde{T} = \theta_1$ and consider the process

$$\tilde{q}(t) = q^*_{u,t}, \quad t \in [0, \tilde{T}] \quad (\tilde{q}(0) = 0)$$

then we may apply theorem 1 of this section to this process. By this theorem $\tilde{q}(t\theta_1)/\sqrt{\theta_1}$ C-converges to a diffusion process. Consequently $\mathbf{P}_{\mathfrak{M}(u)}(q^*_{u,\theta_1} < -\sqrt{\theta_1}) = \mathbf{P}_{\mathfrak{M}(u)}(\tilde{q}(t\theta_1)/\sqrt{\theta_1} < -1)$ converges to some positive limit.

The conditions of theorem 5 of section 3.3 are satisfied and, consequently, theorem 2 is proved. □

THEOREM 3. *Let conditions* (I)–(V) *of theorem* 1 *be satisfied. Then, if the number of service channels is equal to*

$$n = \frac{Ta_e}{\lambda} + D\sqrt{T}, \quad |D| < \infty$$

and requests which find all channels busy form a queue (which are served as channels are freed), then the normalized 'queue'

$$z(t) = \frac{q(tT) - (a_e/\lambda)T}{\sqrt{T}}$$

C-converges to a diffusion process with drift coefficient equal to $\max(-\lambda x, -\lambda D)$ *and diffusion coefficient* $b_{ee} + a_e$.

Proof. The verification of equalities (3.57), (3.58) for $\omega \in \Omega_{u,\theta} \cap \{-N < \zeta(u) < D - \delta\}$ proceeds in precisely the same way as in theorem 2.

Now let $\omega \in \Omega_{u,\theta} \cap \{D + \delta < \zeta(u) < N\}$. In this set the process $\{s_{u,t}, t \geqslant 0\}$ is the same as a Poisson process $\{\Pi(t), t \geqslant 0\}$ with parameter

$$\frac{\lambda n}{T} = a_e + \frac{\lambda D}{\sqrt{T}}$$

until the trajectory $q(u + t) = q(u) + q_{u,t}$ leaves the domain $q < n$, that is, at least until

$$\sup_t |q_{u,t}| < \delta\sqrt{T}.$$

If the latter restriction should not hold (i.e. if $s_{u,t}$ for all t was a Poisson process not depending on $\{e_{u,t}\}$, then we would obviously have

$$\begin{aligned} \mathbf{E}_{\mathfrak{M}(u)} q_{u,\theta} &= \mathbf{E}_{\mathfrak{M}(u)} e_{u,\theta} - \theta\left(a_e + \frac{\lambda D}{\sqrt{T}}\right) \\ \mathbf{E}_{\mathfrak{M}(u)} q_{u,\theta} &= \mathbf{E}_{\mathfrak{M}(u)} (e_{u,\theta} - \mathbf{E}_{\mathfrak{M}(u)} e_{u,\theta})^2 + \theta\left(a_e + \frac{\lambda D}{\sqrt{T}}\right). \end{aligned} \tag{3.61}$$

We will show that this equality is asymptotically preserved even under the above restriction. In fact, by the well-known properties of a Poisson process we will obviously have

$$\mathbf{P}\left(\sup_{t \leqslant \theta} \left|\Pi(t) - t\left(a_e + \frac{\lambda D}{\sqrt{T}}\right)\right| > \delta\sqrt{T}\right) < \frac{\varepsilon\theta}{T}.$$

From here and the similar properties of $e_{u,t}$ (see corollary 1, lemma 2, of section 3.3) we obtain that in $\Omega^e_{u,\theta} \cap \{\zeta(u) > D + \delta\}$

$$\mathbf{P}_{\mathfrak{M}(u)}\left(\sup_{t \leqslant \theta} |q_{u,\theta}| > \delta\sqrt{T}\right) = \mathbf{P}_{\mathfrak{M}(u)}\left(\sup_{t \leqslant \theta} |e_{u,t} - \Pi(t)| > \delta\sqrt{T}\right) < \frac{\varepsilon\theta}{T}.$$

Restricting, as in theorem 2, $q(u)$ to a sufficiently slowly growing sequence $\tilde{N}$ (cf. (3.56)), we find, similarly to (3.60), that

$$\mathbf{E}_{\mathfrak{M}(u)}\left(q^2_{u,\theta}; \sup_{t \leqslant \theta} |q_{u,t}| > \delta\sqrt{T}\right) \leqslant \varepsilon\theta.$$

Here with no difficulty we have preservation (up to terms $o(\theta/\sqrt{T})$ and $o(\theta)$ respectively) of (3.61) for the actual process $q_{u,t}$. Thus, in $\Omega^e_{u,\theta} \cap \{N > \zeta(u) > D + \delta\}$

$$\mathbf{E}_{\mathfrak{M}(u)} q_{u,\theta} = -\frac{\lambda D\theta}{\sqrt{T}} + \frac{r_1\theta}{\sqrt{T}}$$

$$\mathbf{E}_{\mathfrak{M}(u)} q_{u,\theta} = (b_{ee} + a_e)\theta + r_2\theta, \quad |r_i| < \varepsilon.$$

It remains for us to consider the 'transition' domain $|\zeta(u) - D| \leqslant \delta$.

In the same probability space as $q_{u,t}$ it is possible to construct Poisson processes* $\Pi^{\pm}(t)$ with parameters $(\lambda/T)(n \pm 2\delta\sqrt{T})$ such that

$$\Pi^-(t) \leqslant s_{u,t} \leqslant \Pi^+(t)$$

where the right-hand inequality is always valid and the left-hand is satisfied provided

$$\inf_t q(u+t) > n - 2\delta\sqrt{T}.$$

(This is possible, for example, via the introduction of new simulated service lines or the elimination of the old.)

In $\{\zeta(u) > D - \delta\}$ we have

$$\left\{\inf_{t \leqslant \theta} (e_{u,t} - \Pi^+(t)) > -\delta\sqrt{T}\right\} \subset \left\{\inf_{t \leqslant \theta} q_{u,t} > -\delta\sqrt{T}\right\}$$

$$\subset \left\{\inf_{t \leqslant \theta} q(u+t) > n - 2\delta\sqrt{T}\right\}$$

$$\left\{\sup_{t \leqslant \theta} |q_{u,t}| > \delta\sqrt{T}\right\} \subset \left\{\inf_{t \leqslant \theta} (e_{u,t} - \Pi^+(t)) < -\delta\sqrt{T}\right\}$$

$$\cup \left\{\sup_{t \leqslant \theta} (e_{u,t} - \Pi^-(t)) > \delta\sqrt{T}\right\}.$$

Therefore in $\Omega^e_{u,\theta} \cap \{N \geqslant \zeta(u) > D - \delta\}$, as before, we obtain

$$\mathbf{P}_{\mathfrak{M}(u)}\left(\sup_{t \leqslant \theta} |q_{u,t}| > \delta\sqrt{T}\right) < \frac{\varepsilon\theta}{T}.$$

Hence

$$\mathbf{E}_{\mathfrak{M}(u)}\left(q^2_{u,\theta}; \sup_{t \leqslant \theta} |q_{u,t}| > \delta\sqrt{T}\right) < \varepsilon\theta \tag{3.62}$$

which implies, in particular, a condition of type (IV).

Further, for $\omega \in \Omega^e_{u,\theta} \cap \{N \geqslant \zeta(u) > D - \delta\}$ we have

$$\mathbf{E}_{\mathfrak{M}(u)} q_{u,\theta} \geqslant \mathbf{E}_{\mathfrak{M}(u)}(e_{u,\theta} - \Pi^+(\theta)) = -\frac{\lambda D\theta}{\sqrt{T}} - \frac{2\lambda\delta\theta}{\sqrt{T}} + \frac{r_1\theta}{\sqrt{T}}$$

$$\mathbf{E}_{\mathfrak{M}(u)} q_{u,\theta} \leqslant \mathbf{E}_{\mathfrak{M}(u)}\left(e_{u,\theta} - \Pi^-(\theta); \sup_t |q_{u,t}| < \delta\sqrt{T}\right)$$

$$+ \mathbf{E}_{\mathfrak{M}(u)}\left(q_{u,\theta}; \sup_{t \leqslant \theta} |q_{u,t}| > \delta\sqrt{T}\right).$$

* It is useful to note that the conditional distribution, relative to $\mathfrak{M}(u)$, of $s_{u,t}$ at $t = 0$ locally (that is, in the sense of infinitesimal characteristics) behaves like the distribution of a Poisson process with parameter equal to min $(\lambda n/T, (\lambda/T)q(u))$.

The last term here, by (3.62), is estimated by the value $\varepsilon\theta/\delta\sqrt{T}$. Similarly, estimating $\mathbf{E}_{\mathfrak{M}(u)}(e_{u,\theta} - \Pi^{-}(\theta); \sup_{t \leqslant \theta}|q_{u,t}| \geqslant \delta\sqrt{T})$ and taking $\varepsilon < \delta^2$ we obtain

$$\mathbf{E}_{\mathfrak{M}(u)} q_{u,\theta} \leqslant -\frac{\lambda D\theta}{\sqrt{T}} + \frac{4\lambda\delta\theta}{\sqrt{T}}.$$

Since δ is arbitrary in these inequalities for $\mathbf{E}_{\mathfrak{M}(u)} q_{u,t}$, then, comparing these inequalities with the earlier asymptotic representations for $\mathbf{E}_{\mathfrak{M}(u)} q_{u,\theta}$ in $\{\zeta(u) > D + \delta\}$ and $\{\zeta(u) < D - \delta\}$, we conclude that

$$\mathbf{E}_{\mathfrak{M}(u)} q_{u,\theta} = \frac{\lambda\theta}{\sqrt{T}} \max(-\zeta(u), -D) + \frac{r_1\theta}{\sqrt{T}}, \quad |r_1| < \varepsilon.$$

In a similar way we establish that in the corresponding domain

$$\mathbf{E}_{\mathfrak{M}(u)} q_{u,\theta} = \theta(b_{ee} + a_e) + r_2\theta, \quad |r_2| < \varepsilon.$$

(This follows, for example, from $q_{u,\theta} = e_{u,\theta} - s_{u,\theta} = e_{u,\theta} - \Pi(\theta) + p$, where it is easy to establish that $\mathbf{E}_{\mathfrak{M}(u)} p^2 < \varepsilon\theta$.)

It remains only to use theorem 1 of section 3.3.* □

In conclusion of this section we note once again that the exponential nature of the service time is essential for the convergence to a Markov diffusion process. For nonexponential distributions such convergence will not hold (see the theorems of chapter 2).

3.5. Independent input stream and stochastic control of refusals

In this section we clarify how the conditions of theorems 1–6 of section 3.3 are transformed when the input stream $e(t)$ is independent and together with the service process $s(u)$ satisfies the conditions of these theorems, but the refusal process stochastically depends on the queue length (for the definition, see section 3.1). In the previous section the process $s(u)$ was known 'explicitly'; in this section the refusal process $r(u)$ is given explicitly.

We note first of all that, since for an independent input stream the function

$$\mathbf{E}_{\mathfrak{M}(u)} e_{u,\theta} = \mathbf{E}_{\mathfrak{E}(u)} e_{u,\theta}$$

is $\mathfrak{E}(u)$-measurable, then it may depend on $q(u)$ provided $q(u)$ is a deterministic functional of the trajectory $e(t)$ on $[0, u]$. Consequently, if $s(u) + r(u)$ is not a deterministic functional of $e(t)$, $0 \leqslant t \leqslant u$ (or, almost the same, if $\mathfrak{M}(u)$ and $\mathfrak{E}(u)$ do not coincide), then $\mathbf{E}_{\mathfrak{M}(u)} e_{u,\theta}$, for fixed $\mathfrak{E}(u)$, may not depend on $q(u)$. This

* In the proof of theorem 3 it was necessary to estimate the moments of $q_{u,\theta}$ in $|\zeta(u) - D| \leqslant \delta$. If we follow the route mentioned in the remark at the end of section 1.9, then it would be sufficient in $\Omega_{u,\theta} \cap \{|\zeta(u) - D| \leqslant \delta\}$ to estimate the probability of exit from a neighbourhood of the point D, verifying the reflection condition (PRVI) at D.

means, in particular, that under these conditions the coefficients $a_e(z)$ (or $\alpha_e(z)$ in theorems 1–3 of section 3.3) in condition (I) are necessarily constant:

$$a_e(z) = a_e = \text{const.} \quad (\alpha_e(z) = \alpha_e = \text{const.})$$

The same conclusion, obviously, holds for

$$b_{ee}(z) = b_{ee} = \text{const.} \quad (\beta_{ee}(z) = \beta_{ee} = \text{const.})$$

Thus, we consider a system with independent input in which the refusal process $r(u)$ depends stochastically on $q(u)$. The function $1 - p(z)$ of the probability of refusal is supposed known.

To avoid long discussions we restrict ourselves to the case of 'rough' dependence of the probability of refusal on the queue length. More general theorems (of the type of theorems 1–3 of section 3.3), on systems in which the stochastic dependence of $r(u)$ on $q(u)$ is 'more sensitive' to the deviation of $q(u)$, may be obtained quite analogously and the reader may easily reconstruct their proofs and statements.

Suppose first that for the 'point of equilibrium' Q (see section 2.2), defined as the solution of

$$A(Q) = a_e p(Q) - a_s(Q) = 0$$

we have the strict inequality $Q > 0$. As before, let

$$Q(u) = \frac{q(u)}{T}, \quad \Gamma_u^N = \left\{ |Q(u) - Q| < \frac{N}{\sqrt{T}} \right\}.$$

THEOREM 1. (1) *Let each of* $e(u)$ *and* $s(u)$ *satisfy conditions* (I)–(III) *and* (V) *of theorem* 4 *of section* 3.3 *in* $\Omega_{u,\theta} \cap \Gamma_u^N$.

(2) *In* $\Omega_{u,\theta} \cap \Gamma_u^N$ *the 'reduced input process'* $e^*(u) = e(u) - r(u)$ *and the process* $s(u)$ *have the property**

$$\mathbf{E}_{\mathfrak{M}(u)}(e^*_{u,\theta} - \theta a_e p(Q(u)))(s_{u,\theta} - \theta a_s(Q(u))) = r_{e^*,s}\theta, \quad |r_{e^*,s}| < \varepsilon. \tag{3.63}$$

(3) *The homogeneity condition (cf. condition* (V) *of theorem* 1 *of section* 3.4)

$$\mathbf{P}_{\mathfrak{M}(u)}(\Omega(u)\bar{\Omega}(u + \theta)) < \frac{\varepsilon\theta}{T}$$

* On the right-hand side of (3.63) we could also put $b_{e^*,s}(Q(u)) + r_{e^*,s}\theta$ for $b_{e^*,s} \neq 0$. The proof of the theorem would not be changed by this (it is only necessary to add $-2b_{e^*,s}(z)$ to $B(z)$). But, since $b_{e^*,s} = 0$ in all real problems of this type, then we have, for simplicity, restricted ourselves to this special case.

Evidently, (3.63) may be obtained as a corollary of the equality

$$\mathbf{E}_{\mathfrak{M}(u)}(e_{u,\theta} - a_e\theta)(s_{u,\theta} - a_s(Q(u))\theta) = r_{es}\theta, \quad |r_{es}| < \varepsilon.$$

holds, where

$$\Omega(u) = \bigcap_{t \leqslant u} \Omega_{t,\theta}\{e(t) - e(t-0) < \delta\sqrt{T}\}\{s(t) - s(t-0) < \delta\sqrt{T}\}$$

(this set may be taken as $\Omega_{u,\theta}$ from the start).

Then, if $p(z)$ satisfies a Lipschitz condition, the increment $e^*_{u,\theta} = e_{u,\theta} - r_{u,\theta}$ along with $e_{u,\theta}$ and $s_{u,\theta}$ also satisfy conditions (I)–(III) *and* (V) *(theorem* 4 *of section* 3.3), *in which we must put*

$$a_{e^*}(z) = a_e p(z) \quad (= a_e - a_r(z)) \tag{3.64}$$

$$b_{e^*e^*}(z) = b_{ee} p^2(z) + a_e p(z)(1 - p(z))$$

$$(= b_{ee} - 2b_{er}(z) - b_{rr}(z)).$$

If $A(z) = a_e p(z) - a_s(z)$ is continuously differentiable at Q and $B(z) = b_{e^*e^*}(z) + b_{ss}(z)$ is continuous, then theorem 4 *of section* 3.3 *holds.*

Before passing to the proof of this theorem, we make one remark. The function $p(z)$ enters the description of the service process $\mathbf{S}(u)$ only through the component $r(u)$ and plays the role of a parameter which may be different for 'one and the same' components $e(u)$ and $s(u)$. In every case the description of the necessary properties (I)–(III) of $e(u)$ and $s(u)$, as well as the right-hand side in (3.63), is in no way connected with the function p. Under these conditions it is natural to suppose (and to some extent we will use these assumptions) that conditions (1)–(3) of theorem 1 are satisfied for various $p(z)$, say from the class of functions uniformly satisfying a Lipschitz condition. Such an assumption (at least in the form used in the proof) does not limit generality and is introduced for the sake of naturality and brevity in the calculations.

Proof. The last assertion of the theorem will follow immediately from theorem 4 of section 3.3 if we prove the required properties of e^*. Therefore our problem lies in proving the equalities

$$\mathbf{E}_{\mathfrak{M}(u)} e^*_{u,\theta} = \theta a_e p(Q(u)) + r_{e^*} \frac{\theta}{\sqrt{T}}$$

$$\mathbf{E}_{\mathfrak{M}(u)} (e^*_{u,\theta} - \theta a_e p(Q(u))^2 = \theta[b_{ee} p^2(Q(u)) + a_e p(Q(u))(1 - p(Q(u)))]$$

$$+ r_{e^*e^*}\theta$$

$$\mathbf{E}_{\mathfrak{M}(u)}[(e^*_{u,\theta} - \theta a_e p(Q(u)))^2; |e^*_{u,\theta} - \theta a_e p(Q(u))| > \delta\sqrt{T}] < \varepsilon\theta$$

$$|r_{e^*}| < \varepsilon, \quad |r_{e^*e^*}| < \varepsilon. \tag{3.65}$$

Concerning $p(z)$ we may, without loss of generality, suppose that $p(z)$ is constant outside the $\bar{N}/\sqrt{T}$-neighbourhood of Q: $p(z) = p(Q \pm \bar{N}/\sqrt{T})$ for $z \gtrless Q \pm \bar{N}/\sqrt{T}$, where $\bar{N}$ is some increasing sequence which we choose later. This assumption does not limit generality because the probability that $Q(u)$ reaches the boundary of this domain at least once will, as it turns out, converge to zero. We put, as in the previous section,

$$\bar{q}(u) = QT + \min[\max(q(u) - QT, -\bar{N}\sqrt{T}), \bar{N}\sqrt{T}].$$

We have

$$e^*_{u,\theta} = \sum_{j=1}^{e_{u,\theta}} \delta_j \tag{3.66}$$

where $\delta_j = 0$ if the jth request (counted from the epoch u) is refused, and $\delta_j = 1$ otherwise. Let $u + \mu_j$ be the epoch of the appearance of the jth request. Then, by the definition of stochastic dependence of $r(u)$ on the queue (see section 3.1; in the same place see the definition of $\mathfrak{M}^*(\mu_j)$)

$$\begin{aligned}
\mathbf{E}_{\mathfrak{M}(u)} e^*_{u,\theta} &= \mathbf{E}_{\mathfrak{M}(u)} \mathbf{E}_{\mathfrak{M}^*(\mu_1)} \left(\sum_{j=1}^{e_{u,\theta}} \delta_j \right) (I(\mu_1 \leqslant \theta) + I(\mu_1 > \theta)) \\
&= \mathbf{E}_{\mathfrak{M}(u)} \mathbf{E}_{\mathfrak{M}^*(\mu_1)} \delta_1 I(\mu_1 \leqslant \theta) + \mathbf{E}_{\mathfrak{M}(u)} \mathbf{E}_{\mathfrak{M}^*(\mu_2)} I(\mu_1 \leqslant \theta) \sum_{j=2}^{e_{u,\theta}} \delta_j \\
&= \mathbf{E}_{\mathfrak{M}(u)} I(\mu_1 \leqslant \theta) p(Q(u+\mu_1 - 0)) + \mathbf{E}_{\mathfrak{M}(u)} I(\mu_2 \leqslant \theta) \mathbf{E}_{\mathfrak{M}^*(\mu_2)} \delta_2 \\
&\quad + \mathbf{E}_{\mathfrak{M}(u)} \mathbf{E}_{\mathfrak{M}^*(\mu_2)} I(\mu_2 \leqslant \theta) \sum_{j=3}^{e_{u,\theta}} \delta_j \\
&= \cdots \\
&= \mathbf{E}_{\mathfrak{M}(u)} \sum_{j=1}^{\infty} I(\mu_j \leqslant \theta) p(Q(u+\mu_j - 0)) \\
&= \mathbf{E}_{\mathfrak{M}(u)} \sum_{j=1}^{e_{u,\theta}} p(Q(u+\mu_j - 0)).
\end{aligned}$$

Further

$$\begin{aligned}
&|\mathbf{E}_{\mathfrak{M}(u)} e^*_{u,\theta} - a_e p(Q(u))\theta| \\
&\qquad \leqslant \left| \mathbf{E}_{\mathfrak{M}(u)} \sum_{j=1}^{e_{u,\theta}} [p(Q(u+\mu_j - 0)) - p(Q(u))] \right| + p(Q(u)) |\mathbf{E}_{\mathfrak{M}(u)} e_{u,\theta} - a_e \theta|.
\end{aligned}$$

If L is a Lipschitz constant for p, then the first sum in this inequality does not exceed ($\bar{Q}(u) = \bar{q}(u)/T$)

$$\begin{aligned}
&\mathbf{E}_{\mathfrak{M}(u)} L \sup_{i \leqslant e_{u,\theta}} |\bar{Q}(u + \mu_i - 0) - \bar{Q}(u)| e_{u,\theta} \\
&\qquad \leqslant L \sqrt{\mathbf{E}_{\mathfrak{M}(u)} (e_{u,\theta})^2 \mathbf{E}_{\mathfrak{M}(u)} \sup_{v \leqslant \theta} |\bar{Q}(u + v) - \bar{Q}(u)|^2}.
\end{aligned}$$

We will see, thus, that in the corresponding domain

$$|\mathbf{E}_{\mathfrak{M}(u)} e^*_{u,\theta} - a_e p(Q(u))\theta| \leqslant \frac{\varepsilon\theta}{\sqrt{T}} + \frac{c\theta}{T}\sqrt{\mathbf{E}_{\mathfrak{M}(u)} \sup_{v\leqslant\theta} (\bar{q}_{u,v})^2}$$

where $\bar{Q}(u) = \bar{q}(u)/T$, $\bar{q}_{u,v} = \bar{q}(u+v) - \bar{q}(u)$. If we assume that in the required domain

$$\mathbf{E}_{\mathfrak{M}(u)} \sup_{v\leqslant\theta} (\bar{q}_{u,v})^2 < \varepsilon T \tag{3.67}$$

then the first of relations (3.65) will be proved.

Further, for brevity, we put

$$p_i = p(Q(u+\mu_i - 0)), \quad p_0 = p(Q(u)).$$

Then

$$\mathbf{E}_{\mathfrak{M}(u)}(e^*_{u,\theta} - a_e p(Q(u))\theta)^2$$
$$= \mathbf{E}_{\mathfrak{M}(u)}\left(\sum_{j=1}^{e_{u,\theta}} (\delta_j - p_j) + \sum_{j=1}^{e_{u,\theta}} (p_j - p_0) + p_0(e_{u,\theta} - a_e\theta)\right)^2.$$

Here, as before, we find

$$\mathbf{E}_{\mathfrak{M}(u)} \sum_{j=1}^{e_{u,\theta}} (\delta_j - p_j)^2$$
$$= \mathbf{E}_{\mathfrak{M}(u)} \sum_{j=1}^{e_{u,\theta}} \mathbf{E}_{\mathfrak{M}^*(\mu_j)}(\delta_j - p_j)^2 = \mathbf{E}_{\mathfrak{M}(u)} \sum_{j=1}^{e_{u,\theta}} p_j(1-p_j)$$

$$\mathbf{E}_{\mathfrak{M}(u)} \sum_{i<j}^{e_{u,\theta}} (\delta_i - p_i)(\delta_j - p_j)$$
$$= \mathbf{E}_{\mathfrak{M}(u)} \sum_{i<j}^{e_{u,\theta}} \mathbf{E}_{\mathfrak{M}^*(\mu_j)}(\delta_i - p_i)(\delta_j - p_j) = 0$$

$$\mathbf{E}_{\mathfrak{M}(u)}(e_{u,\theta} - a_e\theta) \sum_{j=1}^{e_{u,\theta}} (\delta_j - p_j)$$
$$= \mathbf{E}_{\mathfrak{M}(u)}(e_{u,\theta} - a_e\theta)\mathbf{E}_{(\mathfrak{M}(u),\mathfrak{E}(u+\theta))} \sum_{j=1}^{e_{u,\theta}} (\delta_j - p_j)$$
$$= \mathbf{E}_{\mathfrak{M}(u)}(e_{u,\theta} - a_e\theta) \sum_{j=1}^{e_{u,\theta}} \mathbf{E}_{(\mathfrak{M}(u),\mathfrak{E}(u+\theta))}\mathbf{E}_{(\mathfrak{M}^*(\mu_j),\mathfrak{E}(u+\theta))}(\delta_j - p_j).$$

Since $\mathfrak{E}(\mu_i) \subset \mathfrak{M}^*(\mu_i) \subset \mathfrak{M}(\mu_i)$, then, by the strict independence of input (see (3.4); $\delta_j - p_j$ is measurable relative to $\mathfrak{M}(\mu_j)$)

$$\mathbf{E}_{\mathfrak{M}^*(\mu_j),\mathfrak{E}(u+\theta))}(\delta_j - p_j) = \mathbf{E}_{\mathfrak{M}^*(\mu_j)}(\delta_j - p_j) = 0.$$

Here we have used the theorem of section 3.1 that independence of input implies strict independence.

As a result we obtain

$$\mathbf{E}_{\mathfrak{M}(u)}(e^*_{u,\theta}-a_e p(Q(u))\theta)$$

$$=\mathbf{E}_{\mathfrak{M}(u)}\Bigg[\sum_{i=1}^{e_{u,\theta}} p_i(1-p_i)+p_0^2(e_{u,\theta}-a_e\theta)^2+\sum_{i=1}^{e_{u,\theta}}(p_i-p_0)^2$$

$$+2\sum_{j=1}^{e_{u,\theta}}(\delta_j-p_j)\sum_{i=1}^{e_{u,\theta}}(p_i-p_0)+2p_0(e_{u,\theta}-a_e\theta)\sum_{i=1}^{e_{u,\theta}}(p_i-p_0)\Bigg].$$

Since for $\bar{N}^2=o(T/\theta)$

$$\left|\mathbf{E}_{\mathfrak{M}(u)}\sum_{i=1}^{e_{u,\theta}}(p_i(1-p_i)-p_0(1-p_0))\right|\leqslant\frac{c}{T}\,\mathbf{E}_{\mathfrak{M}(u)}e_{u,\theta}\sup_{v\leqslant\theta}|\bar{q}_{u,v}|\leqslant\frac{c\theta\bar{N}}{\sqrt{T}}<\varepsilon\theta$$

$$\mathbf{E}_{\mathfrak{M}(u)}\sum_{i=1}^{e_{u,\theta}}(p_i-p_0)^2\leqslant\mathbf{E}_{\mathfrak{M}(u)}\frac{L^2}{T^2}\,e_{u,\theta}^2\sup_{v\leqslant\theta}|\bar{q}_{u,v}|^2\leqslant\frac{c\theta^2\bar{N}^2}{T}\leqslant\varepsilon\theta$$

then we obtain on $\Omega_{u,\theta}\cap\Gamma_u^N$

$$\mathbf{E}_{\mathfrak{M}(u)}(e^*_{u,\theta}-a_e p(Q(u))\theta)^2=a_e\theta p_0(1-p_0)+p_0^2 b_{ee}\theta+r_{e^*}(\omega),\quad |r_{e^*}|<\varepsilon.$$

This means, obviously, that the second equality in (3.65) holds and also

$$\mathbf{E}_{\mathfrak{M}(u)}q_{u,\theta}^2=\theta B(Q(u))+r_q\theta,\quad |r_q|<\varepsilon$$

(here we have used (3.63)).

We now use lemma 3 of section 3.3 by which, under our conditions,

$$\mathbf{E}_{\mathfrak{M}(u)}\left(\sup_{t\leqslant\theta}q_{u,t}^2;\Omega(u+\theta)\right)\leqslant N_1\theta$$

for $\omega\in\Omega(u)\cap\Gamma_u^{N-1}$ and sufficiently large N_1. Hence we find that for $\omega\in\Omega(u)\Gamma_u^{N-1}$

$$\mathbf{E}_{\mathfrak{M}(u)}\sup_{t\leqslant\theta}\bar{q}_{u,t}^2\leqslant\bar{N}^2T\,\frac{\varepsilon\theta}{T}+\mathbf{E}_{\mathfrak{M}(u)}\left(\sup_{t\leqslant\theta}\bar{q}_{u,t}^2;\Omega(u+\theta)\right)\leqslant\bar{N}^2\varepsilon\theta+N_1\theta\leqslant\varepsilon T.$$

This proves (3.67) and with it the first equality of (3.65).

It remains for us to verify a condition of type (III), i.e. to estimate

$$\mathbf{E}_{\mathfrak{M}(u)}[(e^*_{u,\theta}-\theta a_e p(Q(u)))^2;|e^*_{u,\theta}-\theta a_e p(Q(u))|>\delta\sqrt{T}].$$

By (3.66)

$$e^*_{u,\theta}=\sum_{j=1}^{a_e\theta}\delta_j+\left(\sum_{j=1}^{e_{u,\theta}}\delta_j-\sum_{j=1}^{a_e\theta}\delta_j\right).$$

It is sufficient for us to show that each of the two terms on the right-hand side of this equality satisfy condition (III) (cf., for example, the proof of theorem 1 of section 3.3). But for the second term this is obvious, since it does not exceed $|e_{u,\theta} - \theta a_e|$ in absolute value. In the first term, by the properties of p, we can write

$$\sum_{j=1}^{a_e\theta} \delta_j^- < \sum_{j=1}^{a_e\theta} \delta_j < \sum_{j=1}^{a_e\theta} \delta_j^+ \tag{3.68}$$

where $\delta_j^\pm$, for fixed $\mathfrak{M}(u)$, are independent for different j,

$$\mathbf{P}_{\mathfrak{M}(u)}(\delta_j^\pm = 1) = p^\pm = p_0 \pm L\frac{\bar{N} + N}{\sqrt{T}}.$$

((3.68) assumes that δ_j, $\delta_j^\pm$ are given in the same probability space. This is not difficult to do. However, this need not be assumed if we take (3.68) to be satisfied in 'distribution'.)

Therefore for the deviation of the first term we have

$$\mathbf{P}_{\mathfrak{M}(u)}\left(\sum_{j=1}^{a_e\theta} \delta_j - \theta a_e p_0 > z\right) \leqslant \mathbf{P}_{\mathfrak{M}(u)}\left(\sum_{j=1}^{a_e\theta} (\delta_j^+ - p^+) + a_e\theta\frac{L(\bar{N}+N)}{\sqrt{T}} > z\right)$$
$$\leqslant \mathbf{P}_{\mathfrak{M}(u)}\left(\sum_{j=1}^{a_e\theta} (\delta_j^+ - p^+) > z - \sqrt{\theta}\right)$$

for sufficiently large T. But, as is well known from the properties of Bernoulli schemes, this probability does not exceed

$$e^{-2(z-\sqrt{\theta})^2/a_e\theta}.$$

A similar inequality is easily obtained for negative deviation of z. From these inequalities it follows that

$$\mathbf{E}_{\mathfrak{M}(u)}\left[\left(\sum_{j=1}^{a_e\theta} \delta_j - \theta a_e p_0\right)^2; \left|\sum_{j=1}^{a_e\theta} \delta_j - \theta a_e p_0\right| > k\sqrt{\theta}\right] < \varepsilon\theta$$

for sufficiently large k, and that, consequently, a condition of type (III) is satisfied.

The theorem for an 'adapted' function p is proved. But by this theorem the probability that the trajectory of $Q(u) - Q$ hits the boundary $\pm N/\sqrt{T}$, where the 'adapting' of p begins, converges to zero. Therefore the theorem is proved even for the original function p. □

In the case $Q = 0$ the theorem on convergence to a diffusion process with reflections differs to the same extent as theorem 5 differs from theorem 4 in section 3.3. Namely, the following theorem holds, the proof of which may be easily supplied by the reader by comparing the proof of theorem 1 and theorems 4, 5 of section 3.3.

THEOREM 2. *Let conditions* (1)–(3) *of theorem* 1 *hold; here conditions* (1) *and* (2) *are valid only in* $\{\zeta(u) = q(u)/\sqrt{T} \geqslant \varepsilon^*\}$ *for any fixed* $\varepsilon^* > 0$.

Further, let the reflection condition be satisfied (*see, for example, condition* (IV) *of theorem 2 of section 3.3). And let, in addition, the functions a_e, $p(z)$, $a_s(z)$, and $b_{ee}(z)$ (for $Q = 0$ it is natural to regard these functions as depending also on T) be such that for $A(z) = a_e p(z) - e_s(z)$ and $B(z) = b_{e^*e^*}(z)$ (see (3.64)) there exist*

$$\lim_{T\to\infty} A(0)\sqrt{T} = A_0, \quad \lim_{T\to\infty} A'(0) = A'_0, \quad \lim_{T\to\infty} B(0) = B_0 > 0$$

and $A'(x)$ and $B(x)$ are equicontinuous from the right at $x = 0$.

Then the result of theorem 5 of section 3.3 is valid.

We consider now the case, important in applications, when $p(z)$ has breaks near Q. For example, let $p_1(z)$ and $p_2(z)$ be two differentiable functions

$$p_2(Q) < p_1(Q) - \alpha, \quad \alpha > 0$$

and let $p(z) = p_T(z)$ be defined by

$$p(z) = \begin{cases} p_1(z), & z \geqslant Q + \dfrac{R}{\sqrt{T}} \\ p_2(z), & z < Q + \dfrac{R}{\sqrt{T}}. \end{cases} \tag{3.69}$$

Related to systems with such functions p are, obviously, systems with refusals for $p_1 = 1, p_2 = 0$. As for systems with refusals, the normalized process $q(u)$ in the systems, considered in this section and having a function $p(z)$ of the form (3.69), will converge to a diffusion process with reflection at the upper boundary R.

We consider here the case when $p_2(u) = 0$, that is, when requests which force $q > QT + R\sqrt{T}$ are refused with probability 1. The general case (3.69) we leave to the reader.*

Thus, let

$$p(u) = \begin{cases} p_1(u), & u \leqslant Q + \dfrac{R}{\sqrt{T}} \\ 0, & u > Q + \dfrac{R}{\sqrt{T}}. \end{cases} \tag{3.70}$$

* Here different approaches are possible. One of them consists of choosing a smooth majorant for $p(u)$ with the help of a suitable function in the necessary form of the form $p_1(Q) + (1/\sqrt{T})f^*((u - Q)\sqrt{T})$, where $f(z)$ does not depend on T and is a sufficiently smooth function increasing for $z > R$. For systems with such $p(u)$ there is, of course, an analogue (or generalization) of theorem 1 in the case of 'fine' dependence of control on the process $\zeta(u) = (q(u) - QT)/\sqrt{T}$ (cf. theorems 1–3 of section 3.3). As already noted such theorems are obtained if we simply repeat the proof of theorem 1 under the new conditions.

In addition, under this approach we also need theorems on convergence of unbounded diffusion processes with strong negative drift in $z > R$ (say with drift coefficient $a_N(z) < N(z - R)$ and $N \to \infty$) to a diffusion process with reflection from the boundary $z = R$. Such a family of theorems may be obtained from theorem 1 of section 1.1.

THEOREM 3. *Suppose that the system satisfies conditions* (1)–(3) *of theorem* 1 *for both the refusal function* (3.70) *and for the function* $p(u) \equiv p_1(u)$. *Then, if* $p_1'(u)$ *is continuous at* Q *and* (3.70) *holds, the process* $(q(tT) - QT)/\sqrt{T}$ *will C-converge to a diffusion with coefficients*

$$zA'(Q) = z(a_e p_1(Q) - a_s(Q))'$$
$$B(Q) = b_{ee} p_1^2(Q) + a_e p_1(Q)(1 - p_1(Q)) + b_{ss}(Q)$$

and reflection from the upper boundary R.

From the remarks made in the footnote it is possible to conclude that the same assertion will hold in case (3.69).

Similarly to theorem 3 we can state a theorem on convergence to a diffusion with two reflecting boundaries (0 and R) in the case when we have the conditions of theorem 2 and (3.70).

Proof of theorem 3. Consider another service process $\mathbf{S}^1(u)$, which is obtained if we take as $p(u)$ a smooth function $p_1(u)$ ($p(u) \equiv p_1(u)$). From theorem 1 it then follows that the process $q^1(u)$, corresponding to $\mathbf{S}^1(u)$, satisfies the conditions of theorem 4 of section 3.3. This means, by corollary 1 of lemma 2 of section 3.3, that on $\Omega_{u,\theta} \cap \Gamma_u^{N-1}$

$$\mathbf{P}_{\mathfrak{M}(u)}\left(\sup_{s\leqslant\theta} |q^1_{u,s}| > \delta\sqrt{T}\right) < \frac{\varepsilon\theta}{T}$$
$$\mathbf{E}_{\mathfrak{M}(u)}\left((q^1_{u,\theta})^2; \sup_{s\leqslant\theta} |q^1_{u,s}| > \delta\sqrt{T}\right) < \varepsilon\theta. \tag{3.71}$$

But from here it follows at once that conditions (I)–(III) will be satisfied equally by $q^1_{u,\theta}$ and the increments $q_{u,\theta}$ of the input process, provided $q(u) < QT + (R - \delta)\sqrt{T}$. To establish this is easy with the help of standard methods which we have already used time and again: we regard $q(u)$ as 'truncated' so that $(q(t) - TQ)/\sqrt{T}$ does not exceed some high level $\bar{N}$ and note that in the domain $\{q(u) < QT + (R - \delta)\sqrt{T}\}$

$$\left\{\sup_{s\leqslant\theta} |q^1_{u,s}| > \delta\sqrt{T}\right\} = \left\{\sup_{s\leqslant\theta} |q_{u,s}| > \delta\sqrt{T}\right\}.$$

From here and (3.71) it then follows that

$$\mathbf{E}_{\mathfrak{M}(u)}(q^2_{u,\theta}; |q_{u,\theta}| > \delta\sqrt{T}) \leqslant \bar{N}^2 T \frac{\varepsilon\theta}{T} = \varepsilon_1\theta \quad \text{for } \varepsilon = \frac{\varepsilon_1}{\bar{N}^2}$$

$$\mathbf{E}_{\mathfrak{M}(u)} q^2_{u,\theta} = \mathbf{E}_{\mathfrak{M}(u)}\left((q^1_{u,\theta})^2; \left|\sup_{s\leqslant\theta} q_{u,s}\right| < \delta\sqrt{T}\right) + \varepsilon_1\theta$$
$$= \mathbf{E}_{\mathfrak{M}(u)}(q^1_{u,\theta})^2 + \varepsilon_2\theta.$$

In a similar way the necessary relations for the first moments $\mathbf{E}_{\mathfrak{M}(u)} q_{u,\theta}$ are established. Thus $q_{u,\theta}$ satisfies conditions (I)–(III) in $\{\zeta(u) < R - \delta\}$ with the same coefficients as $q^1_{u,\theta}$.

To prove the assertion of the theorem it remains for us to verify reflection in theorem 1 of section 1.9. This can be done using again the comparison of $q_{u,t}$ with some other process.

Put $q^+_{u,t} = e^+_{u,t} - s_{u,t}$, where $e^+_{u,t} = \sum_{j=1}^{e_{u,t}} \delta^+_j$, δ^+_j for fixed $\mathfrak{M}(u)$ are independent and and

$$\mathbf{P}_{\mathfrak{M}(u)}(\delta^+_j = 1) = p^+ = p_1(Q) + L\frac{\delta}{\sqrt{T}} \geqslant p_1(Q(u))$$

for

$$|Q(u) - Q| < \frac{\delta}{\sqrt{T}}.$$

The random variables δ^+_j can be defined in the same probability space with the variables δ_j, for which

$$\sum_{j=1}^{e_{u,t}} \delta_j = e^*_{u,t}$$

(for example we may suppose that $\delta_j = \delta^+_j - \delta'_j$, where

$$\mathbf{P}_{\mathfrak{M}^*(\mu_j)}(\delta'_j = 0/\delta^+_j = 1) = p_1(Q(u))(p^+)^{-1}$$

$$\mathbf{E}_{\mathfrak{M}^*(\mu_j)}(\delta'_j = 1/\delta^+_j = 1) = \frac{L(\delta/\sqrt{T}) + p_1(Q) - p_1(Q(u))}{p^+} < \frac{2l\delta}{p^+\sqrt{T}}$$

for the definition of $\mathfrak{M}^*(\mu_j)$ see section 3.1).

Then, obviously, $\delta_j \leqslant \delta^+_j$, $q_{u,t} \leqslant q^+_{u,t}$. At the same time, for $q^+_{u,t}$ in $\{\zeta(u) > R - \delta\}$ we have

$$\mathbf{P}_{\mathfrak{M}(u)}\{q^+_{u,\delta^2 T} < \delta\sqrt{T}\} \geqslant \gamma > 0 \tag{3.72}$$

from which follows the required reflection condition for $q(u)$. Thus we need to prove (3.72). This follows from the convergence of $q^+_{u,t}$, $0 \leqslant t \leqslant \delta^2 T$, after suitable normalization (division by $\delta\sqrt{T}$), to a Wiener process, since $q^+_{u,t}$ satisfies conditions (I)–(III) with constant coefficients.

The verification of these conditions gives rise to no difficulties (they are satisfied by $e^+_{u,\theta}$ and $s_{u,\theta}$). Amongst them the only explanation required, perhaps, is the calculation of the mixed moment

$$\mathbf{E}_{\mathfrak{M}(u)}(s_{u,\theta} - a_s(Q)\theta)(e^+_{u,\theta} - a_e p^+\theta). \tag{3.73}$$

But here

$$e^+_{u,\theta} - a_e p^+\theta = (e^*_{u,\theta} - a_e p_1(Q)\theta) + \left(\sum_{j=1}^{e_{u,\theta}} \delta'_j - a_e\theta\frac{L\delta}{\sqrt{T}}\right).$$

Consequently, from the conditions of the theorem and by the Cauchy–Bunyakovskii inequality, (3.73) does not exceed

$$\varepsilon\theta + \left[\mathbf{E}_{\mathfrak{M}(u)}(s_{u,\theta} - a_s(Q)\theta)^2 \mathbf{E}_{\mathfrak{M}(u)}\left(\sum_{j=1}^{e_{u,\theta}} \delta'_j - a_e\theta\frac{L\delta}{\sqrt{T}}\right)^2\right]^{\frac{1}{2}}.$$

Since $\mathbf{P}_{\mathfrak{M}^*(\mu_j)}(\delta'_j = 1) \leqslant 2L\delta/\sqrt{T}$, then, using the same arguments as were used for the estimate of $\mathbf{E}_{\mathfrak{M}(u)}(e^*_{u,\theta} - a_e p(Q(u))\theta)^2$ in theorem 1, we obtain

$$\mathbf{E}_{\mathfrak{M}(u)}\left(\sum_{j=1}^{e_{u,\theta}} \delta'_j - a_e\theta\frac{L\delta}{\sqrt{T}}\right)^2 < \varepsilon\theta.$$

This means that (3.73) does not exceed $\varepsilon\theta$. The theorem is proved. □

We could now obtain theorem 2 of the preceding section (on systems with refusals and with $n = Ta_e/\lambda + D\sqrt{T}$ exponential service channels) as a corollary of theorem 3.

3.6. Properties of systems with independent output. Loaded systems

Suppose that the conditions of one of the theorems of section 3.3, together with the homogeneity condition in $\Omega_{u,\theta}$ (see (3.41)), are satisfied. For definiteness, let this be the conditions of theorem 1. We assume, in addition, that the systems under discussion have *independent output* (for the definition see section 3.1) in the domain $q \geqslant d$, where $d \leqslant QT - N\sqrt{T}$ for $Q > 0$, any fixed $N > 0$, and all sufficiently large T. We show that under these conditions it is possible to take

$$\alpha_s(z) = \text{const.}, \quad \beta_{ss}(z) = \text{const.}, \quad \beta_{e^*s}(z) = 0.$$

Precisely the same will be the case under the conditions of the other theorems in section 3.3.

We return to the definition of independence of output and as Γ_u take Γ_u^{N-1} and as V_{e^*}, V_s take the sets

$$\begin{aligned} V_{e^*} &= \left\{\sup_{v\leqslant\theta} |e^*_{u,v} - va_{e^*}(Q(u))| \leqslant \delta\sqrt{T}\right\} \in \mathfrak{E}^*(u+\theta) \\ V_s &= \left\{\sup_{v\leqslant\theta} |s_{u,v} - va_s(Q(u))| \leqslant \delta\sqrt{T}\right\} \in \mathfrak{S}(u+\theta). \end{aligned} \tag{3.74}$$

For these sets Γ_u, V_{e^*}, V_s condition (3.8) will be satisfied:

$$\begin{aligned} \Gamma_u V_{e^*} V_s &\subset \Gamma_u^{N-1}\left\{\sup_{v\leqslant\theta} |q_{u,\theta} - vA(Q(u))| \leqslant 2\delta\sqrt{T}\right\} \\ &\subset \left\{\inf_{u\leqslant v\leqslant u+\theta} q(v) > QT - (N-1)\sqrt{T} - 2\delta\sqrt{T}\right\} \end{aligned}$$

$$\subset \left\{ \inf_{u \leqslant v \leqslant u+\theta} q(v) > QT - N\sqrt{T} \right\} \subset \left\{ \inf_{u \leqslant v \leqslant u+\theta} q(v) > d \right\}.$$

We now consider the moments of $s_{u,\theta}$, for example, $\mathbf{E}_{\mathfrak{M}(u)} s_{u,\theta}^2$. On $\Omega_{u,\theta} \cap \Gamma_u^{N-1}$ we will have, by corollary 1 of lemma 2 of section 3.3

$$\mathbf{E}_{\mathfrak{M}(u)} s_{u,\theta}^2 = \mathbf{E}_{\mathfrak{M}(u)} (s_{u,\theta}^2; V_{e*} V_s) + \varepsilon\theta.$$

But here

$$I(\Gamma_u^{N-1}) \mathbf{E}_{\mathfrak{M}(u)} (s_{u,\theta}^2; V_{e*} V_s) = \mathbf{E}_{\mathfrak{M}(u)} I(\Gamma_u V_{e*}) \mathbf{E}_{(\mathfrak{M}(u) \mathfrak{E}^*(u+\theta))} (s_{u,\theta}; V_s). \qquad (3.75)$$

Using the independence of output in the form (3.9), applicable to random variables, we obtain that the right-hand side of (3.75) is

$$\mathbf{E}_{\mathfrak{M}(u)} I(\Gamma_u V_{e*}) \mathbf{E}_{\mathfrak{S}(u)} (s_{u,\theta}^2; V_s)$$

and, consequently, on $\Omega_{u,\theta} \cap \Gamma_u^{N-1}$

$$\begin{aligned} \mathbf{E}_{\mathfrak{M}(u)} s_{u,\theta}^2 &= r_s \theta + \mathbf{E}_{\mathfrak{S}(u)} (s_{u,\theta}^2; V_s) \mathbf{P}_{\mathfrak{M}(u)} V_{e*} \\ &= r_{ss} \theta + \mathbf{E}_{\mathfrak{S}(u)} s_{u,\theta}^2; \quad |r_s| < \varepsilon, \quad |r_{ss}| < \varepsilon. \end{aligned} \qquad (3.76)$$

Thus the principal part of $\mathbf{E}_{\mathfrak{S}(u)} s_{u,\theta}^2$ is an $\mathfrak{S}(u)$-measurable function. If $q(u)$ (or $\zeta(u) = (q(u) - QT)/T)$ is not measurable with respect to $\mathfrak{S}(u)$ (measurability holds, for example, for a deterministic $e^*(u)$), then this is possible only when $\beta_{ss}(z) = \beta_{ss} = \text{const.}$ in condition (II).

It is possible to make this not quite rigorous argument precise and to state the following

THEOREM 1. *Let the conditions of theorem 4 of section 3.3 be satisfied except, possibly, condition (II) on mixed moments. Assume that for each two points z_1, z_2 from the interval $(-N+1, N-1)$ there are $\mu = o(1)$ and a set $U_s \in \mathfrak{S}(u)$ of positive probability such that on U_s*

$$\mathbf{P}_{\mathfrak{S}(u)}([z_i] V_{e*}) > 0$$

where

$$[z_i] = \{z_i - \mu \leqslant \zeta(u) \leqslant z_i + \mu\} \Omega_{u,\theta} \in \mathfrak{M}(u).$$

Then, if the output system is independent in the domain $q \geqslant d$, $d \geqslant QT - N\sqrt{T}$, and the functions $\alpha_s(z)$ and $\beta_{ss}(z)$ are continuous, we can take

$$\alpha_s(z) = \text{const.}, \quad \beta_{ss}(z) = \text{const.}, \quad \beta_{e^*s} = 0.$$

Proof. By the independence of output

$$\mathbf{E}_{\mathfrak{S}(u)} (s_{u,\theta}^2; [z_i] V_{e*} V_s) = \mathbf{P}_{\mathfrak{S}(u)} [z_i] V_{e*} . \mathbf{E}_{\mathfrak{S}(u)} (s_{u,\theta}^2; V_s).$$

Consequently, on U_s, the ratio

$$\frac{\mathbf{E}_{\mathfrak{S}(u)}(s_{u,\theta}^2; [z_i]V_{e*}V_s)}{\mathbf{P}_{\mathfrak{S}(u)}[z_i]V_{e*}} \tag{3.77}$$

does not depend on i.

Consider the numerator in (3.77), i.e.

$$\begin{aligned} \mathbf{E}_{\mathfrak{S}(u)}(s_{u,\theta}; [z_i]V_{e*}V_s) &= \mathbf{E}_{\mathfrak{S}(u)}I([z_i]V_{e*}) \\ \mathbf{E}_{(\mathfrak{M}(u)\mathfrak{E}^*(u+\theta))}[s_{u,\theta}^2 &- \theta\beta_{ss}(\zeta(u)); V_s] + \theta\mathbf{E}_{\mathfrak{S}(u)}I([z_i]V_{e*}V_s)\beta_{ss}(\zeta(u)). \end{aligned} \tag{3.78}$$

Here, by the independence, the conditional expectation $\mathbf{E}_{(\mathfrak{M}(u)\mathfrak{E}^*(u+\theta))}$ in the first term may be replaced by $\mathbf{E}_{\mathfrak{M}(u)}$ (actually, even by $\mathbf{E}_{\mathfrak{S}(u)}$). Next, using corollary 1 of lemma 2 of section 3.3, we obtain a uniform estimate $o(\theta)$ for this conditional expectation. This means that the first term in (3.78) is equal to

$$o(\theta)\mathbf{P}_{\mathfrak{S}(u)}[z_i]V_{e*}.$$

Since $1 - \mathbf{P}_{\mathfrak{M}(u)}V_s \leqslant \varepsilon\theta/T$ on z_i, then, by continuity of β_{ss}, the second term in (3.78) is equal to

$$\begin{aligned} \theta\mathbf{E}_{\mathfrak{S}(u)}I([z_i]V_{e*})\beta_{ss}(\zeta(u))V_s &= \theta\mathbf{E}_{\mathfrak{S}(u)}I([z_i]V_{e*})\beta_{ss}(\zeta(u))(1 + o(1)) \\ &= \theta\beta_{ss}(z_i)\mathbf{P}_{\mathfrak{S}(u)}[z_i]V_{e*}(1 + o(1)) \end{aligned}$$

where the estimate $o(1)$ is uniform in ω.

From these expressions we form the sum (3.78) and put it in the ratio (3.77). Then this ratio takes the form

$$o(\theta) + \beta_{ss}(z_i) \tag{3.79}$$

where the estimate $o(\theta)$ is uniform in ω.

Assume now that the assertion of the theorem for β_{ss} is false. That is, there exist z_1, z_2 from the interval $(-N+1, N-1)$ and $c > 0$ such that

$$|\beta_{ss}(z_i) - \beta_{ss}(z_2)| > c.$$

We then obtain, obviously, a contradiction to the fact that (3.77) (or (3.79)) does not depend on i.

The constancy of $\alpha_s(z)$ is proved similarly.

Further, for mixed moments on $\Omega_{u,\theta}\Gamma_u^{N-1}$ we have (see again corollary 1 of lemma 2 of section 3.3; in subsequent formulae $x_{u,\theta}^0 = x_{u,\theta} - \alpha_x(\zeta(u))$)

$$\mathbf{E}_{\mathfrak{M}(u)}s_{u,\theta}^0 e_{u,\theta}^0 = o(\theta) + \mathbf{E}_{\mathfrak{M}(u)}e_{u,\theta}^{*0}I(\Gamma_u^{N-1}V_{e*})\mathbf{E}_{(\mathfrak{M}(u)\mathfrak{E}^*(u+\theta))}s_{u,\theta}^0 I(V_s) \tag{3.80}$$

where $\mathbf{E}_{(\mathfrak{M}(u)\mathfrak{S}^*(u+\theta))}$, by the independence of output, may be replaced by $\mathbf{E}_{\mathfrak{M}(u)}$. But on $\Omega_{u,\theta}\Gamma_u^{N-1}$ this newly obtained expectation is uniformly $o(\theta/\sqrt{T})$; substituting this in (3.80) we obtain

$$\mathbf{E}_{\mathfrak{M}(u)}s_{u,\theta}^0 e_{u,\theta}^{*0} = o(\theta) + o\left(\frac{\theta^2}{T}\right) = o(\theta). \quad \square$$

Remark 1. If, for example, in the conditions of theorem 4 of section 3.3 $a_s(z)$ is differentiable, then we have the same situation as in theorem 1 for $\alpha_x(z) = a'_s(Q)z$. From the independence of output it will then follow that

$$\alpha_x(z) = a'_s(Q) = 0.$$

Remark 2. Since $\zeta(u)$, like $q(u)$, is a discrete random variable, then as the sets $[z_i]$ in theorem 1 we can take the sets generated by one point (i.e. take $\mu = 0$). Then the continuity of α_s and β_{ss} becomes unnecessary. A certain inconvenience in this approach is that the points z_i here are 'moving', i.e. dependent on T.

We now consider the so-called *loaded systems*. By this is usually meant service systems with queues and homogeneous input and output streams (i.e. with independent inputs and outputs), for which $r(t) \equiv 0$, and the intensity of the input stream is close to the intensity of the output stream. In the language of condition (I) of the theorems of section 3.3 this means that the coefficients α_e and α_s (or a_e and a_s; in our case these coefficients are necessarily constant) are close to each other. Thus, to apply the theorems of section 3.3 to the study of loaded systems we must assume that the function $A(Q) = a_e(Q) - a_s(Q)$ (see section 3.2) does not depend on Q:

$$A(Q) = A \to 0$$

as $T \to \infty$.

The fact that $A(Q)$ is defined allows us to consider the system as a system with rough dependence of control on the queue length. If the initial value $q(0)$ turns out to be in the $\sqrt{T}$-neighbourhood of 0, then we may use theorem 5 of section 3.3. But if $q(0)$ is comparable to T, then it is better to apply theorem 1 of section 3.3 (together with remark 2). Here, naturally, the limit distribution of the queue length for loaded systems will be different depending on where $q(0)$ is situated.

Below we give two corollaries of the theorems of section 3.3 describing the behaviour of loaded systems.

Let

$$\zeta(u) = \frac{q(u)}{\sqrt{T}}, \quad r(u) \equiv 0.$$

THEOREM 2. *We consider a system with independent output in the domain $q \geq d = o(\sqrt{T})$, for which conditions* (I)–(III) *hold in the following form: on $\Omega_{u,\theta} \cap \{\varepsilon^* < \zeta(u) < N\}$ (for the definition of $\Omega_{u,\theta}$, N, and ε^* see, for example, section* 3.3).

$$\text{(I)}\quad \mathbf{E}_{\mathfrak{M}(u)} x_{u,\theta} = \theta a_x + \frac{\theta}{\sqrt{T}} r_x(\omega)$$

$$\text{(II)}\quad \mathbf{E}_{\mathfrak{M}(u)} (x_{u,\theta} - \theta a_x)^2 = \theta b_{xx} + \theta r_{xx}$$

$$\text{(III)}\quad \mathbf{E}_{\mathfrak{M}(u)} [(x_{u,\theta} - \theta a_x)^2; |x_{u,\theta} - \theta a_x| > \delta\sqrt{T}] < \varepsilon\theta$$

for any $\delta > 0$. *Here* $|r_x| < \varepsilon$, $|r_{xx}| < \varepsilon$, x *takes the values* e, s, *and the numbers* a_x *and* r_{xx}, *in general, depend on* T.

In addition, let condition (V) *of theorem* 1 *of section* 3.3, *on the absence of large jumps, and condition* (VI) *of theorem* 2 *of section* 3.3, *on the positivity of the probability of reflection from the domain* $\{\zeta \leqslant \varepsilon^*\}$, *both hold.*

Then, if

$$A\sqrt{T} = (a_e - a_s)\sqrt{T} \to A_0, \quad b_{ee} + b_{ss} \to B_0 > 0 \tag{3.81}$$

and the distribution of $\zeta(0) = q(0)/\sqrt{T}$ *converges to the distribution* p_0, *the process* $z(t) = q(tT)/\sqrt{T}$ *C-converges to a diffusion process* $w_R(t)$ *with coefficients* (A_0, B_0) *and reflection from the zero boundary.*

If, under the conditions of this theorem, we bound the length of the queue by $R\sqrt{T}$ (requests finding such a queue are refused) then, obviously, we obtain convergence of $q(tT)/\sqrt{T}$ to a diffusion process with coefficients (A_0, B_0) and with two reflecting boundaries 0 and R.

THEOREM 3. *Consider a system with independent output in the domain* $q \geqslant d = Q_0 T - N_0\sqrt{T}$, *for any fixed* $N_0 > 0$ *and some* $Q_0 > 0$. *Further, let conditions* (I)–(III) *of theorem* 2 *be satisfied in the domain* $\Omega_{u,\theta} \cap \{|\zeta(u)| < N\}$ *where* $\zeta(u)$ *denotes*

$$\zeta(u) = \frac{q(u) - Q_0 T}{\sqrt{T}}.$$

In addition, let condition (V) *of theorem* 1 *of section* 3.3, *on the absence of large jumps, hold.*

Then, if (3.81) *is satisfied, and the distribution of* $\zeta(0) = (q(0) - Q_0 T)/\sqrt{T}$, *for some* $Q_0 > 0$, *converges to the distribution* p_0, *the process* $z(t) = (q(tT) - Q_0 T)/\sqrt{T}$ *C-converges to an unbounded diffusion process with coefficients* (A_0, B_0).

In section 3.3 we have mentioned explicit formulae for the distribution of the limit processes, which arise in the conditions of theorems 2 and 3. From these formulae it follows that the stationary distribution for the limit process in theorem 2 exists only for $A_0 < 0$. Namely, from (3.34) it follows that

$$\lim_{v \to \infty} \mathbf{P}(w_R(v) > y) = e^{2A_0 y/B_0}. \tag{3.82}$$

Thus, for large v and T, the distribution of $z(v) = q(vT)/\sqrt{T}$ is close to (3.82). Since $A = a_e - a_s$ and $T^{-\frac{1}{2}}$ have the same order of smallness, this fact is often used only in the following form: for values $t \gg A^{-2}$

$$\mathbf{P}\left(q(t) > \frac{y}{|A|}\right) \approx e^{-2y/B_0}, \quad A < 0.$$

We now discuss examples of loaded systems. The most graphic and widespread is, apparently, an example of a system $\langle G_I, G_I, G_I, G_I \rangle$, in the notation of Borovkov (1976a). Example 1 of the introductory part of this chapter is devoted to the description of this system. We recommend the reader to look again at this example, so that we may use the notation there without further explanation. As already noted, this system has independent output in the domain $q \geqslant 1$.

The input stream $e(t)$ of this system is a generalized renewal process. As we saw in section 1.11, $\mathbf{E}e(t) \sim a_e t$ as $t \to \infty$, where $a_e = \mathbf{E}\nu^e/\mathbf{E}\tau^e$.

Service in systems $\langle G_I, G_I, G_I, G_I \rangle$ also occurs via a generalized renewal process

$$s^0(t) = \sum_{j=1}^{\eta_s(t)} \nu_j^s$$

(more precisely, with the help of 'pieces' of this process), so that $\mathbf{E}s^0(t) \sim a_s t$ as $t \to \infty$, where $a_s = \mathbf{E}\nu^s/\mathbf{E}\tau^s$.

If during the course of time t service was not broken off (the system remained busy), then we would also have $\mathbf{E}s(t) \sim a_s t$. Thus, in our case

$$A = a_e - a_s = \frac{\mathbf{E}\nu^e}{\mathbf{E}\tau^e} - \frac{\mathbf{E}\nu^s}{\mathbf{E}\tau^s}.$$

This equality means that for the application of theorems 2 and 3 we must consider the controlling sequence $\{\tau_j^e, \nu_j^e, \tau_j^s, \nu_j^s\}$ for our system as a series array, that is, depending on a parameter T.

Thus, let the distributions

$$\mathbf{P}(\tau_j^x < z) = F_x(z), \quad \mathbf{P}(\nu_j^x < z) = H_x(z)$$

(here and later x takes the values e, s) depend on T so that

$$A = \frac{\mathbf{E}\nu^e}{\mathbf{E}\tau^e} - \frac{\mathbf{E}\nu^s}{\mathbf{E}\tau^e} \sim \frac{A_0}{\sqrt{T}}.$$

(The index T, indicating this dependence, will be omitted.)

Suppose that ($x = e, s$)

$$\mathbf{E}\tau^x \geqslant c_x > 0 \tag{3.83}$$

$$\mathbf{D}\tau^x \geqslant c_x > 0, \quad \mathbf{E}[(\tau^x)^2; \tau^x \geqslant N)] \leqslant \varepsilon_x(N) \tag{3.84}$$

$$\mathbf{E}[(\nu^x)^2; |\nu^x| \geqslant N] < \varepsilon_x(N) \tag{3.85}$$

$$\sigma_x^2 = \frac{\mathbf{D}\nu^x}{\mathbf{E}\tau^x} + \frac{\mathbf{D}\tau^x(\mathbf{E}\nu^x)^2}{(\mathbf{E}\tau^x)^3} \geqslant c_x > 0 \tag{3.86}$$

$$\sigma_e^2 + \sigma_s^2 \to \sigma^2 > 0 \tag{3.87}$$

where c_x and $\varepsilon_x(N)$ do not depend on T, $\varepsilon_x(N) \to 0$ as $N \to \infty$. It is clear that for, say, the second inequality in (3.84) to hold, it is sufficient that $\sup_T \mathbf{E}|\tau^x|^{2+\delta} < \infty$ for some $\delta > 0$ or (more generally) that there exists a function $\psi(t)\uparrow\infty$ as $t \to \infty$ such that $\sup_T \mathbf{E}(\tau^x)^2\psi(\tau^x) < \infty$.

In theorem 2 of section 1.11 it was shown that when (3.83)–(3.86) hold each of the processes $e(t)$ and $s^0(t)$ satisfy conditions (I)–(III) of theorem 2 with sets $\Omega^x_{u,\theta} = \{\chi^x(u) \leqslant g\}$, where $\chi^x(u)$ is the size of the 'overshoot' (or 'excess') through the level u in the random walk with jumps $\tau^x_1, \tau^x_2, \ldots$:

$$\chi^x(u) = \tau^x_{\eta_x(u)} - u, \quad \eta_x(u) = \min\{k: \tau^x_1 + \cdots + \tau^x_k > u\}.$$

The number g can be chosen as the solution of the equation $g = \sqrt{T}\{\mathbf{E}[(\tau^x)^2;\ \tau^x > g]\}^{\frac{1}{3}}$. The σ-algebras $\mathfrak{S}(u)$ and $\mathfrak{S}(u)$ will coincide, respectively, with the σ-algebras generated by $\gamma^e(u)$ and $\tilde{\gamma}^s(u) = \inf\{v \geqslant 0: s(u - v + 0) - s(u - v - 0) > 0) > 0\}$;

$$\mathfrak{M}(u) = (\mathfrak{E}(u)\mathfrak{S}(u)).$$

In addition, in the same place, it was shown that $e(t)$ and $s^0(t)$ satisfy the condition on the absence of large jumps (condition (V) of theorem 1 of section 3.3. See the statement of theorem 2). Here, from the proof of theorem 2 of section 1.11, it follows that condition (V) and the sets $\Omega^x_{u,\theta}$ will have the homogeneity property (see (3.41)).

We now show that even the process $q(t) = e(t) - s(t)$ satisfies conditions (I)–(III). For this process, as before, without loss of generality, we can take the 'truncated' process for which values larger than $\bar{N}\sqrt{T}$ are replaced by $\bar{N}\sqrt{T}$, where $\bar{N} \to \infty$ as $T \to \infty$. If we establish the convergence of $q(tT)/\sqrt{T}$ to a diffusion process for this process, then we obtain that the trajectories of the initial and truncated processes coincide with probability approaching 1.

Consider, for example, condition (II) for the truncated process $q(t)$. On $\Omega_{u,\theta} \cap \{2\delta\sqrt{T} < q(u) < (N-1)\sqrt{T}\}$

$$\begin{aligned}\mathbf{E}_{\mathfrak{M}(u)} q^2_{u,\theta} &= \mathbf{E}_{\mathfrak{M}(u)}(q^2_{u,\theta};\ V_e V_s) + \mathbf{E}_{\mathfrak{M}(u)}(q^2_{u,\theta};\ \overline{(V_e V_s)}) \\ &= \mathbf{E}_{\mathfrak{M}(u)}((e_{u,\theta} - s^0_{u,\theta})^2;\ V_e V_s) + \bar{N}^2 T \mathbf{P}_{\mathfrak{M}(u)}(\bar{V}_e + \bar{V}_s).\end{aligned}$$

By corollary 1 of lemma 2 of section 3.3, this expression is equal to $o(\theta) + \mathbf{E}_{\mathfrak{M}(u)}(e_{u,\theta} - s^0_{u,\theta})^2 + \bar{N}^2 T o(\theta/T)$, where the estimates are uniform in ω. Hence, by the independence of $e_{u,\theta}$ and $s^0_{u,\theta}$ and for suitable choice of $\bar{N}$, we obtain

$$\mathbf{E}_{\mathfrak{M}(u)} q^2_{u,\theta} = \theta(\sigma^2_e + \sigma^2_s) + o(\theta).$$

Conditions (I) and (III) are verified similarly.

We note that here we deviate somewhat from the direct verification of conditions (I)–(III) of theorem 2, since in our case it is simpler to directly verify these conditions for $q(t)$ rather than for $e(t)$ and $s(t)$ (the difficulty lies in the verification for $s(t)$).

It remains for us to verify the reflection condition which is that in $\Omega_{u,\theta} \cap \{q(u) < \delta\sqrt{T}\}$

$$\mathbf{P}_{\mathfrak{M}(u)}(q_{u,\delta^2 T} > \delta\sqrt{T}) > p > 0.$$

This condition is obviously satisfied, since for $\theta_1 = \delta^2 T$, $s_{u,\theta_1} \leqslant s^0_{u,\theta_1}$, where s^0_{u,θ_1} is obtained from the trajectory $s_{u,t}$ by the suppression of idle intervals (that is, intervals when the service system is not operating), so that s^0_{u,θ_1}, like $s^0(t)$, is the increment of a generalized renewal process. But since the distribution of $(e_{u,\theta_1} - s^0_{u,\theta_1})/\sqrt{\theta_1}$ converges to a normal distribution with parameters (A_0, σ^2), then

$$\mathbf{P}_{\mathfrak{M}(u)}\{q_{u,\theta} > \delta\sqrt{T}\} \geqslant \mathbf{P}_{\mathfrak{M}(u)}\left\{\frac{e_{u,\theta_1} - s^0_{u,\theta_1}}{\sqrt{\theta_1}} > 1\right\} \to 1 - \Phi\left(\frac{1 - A_0}{\sigma}\right) > 0.$$

From all this it follows that a system $\langle G_I, G_I, G_I, G_I \rangle$, in a loaded state, satisfies the conditions of theorem 2 and, consequently, in accordance with this, the assertion of the theorem holds. That is, $q(tT)/\sqrt{T}$ will C-converge to a diffusion process with coefficients (A_0, σ^2) and with reflection from the zero boundary. □

This example is easily generalized to *multichannel* systems, when the input stream $e(t)$ is formed from the sum of a finite number l of generalized renewal processes (in general, distinct) of the same type as considered above, and service occurs in $m \geqslant 1$ channels, each of which is also controlled by a generalized renewal process. Such a system will have independent output in the domain $q > m$. The role of A_0 here is played by

$$A_0 = \lim_{T\to\infty} \left(\sum_{i=1}^{l} a_{ei} - \sum_{i=1}^{m} a_{si} \right) \sqrt{T}$$

where a_{ei} and a_{sj} are the means of a_e and a_s, introduced earlier, respectively for the ith input and jth output channels. The parameter σ^2 will be the limit of the sum $\sum_{i=1}^{l} \sigma^2_{ei} + \sum_{j=1}^{m} \sigma^2_{sj}$, with similar understandings regarding notation. Here, under conditions (3.83)–(3.87) on the input and output channels, relative to $q(tT)/\sqrt{T}$, the same result as in the single-channel case will hold. We note that for the final assertion condition (3.86) is superfluous (even for single-channel systems). It is enough to require only the positivity of the limit of the sum of the values σ^2 (see (3.87)). The truth of this remark the reader may verify for himself (in fact, in section 1.11, from where (3.83) is adopted, convergence to the limit process was proved for each of the components $(e(tT) - a_e tT)/\sqrt{T}$ and $s(tT) - a_s tT/\sqrt{T}$).

If, in these examples, the queue is bounded by the level $D\sqrt{T}$ (for example, requests finding a queue greater than this value are refused), then, in this case, according to theorem 3, we obtain convergence to a diffusion process with the same coefficients but with reflection from the boundaries 0 and D.

We may obtain a large number of other examples of service systems which satisfy the above theorems, if as input process $e(t)$ and process $s^0(t)$, controlling the service process $s(t)$ ($s(t)$ differs from $s^0(t)$ in that its trajectories contain intervals of constancy corresponding to the idle periods of the system), we consider the various processes mentioned as examples in sections 1.11 and 1.12, including input and output processes constructed with the help of independent stationarily connected random variables, etc. As shown in chapter 1, all of these processes (see sections 1.11 and 1.12) satisfy conditions of the type (I)–(III) and (V) of theorems 1 and 2 of section 1.5 (and, consequently, conditions of the same type in the theorems of section 3.3).

One of these examples is considered in more detail in the next section.

3.7. A numerical example

In order to some extent to illustrate the use of these results and their effectiveness for computation in real service systems we consider the following concrete problem.

In a certain communication system, through independent integer-valued intervals of time $\tau_1^e, \tau_2^e, \ldots$

$$\mathbf{P}(\tau_j^e = 1) = \tfrac{1}{3}, \quad \mathbf{P}(\tau_j^e = k+1) = \frac{1}{3\,.\,2^{k-1}}, \quad k = 1, 2, \ldots$$

$$(\mathbf{E}\tau^e = \tfrac{7}{3}, \mathbf{E}(\tau^e)^2 = 7)$$

communications (requests) are received in groups of sizes

$$\nu_1^e = \tau_1^e + \nu_1, \quad \nu_2^e = \tau_2^e + \nu_2, \ldots$$

where the ν_j are independent (among themselves and from $\{\tau_j^e\}$),

$$\mathbf{P}(\nu_j = -1) = \mathbf{P}(\nu_j = 0) = \mathbf{P}(\nu_j = 1) = \tfrac{1}{3}$$

$$(\mathbf{E}\nu = 0, \mathbf{E}\nu^2 = \tfrac{2}{3}).$$

The processing of communications (service of requests) is carried out by two independent mechanisms (two channels); here the transmission of the kth communication takes time $\tau_k^s = \frac{1}{2} + \tau_k$ (that is, the duration of service is equal to the length of the communication), where $\{\tau_k\}$ does not dependent on $\{\tau_j^e\}$ and $\{\nu_j\}$ and is made up of independent variables

$$\mathbf{P}(\tau_k > t) = \mathrm{e}^{-\frac{3}{4}t} \quad (\mathbf{E}\tau^s = \tfrac{11}{6}, \mathbf{D}\tau^s = \tfrac{16}{9}).$$

If the number of communications accumulating in the queue reaches 20, then communications joining at this time go to another system (to this system they are refused).

Obviously this is a system with independent output in the domain $q \geqslant 2$.

Suppose that the queue at $t = 0$ is $q(0) = 0$. Our interest is in the distribution of $q(u)$ for the value $u = 600$, the stationary distribution of $q(u)$, and also the probability that $q(u)$, in the time interval from 0 to 600, reaches the level 20 at least once.

The input stream $e(t)$ here (in the notation of example 1) is

$$e(t) = \sum_{j=1}^{\eta_e(t)} v_j^e = t + \chi(t) + \sum_{j=1}^{\eta_e(t)} v_j \tag{3.88}$$

where

$$\eta(t) = \min(k: \tau_1^e + \cdots + \tau_k^e > t)$$

$$\chi(t) = \tau_1^e + \cdots + \tau_{\eta(t)}^e - t$$

(so that the summation in (3.88) is up to $\eta_e(t)$ (which is more convenient) and not up to $\eta_e(t) - 1$, we suppose that the first group of requests join the system at time 0, not τ_1^e). As remarked in subsection 1.11.4, such a process $e(t)$ satisfies all the necessary conditions of theorems 2 and 3 of the previous section; here a_e and σ_e^2 (the coefficients of θ in the expressions for $\mathbf{E}_{\mathfrak{M}(u)} e_{u,\theta}$ and $\mathbf{E}_{\mathfrak{M}(u)}(e_{u,\theta} - a_e\theta)$ respectively) will be

$$a_e = \frac{\mathbf{E}\tau^e + \mathbf{E}v}{\mathbf{E}\tau^e} = 1, \quad \sigma_e^2 = \frac{\mathbf{D}v}{\mathbf{E}\tau^e} + \frac{(\mathbf{E}v)^2\mathbf{D}\tau^e}{(\mathbf{E}\tau^e)^3} = \frac{\mathbf{D}v}{\mathbf{E}\tau^e} = \frac{2}{7}.$$

The service process is controlled by two renewal processes $\eta_s^1(t)$ and $\eta_s^2(t)$, generated by two independent sequences of random variables distributed as $\{\tau_k^s\}$. The service processes $s(t)$ considered at the end of the previous section were of the same type. They also satisfy all the conditions of theorems 2 and 3 of section 3.6; here the coefficients a_s and σ_s^2 are

$$a_s = \frac{2}{\mathbf{E}\tau^s} = \frac{12}{11}, \quad \sigma_s^2 = 2\frac{\mathbf{D}\tau^s}{(\mathbf{E}\tau^s)^3} = \frac{2 . 16(6)}{9 . (11)^3} = \frac{32 . 24}{(11)^3}.$$

We have obtained that in our case the number A is

$$A = a_e - a_s = -\tfrac{1}{11}.$$

We try to consider our system as an element of a sequence of systems, for which $A \to 0$. The queue length for us is bounded by the value 20. Therefore we must use a theorem on loaded systems of the type of theorem 3 of section 3.3 (on convergence to a diffusion with two boundaries) (see also the remark to theorem 2 of section 3.6). Since there is some arbitrariness in the choice of parameter T, then we can, without loss of generality, suppose that the second reflecting boundary is at the level $R = 1$. Then, since $R\sqrt{T} = 20$, we obtain $T = 400$.

Consequently, using the theorem mentioned, the distribution of

$$z(t) = \frac{q(tT)}{\sqrt{T}} = \frac{q(400t)}{20} \tag{3.89}$$

is close (if $T = 400$ is 'sufficiently large') to the distribution of a diffusion process $w_{RR}(t)$ with initial value $w_{RR}(0) = 0$, reflection from the boundaries 0 and 1, and with coefficients

$$A_0 = A\sqrt{T} = -\tfrac{20}{11} \approx -1.82$$

and

$$\sigma^2 = \sigma_e^2 + \sigma_s^2 = \frac{2}{7} + \frac{24.32}{11^3} \approx 0.867.$$

In section 3.3, in the remark to theorem 6, there is given an explicit form for the transition for such processes. Putting $\alpha = A_0/\sigma^2$, we obtain the following expression for the density $f_t(y)$ of the distribution of $w_{RR}(t)$ $(w_{RR}(0) = 0, R = 1)$:

$$f_t(y) = \frac{2\alpha\, e^{2\alpha y}}{e^{2\alpha} - 1} + 2\exp\left\{-\frac{\alpha^2\sigma^2 t}{2} + \alpha y\right\} \sum_{k=1}^{\infty} e^{-(t\sigma^2/2)(\pi k)^2}$$

$$\times \frac{\pi k}{\alpha^2 + (\pi k)^2}(\pi k \cos \pi k y + \alpha \sin \pi k y).$$

Integrating this expression we arrive at the distribution function

$$F_t(y) = \frac{e^{2\alpha y} - 1}{e^{2\alpha} - 1} + 2\exp\left\{-\frac{\alpha^2\sigma^2 t}{2} + \alpha y\right\} \sum_{k=1}^{\infty} e^{-(t\sigma^2/2)(\pi k)^2}$$

$$\times \frac{\pi k}{\alpha^2 + (\pi k)^2} \sin \pi k y.$$

Hence it is clear, in particular, that as $t \to \infty$ $w_{RR}(t)$ has stationary distribution

$$\frac{e^{2\alpha y} - 1}{e^{2\alpha} - 1}$$

(a truncated exponential distribution) and that the speed of convergence to this distribution is very high. (The correction term does not exceed $\exp\{-\tfrac{1}{2}(\alpha^2 + \pi^2)\sigma^2 t\}$.)

In our case ($\alpha = A_0/\sigma^2 \approx -2.10$) this works out at

$$\exp\left\{-\frac{14.3 \cdot 0.867}{2} t\right\} = \exp(-6.20t).$$

Consequently we may anticipate that the distribution of $q(600)$, corresponding to the value $t = 1.5$ (see (3.89)), will be close to stationary.

We note further that the limit distribution calls for a nice description of the 'regular' domain of q, where the conditions of the limit theorems act on proper behaviour of the moments of the increments. Therefore it is difficult to expect that the approximation of the limit distribution will be good for the values $q = 0$ and $q = 1$, where the functioning laws of the system, acting in $q \geqslant 2$, are broken. From this point of view we may join the states $q = 0$, $q = 1$, and $q = 2$ into one and consider a new variable $\tilde{q}(u) = 1$, if $q(u) = 0$, 1, or 2 and $\tilde{q}(u) = q(u) - 1$ if $q(u) > 2$, and approximate this new variable $\tilde{q}$ by the process w_{RR}. The variable $\tilde{q}$

takes values from 1 up to 19 and the normalized value $\tilde{z}(t) = \tilde{q}(tT)/T = \tilde{q}(400t)/20$ will be in $[0.05, 0.95]$. Consequently we must 'contract' the limit distribution from $[0, 1]$ to this domain. But the value $\tilde{q}(t) = 1$ is taken with positive probability and it is natural to put in correspondence not $F_t(0) = 0$ but the quantity $F_t(y_0)$, for some positive value y_0. There is certainly some arbitrariness in the choice of y_0 which can be removed in various ways. The arbitrariness is also closely connected with the fact that we have to approximate a discrete distribution concentrated on a relatively small number of points by a continuous distribution. We consider here the version when $y_0 = 1/19$ and the distribution function $\mathbf{P}(\tilde{z}(t) \leqslant y)$ is approximated by*

$$\tilde{F}_t(y) = \begin{cases} F_t(20y/19) & 0.05 \leqslant y \leqslant 0.95 \\ 0 & y < 0.05. \end{cases} \tag{3.90}$$

Below we give a table comparing the empirical distribution of $\tilde{z}(t) = \tilde{q}(400t)/20$ for $t = 1.5$ (i.e. for non-normalized time $u = 600$) with the distribution $\tilde{F}_t(y)$ for the value $t = \infty$. The empirical distribution function is obtained for $n = 400$ observations so that the maximal standard deviation of the empirical distribution function is

$$\sqrt{\frac{1}{4.400}} = \frac{1}{40} = 0.025.$$

The symbol $\mathbf{P}_n$ denotes the empirical probability. For example, $\mathbf{P}_n(\tilde{q}(600) \leqslant k)$ is the relative frequency of the event $\tilde{q}(600) \leqslant k$ in n observations.

The table compares empirical values $P_{n,k} = \mathbf{P}_n(\tilde{q}(600) \leqslant k) = \mathbf{P}_n(\tilde{z}(1.5) \leqslant k/20)$ ($n = 400$) with the values $\pi_k = \tilde{F}_\infty(k/20) = F_\infty(k/19)$ in the limit of the distribution function (3.90).

k	$P_{n,k}$	π_k	k	$P_{n,k}$	π_k	k	$P_{n,k}$	π_k
1	0.25	0.20	8	0.80	0.84	14	0.96	0.97
2	0.37	0.36	9	0.86	0.87	15	0.98	0.98
3	0.46	0.49	10	0.88	0.90	16	0.99	0.98
4	0.54	0.59	11	0.91	0.92	17	0.99	0.99
5	0.65	0.68	12	0.93	0.94	18	1.00	1.00
6	0.68	0.74	13	0.95	0.96	19	1.00	1.00
7	0.75	0.80						

The probability that $\max_{0 \leqslant u \leqslant 1.5} q(uT) < 20$ can be found approximately by the 'principle of invariance' and the results of Kac (1947), where random walks with two boundaries, one reflecting ($q = 0$) and one absorbing ($q = 20$), are studied.

* Here as distribution function of ξ we take $\mathbf{P}(\xi \leqslant y)$, which includes the probability of the value y.

CHAPTER 4

Stability Theorems

As already mentioned in the introduction it is natural to associate with asymptotic methods in queuing theory also stability (or continuity) theorems for various types of systems. These theorems are also limit theorems for random processes, but, in contrast to the results of chapters 2 and 3, they are not of a 'collective' nature. Stability theorems provide conditions, in a given service system, under which small deviations in the distribution of a given concrete controlling sequence lead to small deviations in the distribution of stationary (or prestationary) characteristics of the system which are under discussion (for example, stationary waiting time, probability of refusal, etc.). In other words, we give conditions here under which closeness in some sense or other of the controlling sequences leads to closeness of the characteristics being studied. Theorems of this type allow one to use different kinds of approximation of individual systems with the help of others.

Amongst the first works on stability of queuing systems, apparently, is the work of Rossberg (1965) and the works of Gnedenko (1970) and Franken (1970). Somewhat later, articles of Kennedy (1972), Borovkov (1972a, e), Stoyan (1972), and Kalashnikov and Tsitsiashvili (1972) appeared; all of these contain various formulations and methods in the stability problem. Now the main directions of research may, evidently, be regarded as developed, however, the number of papers on stability has become so large that an account of their contents within the limits of one book is impossible. We note here only the main directions of research in this area, having in view first of all the stability of the *stationary* characteristics of the service system.

The works of Franken (1970, 1978) and Arndt (1978) use the theory and technique of random marked point processes. A systematic exposition of this approach can be found in Franken *et al.* (1981).

Zolotarev (1975b, 1976) considered the problem of stability as a problem of continuity arising for mappings of one metric space to another (in our case—spaces connected with the controlling sequence to a space connected with the characteristic of the system being studied).

Kalashnikov, for the analysis of stability problems, applied the so-called method of trial functions, closely connected with the classical Lyapunov method for investigation of stability for differential equations. Along with the papers by Kalashnikov and Tsitsiashvili (1972) and Kalashnikov (1977), the book by Kalashnikov (1978) is devoted to an exposition of this method.

In the papers of D. Stoyan (1972) and H. Stoyan (1973) (see also Delbrouch, 1973; Kac, 1947) the problems of stability are connected with certain order relations for controlling sequences.

We mention also the work of Kennedy (1972), Whitt (1974b), and many other authors (see Cohen, 1973; Iglehart, 1973; Whitt, 1974a) who have considered stability of systems only on finite intervals of time (that is, stability of prestationary characteristics). This problem is simpler by comparison with one for stationary characteristics because in the first case there is usually known an explicit form of the mappings which determine the characteristics under discussion as a function of the controlling sequences. (In a number of these papers (for example Whitt, 1974b) closeness of service processes in $D(0, \infty)$ is discussed, but with the topology of convergence in $D(0, t)$ (with the usual Skorokhod topology) for any fixed t (see Lindvall, 1973; Stone, 1963; Whitt, 1970).)

For some service systems it is also possible to use the approach of the analysis of explicit formulae or equations which describe the behaviour of the stationary characteristics (see, for example, Rossberg, 1965; Borovkov, 1976a). However, such an analysis can be carried out only for a very limited circle of service systems. At the same time numerical estimates of the speed of convergence, obtained in this way, turn out to be very precise (see section 4.3).

In this chapter we set forth the approach which seems to us to be very natural and which applied to many types of system, could be called the 'renewal method'. This method is first illustrated (sometimes implicitly) in a series of examples in sections 4.2–4.5 and then stated in general form in section 4.6. It must be noted that for systems with an infinite number of service channels we use a direct analysis of the indicator representation and the title 'renewal method' is not justified. All the same for convenience and simplicity of exposition we will use this terminology for the approach presented here.

The foundation of this approach was laid in Borovkov (1972d) (see also Sakhanenko, 1974) in which were studied functionals of processes and sequences given on the whole axis (see also the works of the author at that time (Borovkov, 1972a, e; 1976a), devoted directly to stability theorems for service processes). These functionals have, roughly speaking, the property that on sets A_N, with $\mathbf{P}(A_N)\to 1$ ad $N\to\infty$, they depend only on a finite number N of elements of the controlling sequence, where the events A_N are defined by the behaviour of the 'tails' of these sequences beginning at N. But it is just these functionals which are the stationary characteristics of the service system (in the presence of 'renewals'); the typical, and simplest, of which is the supremum $\sup_{k\geqslant 0} X_k$, where X_k is the cumulative sum of the elements of the controlling sequence with negative mean.

In section 4.1 we give subsidiary results concerning the stability of the distribution of the supremum of a sum of stationarily connected variables and the supremum of a process with stationary increments. Immediate corollaries of these results are theorems on stability of stationary waiting time for single-channel systems with queues and systems with autonomous service, presented in section 4.2. The results of these sections are in essence final.

In section 4.4 we consider stability theorems for single-channel systems with a bounded queue (bounded waiting time).

Section 4.5 is devoted to systems with an infinite number of service channels.

In section 4.6 we give a general method of obtaining ergodic theorems and stability theorems, called in this book the 'renewal method'. In section 4.7 this method is applied to the investigation of multichannel systems with queues and with refusals.

To compare the renewal method with other approaches is fairly difficult (sometimes the comparison seems artificial, since essentially the same operations are being carried out), and yet on the basis of the results obtained it seems to us at present to be the most preferable. In addition, the existence of renewing events is a necessary condition for ergodicity (see theorem 1 of section 4.6), if by convergence to stationary distribution is meant strong convergence, defined in (4.58) (or 4.58a)). Connected with these facts is the situation that often the results obtained by the renewal method turn out to be unimprovable. Some exceptions are, perhaps, the qualitative questions of ergodicity and stability for multichannel systems (section 4.7), where our specially constructed renewing events are rather simple and, apparently, 'rough'. Most likely this explains the fact that the conditions obtained for ergodicity and stability contain a condition explicitly connected with the method (more precisely, with the particular construction of renewing events)—this is the unboundedness condition on τ_j^e in theorem 3 of section 4.7.

For brevity in the above mentioned sections of this chapter we have considered principally only *qualitative* questions of stability of *stationary* characteristics of service systems. However, the renewal method gives the same success in the study of stability of *prestationary* characteristics (as already remarked, these are noticeably simpler), and also in obtaining *quantitative* estimates of the speed of convergence. The estimates obtained in this way turn out in a well-known sense to be unimprovable or close to unimprovable. Both of these circumstances are illustrated in section 4.3, in which, for single-channel systems, estimates of the speed of convergence are obtained both for stationary and prestationary characteristics, and also at the end of section 4.7, where the results of Akhmarov (1979a, b, c, d), which have used the renewal method, are stated. These estimates are either unimprovable or close to it.

To the merits of the renewal method noted above, we must also add the qualities connected with the methodological side of the matter and, in particular, that it gives us a common approach to the proof of stability and ergodic theorems which, in essence, are closely connected.

The results of sections 4.1, 4.3, 4.4, 4.5, 4.6, and 4.7 are contained completely, or partially, in the works of Borovkov (1972b, d, e, 1975a, 1976a, 1977b, 1978a, b).

4.1. Subsidiary results on the distribution of the maximum of sequences of sums of stationarily related variables

Consider a series of stationary sequences $\xi^{(r)} = \{\xi_k^{(r)};\ -\infty < k < \infty\}, r = 1, 2, \ldots$ and a stationary sequence $\xi = \{\xi_k;\ -\infty < k < \infty\}$. We denote

$$Y_{m,k}^{(r)} = \sum_{i=m-k}^{m-1} \xi_i^{(r)}, \quad Y_{m,k} = \sum_{i=m-k}^{m-1} \xi_i$$

$$Y_{m,0}^{(r)} = Y_{m,0} = 0, \quad Y_m^{(r)} = \sup_{k \geqslant 0} Y_{m,k}^{(r)}, \quad Y(m) = \sup_{k \geqslant 0} Y_{m,k}$$

(we consider here the maximum of sequences of sums, which extend *to the left*, i.e. a sum of the form $\zeta_{m-1} + \zeta_{m-2} + \cdots$, since it is just such sums which arise in the problems in queuing theory; see Borovkov, 1976a).

THEOREM 1. *Let the following conditions be satisfied.*

(I) *The sequence ξ is ergodic,* $\mathbf{E}\xi_1 < 0$.

(II) *The finite-dimensional distributions of $\xi^{(r)}$ converge weakly, as $r \to \infty$, to the finite-dimensional distributions of ξ.*

(III) $\mathbf{E}(\xi_1^{(r)}; \xi_1^{(r)} \geqslant 0) \to \mathbf{E}(\xi_1; \xi_1 \geqslant 0) < \infty$.

Then the finite-dimensional distributions of the sequence $\{Y^{(r)}(n);\ -\infty < n < \infty\}$ converge weakly to the finite-dimensional distributions of $\{Y(n);\ -\infty < n < \infty\}$.

Remark. Conditions (I)–(III) imply convergence, as $r \to \infty$, of the distributions of a whole class of functionals on $\{Y_{0,k}^{(r)};\ k \geqslant 0\}$ (the so-called V-continuous functionals, see Borovkov (1972d) and Sakhanenko (1974)). Condition (III) is a compactness condition on the sequence of distributions of these functionals.

If condition (III) is not satisfied, then the result of the theorem, as is easy to verify, will not hold. The condition

$$\mathbf{E}\xi_1^{(r)} \to \mathbf{E}\xi_1$$

is also insufficient for the convergence of the distributions of $Y^{(r)}(n)$. The following example shows this. Let the sequence $\{\xi_k^{(r)}\}$ $r = 1, 2, \ldots$ be given on one probability space and be made up of independent variables $\xi_k^{(r)}$ taking four values $-2r, -1, 1, 2r$ with probabilities

$$\frac{1}{r}, \quad \frac{3}{4} - \frac{1}{r}, \quad \frac{1}{4} - \frac{1}{r}, \quad \frac{1}{r}$$

respectively. In this case

$$\mathbf{E}\xi_1^{(r)} = -\tfrac{1}{2} = \mathbf{E}\xi_1, \quad \mathbf{P}(Y(0) < \infty) = 1.$$

At the same time $\liminf_{r\to\infty} \mathbf{P}(Y^{(r)}(0) \geqslant r) > 0$. Indeed, let A_r be the event that among the r values $\xi^{(r)}_{-1}, \ldots, \xi^{(r)}_{-r}$ the value $-2r$ does not appear and the value $2r$ appears at least once. Then

$$A_r \subset \{Y^{(r)}(0) \geqslant r\}$$

$$\mathbf{P}(A_r) = \left(1 - \frac{1}{r}\right)^r \left(1 - \left(1 - \frac{1}{r}\right)^r\right) \to \mathrm{e}^{-1}(1 - \mathrm{e}^{-1}) > 0.$$

There are some additional remarks to theorem 1 at the end of the section.

We move on to the proof of the theorem. For this we need two auxiliary propositions.

LEMMA 1. *Let ξ be an arbitrary stationary sequence, $Y = Y(0)$. Then for any $c > 0$*

$$\mathbf{P}(Y > 0) \leqslant \frac{1}{c}\mathbf{E}(\xi_1; \xi_1 \geqslant 0) + \mathbf{P}(\xi_1 > -c).$$

Proof. We use the inequality (Doob, 1953, p. 418)

$$\mathbf{E}(\xi_{-1}; Y > 0) \geqslant 0 \tag{4.1}$$

valid under the conditions of the theorem and $\mathbf{E}|\xi_1| < \infty$.

First consider the restricted class of sequences for which $\mathbf{E}|\xi_{-1}| < \infty$ and $\mathbf{P}(\xi_{-1} \in (-c, 0)) = 0$ for some $c > 0$. Then, by (4.1),

$$\mathbf{E}(\xi_{-1}; \xi_{-1} \geqslant 0, Y > 0) \geqslant -\mathbf{E}(\xi_{-1}; \xi_{-1} \leqslant -c, Y > 0)$$
$$\geqslant c\mathbf{P}(\xi_{-1} \leqslant -c, Y > 0).$$

Consequently

$$\mathbf{P}(Y > 0) = \mathbf{P}(\xi_{-1} \geqslant 0, Y > 0) + \mathbf{P}(\xi_{-1} \leqslant -c, Y > 0)$$
$$\leqslant \mathbf{P}(\xi_{-1} \geqslant 0) + \frac{1}{c}\mathbf{E}(\xi_{-1}; \xi_{-1} \geqslant 0). \tag{4.2}$$

Now let the probability of a hit in the interval $(-c, 0)$ be positive. We introduce the random variable

$$\xi_j^* = \begin{cases} \xi_j & \text{if } \xi_j \notin (-c, 0) \\ 0 & \text{if } \xi_j \in (-c, 0). \end{cases}$$

Then, using (4.2) and putting $Y_k^* = \sum_{j=1}^k \xi^*_{-j}$, $Y^* = \sup_{k \geqslant 0} Y_k^*$ we may write

$$\mathbf{P}(Y > 0) \leqslant \mathbf{P}(Y^* > 0) \leqslant \mathbf{P}(\xi^*_{-1} \geqslant 0) + \frac{1}{c}\mathbf{E}(\xi^*_{-1}; \xi^*_{-1} \geqslant 0)$$

$$= \mathbf{P}(\xi_{-1} > -c) + \frac{1}{c}\mathbf{E}(\xi_{-1}; \xi_{-1} \geqslant 0).$$

It remains to remove the restriction $\mathbf{E}\xi_{-1} > -\infty$. This is done by a device quite analogous to that just used. We need to introduce the random variable $\xi_j^{**} = \max(\xi_j, -A)$ for $A > 0$, and use $\mathbf{P}(Y > 0) \leqslant \mathbf{P}(Y^{**} > 0)$ (with the obvious meaning for Y^{**}). The lemma is proved. □

In what follows we will put

$$x^+ = (x)^+ = \max(x, 0).$$

LEMMA 2. *If $(\xi_1^{(r)}, \ldots, \xi_L^{(r)})$ is a vector whose distribution as $r \to \infty$ weakly converges to the distribution of $(\xi_1, \ldots, \xi_L)$ and*

$$\mathbf{E}(\xi_j^{(r)})^+ \to \mathbf{E}(\xi_j)^+, \quad j = 1, \ldots, L \tag{4.3}$$

then, for the sum $Y_L^{(r)} = \sum_{j=1}^L \xi_j^{(r)}$ we have

$$\mathbf{E}(Y_L^{(r)})^+ \to \mathbf{E}(Y_L)^+.$$

Proof. Obviously it is sufficient to verify the lemma for $L = 2$. Since

$$\liminf_{r\to\infty} \mathbf{E}(Y_2^{(r)})^+ \geqslant \mathbf{E}(Y_2)^+$$

then it is necessary for us to show that

$$\limsup_{r\to\infty} \mathbf{E}(Y_2^{(r)})^+ \leqslant \mathbf{E}(Y_2)^+. \tag{4.4}$$

Denoting the distribution functions of $(\xi_1^{(r)}, \xi_2^{(r)})$ and (ξ_1, ξ_2) by $F^{(r)}(x, y)$ and $F(x, y)$, respectively, we have

$$\mathbf{E}(Y_2^{(r)})^+ - \mathbf{E}(Y_2)^+ = \int_{x+y\geqslant 0} (x + y)\, \mathrm{d}(F^{(r)}(x, y) - F(x, y)).$$

Using the convergence $F^{(r)} \Rightarrow F$, we obtain for any fixed point of continuity of $F(x, \infty)$:

$$\limsup_{r\to\infty} \int_{x+y\geqslant 0} x\, \mathrm{d}(F^{(r)}(x, y) - F(x, y))$$

$$\leqslant \limsup_{r\to\infty} \int_{\substack{0\leqslant x\leqslant M\\ x+y>0}} + \limsup_{r\to\infty} \int_{\substack{x>M\\ x+y>0}} \leqslant \limsup_{r\to\infty} \int_{x>M} x\, \mathrm{d}F^{(r)}(x, y)$$

$$= \limsup_{r\to\infty} \mathbf{E}(\xi_1^{(r)}; \xi_1^{(r)} > M) \equiv \varepsilon(M).$$

But the uniform, with respect to r, convergence $\mathbf{E}(\xi_1^{(r)}; \xi_1^{(r)} > M) \to 0$ as $M \to \infty$ is, as is easy to see, a necessary and sufficient condition for (4.3). Therefore

$\varepsilon(M) \to 0$ as $M \to \infty$ and, consequently,

$$\limsup_{r\to\infty} \int_{x+y\geqslant 0} x \,\mathrm{d}(F^{(r)}(x, y) - F(x, y)) \leqslant 0$$

$$\limsup_{r\to\infty} \int_{x+y\geqslant 0} y \,\mathrm{d}(F^{(r)}(x, y) - F(x, y)) \leqslant 0.$$

(4.4), and with it lemma 2, is proved. □

Proof of the theorem. Denote $\Psi(N) = \mathbf{P}(\sup_{k>N} Y_{m,k} \geqslant 0)$ and assume that

$$\Phi(N) = \limsup_{r\to\infty} \mathbf{P}(\sup Y^{(r)}_{m,k} \geqslant 0) \to 0 \tag{4.5}$$

as $N \to \infty$ (by stationarity, this probability does not depend on m). By condition (4.1) and the strong law of large numbers $\Psi(N) \to 0$ as $N \to \infty$ and for a given $\varepsilon > 0$ we may choose N so that $\Psi(N) < \varepsilon$, $\Phi(N) < \varepsilon$. Now let $(x_1, \ldots, x_l)$ be a point of continuity of the joint distribution of $\bar{Y}_{m_1,N}, \ldots, \bar{Y}_{m_l,N}$ for the chosen N, where

$$\bar{Y}_{m,N} = \max_{0\leqslant k\leqslant N} Y_{m,k}.$$

Then, for the probability of $A^{(r)} = \{Y^{(r)}(m_1) \geqslant x_1, \ldots, Y^{(r)}(m_l) \geqslant x_l\}$, we will have, by condition (II),

$$\begin{aligned}
\limsup_{r\to\infty} \mathbf{P}(A^{(r)}) &\leqslant l\varepsilon + \limsup_{r\to\infty} \mathbf{P}\left(A^{(r)}; \sup_{k>N} Y^{(r)}_{m_1,k} < 0, \ldots, \sup_{k>N} Y^{(r)}_{m_l,k} < 0\right) \\
&\leqslant l\varepsilon + \limsup_{r\to\infty} \mathbf{P}(\bar{Y}^{(r)}_{m_1,N} \geqslant x_1, \ldots, \bar{Y}^{(r)}_{m_l,N} \geqslant x_l) \\
&= l\varepsilon + \mathbf{P}(\bar{Y}_{m_1,N} \geqslant x_1, \ldots, \bar{Y}_{m_l,N} \geqslant x_l) \\
&\leqslant 2l\varepsilon + \mathbf{P}(Y(m_1) \geqslant x_1, \ldots, Y(m_l) \geqslant x_l).
\end{aligned} \tag{4.6}$$

Obtaining, in a similar way, the converse inequality for lim inf we establish thereby, by the arbitrariness of $\varepsilon > 0$, the required convergence of the finite-dimensional distributions of the sequence $\{Y^{(r)}(n);\ -\infty < n < \infty\}$.

Thus for the proof of the theorem we must show that (4.5) holds. As already noted, the distribution of $\{Y_{m,k};\ k \geqslant 0\}$ does not depend on m. Therefore we will assume for simplicity that $m = 0$ and denote

$$Y^{(r)}_k = Y^{(r)}_{0,k} = \xi^{(r)}_{-1} + \cdots + \xi^{(r)}_{-k}$$

$$Y_k = Y_{0,k} = \xi_{-1} + \cdots + \xi_{-k}, \quad Y = Y(0) = \sup_{k\geqslant 0} Y_k.$$

For arbitrary $\beta > 0$ and integer $L > 0$ we have

$$\mathbf{P}\Big(\sup_{k\geqslant N} Y_k^{(r)}\geqslant 0\Big)\leqslant \mathbf{P}(Y_N^{(r)}\geqslant -\beta N)+\mathbf{P}\Big(Y_N^{(r)}<\beta N, \sup_{k\geqslant N} Y_k^{(r)}\geqslant 0\Big)$$

$$\leqslant \mathbf{P}\Big(Y_N^{(r)}<-\beta N; \sup_{k\geqslant N} Y_k^{(r)}\geqslant 0; \bigcap_{k=0}^{\infty}\{Y_{N+kL}^{(r)}-Y_N^{(r)}\leqslant -\beta kL\}\Big)$$

$$+\mathbf{P}\Big(\bigcup_{k=0}^{\infty}\{Y_{N+kL}^{(r)}-Y_N^{(r)}>-\beta kL\}\Big)+\mathbf{P}(Y_N^{(r)}\geqslant -\beta N).$$

We denote the terms on the right-hand side of the last inequality, according to their order of appearance, by I_1, I_2, and I_3.

Then, by stationarity of $\xi^{(r)}$ and lemma 1 (for $c = \beta L$),

$$I_2=\mathbf{P}\Big(\sup_{k\geqslant 0}(Y_{kL}^{(r)}+\beta kL)>0\Big)\leqslant \mathbf{P}\Big(\frac{Y_L^{(r)}}{L}>-2\beta\Big)+\frac{1}{\beta}\mathbf{E}\Big(\frac{Y_L^{(r)}+\beta L}{L}\Big)^{+}$$

$$I_1\leqslant \mathbf{P}\Big(\bigcup_{j=0}^{\infty}\bigcup_{k=jL+1}^{(j+1)L}\Big\{\xi_k^{(r)}>\frac{N\beta+jL\beta}{L}\Big\}\Big)\leqslant L\sum_{j=0}^{\infty}\mathbf{P}\Big(\xi_1^{(r)}>j\beta+\frac{N\beta}{L}\Big)$$

$$\leqslant \frac{L}{\beta}\int_{(N\beta/L)-\beta}^{\infty} P^{(r)}(t)\,dt$$

where $P^{(r)}(t) = \mathbf{P}(\xi_1^{(r)} > t)$. The convergence $\mathbf{E}(\xi_1^{(r)}; \xi_1^{(r)} > 0) \to \mathbf{E}(\xi_1; \xi_1 > 0)$ implies uniform, with respect to r, convergence of $\phi^{(r)}(u) = \int_u^\infty P^{(r)}(t)\,dt$ as $u \to \infty$. Therefore there is a function $\phi(u) \to 0$ as $u \to \infty$ such that

$$I_1 \leqslant \frac{L}{\beta}\phi\Big(\frac{N\beta}{L}\Big).$$

Since, for each L, $L\phi(N/L) \to 0$ as $N \to \infty$, then it is possible to give a function $M = M(L)$, unboundedly increasing as $L \to \infty$, such that $L\phi(M(L)/L) \to 0$ as $L \to \infty$. Henceforth we will assume that L and N are connected by $N = (1/\beta)M(L)$.

Thus, for arbitrary N (or L) and β we have

$$\mathbf{P}\Big(\sup_{k\geqslant N} Y_k^{(r)} \geqslant 0\Big) \leqslant \mathbf{P}\Big(\frac{Y_N^{(r)}}{N} \geqslant -\beta\Big) + \mathbf{P}\Big(\frac{Y_L^{(r)}}{L} > -2\beta\Big)$$

$$+ \frac{1}{\beta}\mathbf{E}\Big(\frac{Y_L^{(r)} + L}{L}\Big)^{+} + \frac{L}{\beta}\phi\Big(\frac{N\beta}{L}\Big).$$

Consequently, for those β (an everywhere dense set), which are points of continuity of the distributions of $-Y_N/N$ and Y_L/L, we obtain by lemma 2

$$\limsup_{r\to\infty} \mathbf{P}\Big(\sup_{k\geqslant N} Y_k^{(r)} \geqslant 0\Big) \leqslant A(N, \beta) + \frac{L}{\beta}\phi\Big(\frac{N\beta}{L}\Big) \qquad (4.7)$$

where

$$A(N,\beta) = \mathbf{P}\left(\frac{Y_N}{N} \geqslant -\beta\right) + \mathbf{P}\left(\frac{Y_L}{L} > -2\beta\right) + \frac{1}{\beta}\mathbf{M}\left(\frac{Y_L + \beta L}{L}\right)^+.$$

We will show that by the choice of N and β the right-hand side of (4.7) may be made arbitrarily small. Given an arbitrary $\varepsilon > 0$ consider the value $A(N, \beta)$.

For $\beta = -a/3$, where $a = \mathbf{E}\xi_k < 0$, the terms of $A(N, \beta)$ satisfy

$$\frac{1}{\beta}\mathbf{E}\left(\frac{Y_L + \beta L}{L}\right)^+ \leqslant \frac{1}{\beta}\mathbf{E}\left(\frac{Y_L}{L} - a; \frac{Y_L}{L} > \frac{a}{2}\right)$$

$$\mathbf{P}\left(\frac{Y_L}{L} > -2\beta\right) \leqslant \mathbf{P}\left(\frac{Y_L}{L} > \tfrac{2}{3}a\right)$$

so that

$$A(N,\beta) \leqslant \mathbf{P}\left(\frac{Y_N}{N} \geqslant \frac{a}{3}\right) + \mathbf{P}\left(\frac{Y_L}{L} > \frac{2a}{3}\right) + \frac{1}{\beta}\mathbf{E}\left|\frac{Y_L}{L} - a\right|.$$

Since, by the strong law of large numbers, $\mathbf{E}|Y_L/L - a| \to 0$ as $L \to \infty$ (Y_L/L L_1-converges to a), then we may now choose N (and, consequently, L) so large that

$$A(N,\beta) \leqslant \frac{\varepsilon}{2}, \quad \frac{L}{\beta}\phi\left(\frac{N\beta}{L}\right) \leqslant \frac{\varepsilon}{2}.$$

Thus we have shown that, by the choice of N, the left-hand side of (4.7) can be made arbitrarily small. (4.5), and with it the theorem, is proved. □

We now consider condition (III) more carefully. We first give an example showing that condition (III) is not necessary for the convergence of the distributions of $Y^{(r)}(n)$ (convergence of the finite-dimensional distributions of $\xi^{(r)}$ is assumed, as is $\mathbf{E}\xi_1 < 0$). Let $\xi_k^{(r)}$ be independent

$$\xi_k^{(r)} = \begin{cases} r^2 - 1 & \text{with probability } r^{-2} \\ -1 & \text{with probability } 1 - r^{-2} - r - 1 \\ -r^2 - 1 & \text{with probability } r - 1. \end{cases}$$

The finite-dimensional distributions of this sequence converge, obviously, to the finite-dimensional distributions of the sequence $\xi = (-1, -1, -1, \ldots)$. Here the distributions of $Y^{(r)}(n)$ also converge. In fact, since $\mathbf{P}(Y(n) = 0) = 1$ for any n, then it is sufficient to show that $\mathbf{P}(Y^{(r)}(n) > 0) \to 0$ for any n and for $r \to \infty$. We have

$$\mathbf{P}(Y^{(r)}(0) > 0) \leqslant \mathbf{P}(Y_*^{(r)}(0) > 0) = \mathbf{P}(Y_{**}^{(r)}(0) > 0) \tag{4.8}$$

where $Y_*^{(r)}$ and $Y_{**}^{(r)}$ respectively are constructed relative to sums of the random variables $\xi_k^{(r)*} = \xi_k^{(r)} + 1$ and

$$\xi_k^{(r)**} = \begin{cases} 1 & \text{with probability } \dfrac{r^{-2}}{r^{-1}+r^{-2}} = (r+1)^{-1} \\ -1 & \text{with probability } \dfrac{r^{-1}}{r^{-1}+r^{-2}} = r(r+1)^{-1} \end{cases}$$

($\xi_k^{(r)**}$ is obtained from $\xi^{(r)*}$ by rejecting zero elements). By lemma 1 the right-hand side in (4.8) does not exceed

$$\mathbf{P}(\xi_1^{(r)**} > -1) + \mathbf{E}(\xi_1^{(r)**})^+ = (r+1)^{-1} + (r+1)^{-1} \to 0$$

as $r \to \infty$. Thus $\mathbf{P}(Y^{(r)}(n) = 0) \to \mathbf{P}(Y(n) = 0) = 1$. At the same time condition (III) of theorem 1 is not satisfied since

$$\mathbf{E}(\xi_1^{(r)})^+ = 1 - r^{-2} \to 1 \neq \mathbf{E}(\xi_1)^+ = 0.$$

We note now that condition (III), although not necessary, is in some sense close to it. Namely, we will prove the following result.

Let conditions (I), (II) *of theorem* 1 *hold, in which it is assumed in addition that the* $\xi_j^{(r)}$ *are independent and*

$$\mathbf{E}(\xi_1^{(r)})^- \to \mathbf{E}\xi_1^-$$

where $x^- = (x)^- = \min(0, x)$. *Then the weak convergence of the distributions of* $Y^{(r)} = Y^{(r)}(0)$ *to the distribution of* $Y = Y(0)$ *implies condition* (III).*

Proof. As is well known (see Borovkov, 1976a)

$$\mathbf{E}\,\mathrm{e}^{\mathrm{i}\lambda Y} = \exp\left\{\sum_{k=1}^{\infty}\int_0^{\infty}(\mathrm{e}^{\mathrm{i}\lambda x}-1)\frac{\mathrm{d}\mathbf{P}(Y_k < x)}{k}\right\}$$

where $Y_k = Y_{0,k}$ is a sum of k independent variables $\xi_{-1}, \ldots, \xi_{-k}$ (for brevity, we will also write $Y_k^{(r)} = Y_{0,k}^{(r)}$). This means that the distributions of $Y^{(r)}$ and Y are infinitely divisible and a necessary and sufficient condition for convergence of their distributions is weak convergence of the spectral functions

$$\Phi^{(r)}(x) = \sum_{k=1}^{\infty}\frac{1}{k}\mathbf{P}(Y_k^{(r)} > x) \Rightarrow \Phi(x) = \sum_{k=1}^{\infty}\frac{1}{k}\mathbf{P}(Y_k > x) \quad x > 0.$$

For this, in turn, it is necessary and sufficient that the series

$$\sum_{k=1}^{\infty}\frac{1}{k}\mathbf{P}(Y_k^{(r)} > x)$$

converge uniformly with respect to r at each point of continuity of $\Phi(x)$. We now choose as $b > 0$ a point of continuity of $\Phi(x)$ and of $\mathbf{P}(Y_k > kx)$, and form the sum

* This assertion was proved by A. A. Mogul'skii.

of the values $\xi_k^b = \xi_k - b$:

$$Y_k^b = Y_k - kb, \quad Y_k^{(r)b} = Y_k^{(r)} - kb.$$

Then, putting $Y^b = \sup_{k \geqslant 0} Y_k^b$, $Y^{(r)b} = \sup_{k \geqslant 0} Y_k^{(r)b}$, on the basis of the above and in view of the convergence of the distributions of $Y^{(r)}$ and Y we obtain

$$\mathbf{P}(Y^{(r)b}=0) = \exp\left\{-\sum_{k=1}^{\infty} \frac{1}{k} \mathbf{P}(Y_k^{(r)b}>0)\right\} = \exp\left\{-\sum_{k=1}^{\infty} \frac{1}{k} \mathbf{P}(Y_k^{(r)}>kb)\right\}$$

$$\rightarrow \exp\left\{-\sum_{k=1}^{\infty} \frac{1}{k} \mathbf{P}(Y_k>kb)\right\} = \mathbf{P}(Y^b=0).$$

If we denote by χ the first nonpositive sum amongst $Y_1, Y_2, \ldots$, then for $\mathbf{E}\xi_1 < 0$, from Borovkov (1976a, section 16), we find

$$\mathbf{E}\chi \,.\, \mathbf{P}(Y = 0) = \mathbf{E}\xi.$$

Therefore with the obvious agreements concerning notation, we obtain

$$\frac{\mathbf{E}\xi^{(r)b}}{\mathbf{E}\chi^{(r)b}} \rightarrow \frac{\mathbf{E}\xi^b}{\mathbf{E}\chi^b}.$$

In addition, by the condition and the inequality $\chi^b \leqslant 0$

$$\liminf_{r\to\infty} \mathbf{E}\xi^{(r)b} \geqslant \mathbf{E}\xi^b$$

$$\limsup_{r\to\infty} \mathbf{E}\chi^{(r)b} \leqslant \mathbf{E}\chi^b$$

$$\limsup_{r\to\infty} \mathbf{E}\xi^{(r)b} \leqslant \frac{\mathbf{E}\xi^b}{\mathbf{E}\chi^b} \limsup_{r\to\infty} \mathbf{E}\chi^{(r)b} \leqslant \mathbf{E}\xi^b.$$

This means that

$$\mathbf{E}\xi^{(r)b} \rightarrow \mathbf{E}\xi^b, \quad \mathbf{E}\xi^{(r)} \rightarrow \mathbf{E}\xi, \quad \mathbf{E}(\xi^{(r)})^+ \rightarrow \mathbf{E}\xi^+. \qquad \square$$

We pass on now to a corollary of theorem 1 for the case when $\xi_k^{(r)} = \xi_{k+}^{(r)} - \xi_{k-}^{(r)}$, $\xi_k = \xi_{k+} - \xi_{k-}$, where the sequence $\{\xi_{k-}^{(r)}, \xi_{k+}^{(r)}\}$ is stationary and $\xi_k^{(r)} \geqslant 0$.

THEOREM 2. *Suppose that*

(I) *The sequence* $(\xi_-, \xi_+) = \{\xi_{k-}, \xi_{k+};\ -\infty < k < \infty\}$ *is ergodic and* $\mathbf{E}\xi_1 < 0$.

(II) *The finite-dimensional distributions of* $(\xi_-^{(r)}, \xi_+^{(r)})$ *converge, as* $r \rightarrow \infty$, *to the distributions of* (ξ_-, ξ_+).

(III) $\mathbf{E}\xi_{1+}^{(r)} \rightarrow \mathbf{E}\xi_{1+} < \infty$.

Then theorem 1 holds.

The proof of theorem 2 consists of verifying the conditions of theorem 1. Conditions (I) and (II) of theorem 1, obviously, are satisfied. Condition (III) follows from lemma 2 and condition (III) of theorem 2. □

Further discussions of stability questions when $\{\xi_k^{(r)};\ -\infty < k < \infty\}$ is a sequence of independent variables (the condition for uniform convergence of the distributions of $Y^{(r)}(n)$, an estimate of the speed of convergence, and so on) can be found in Borovkov (1976a).

We will now prove the analogue of theorem 1 for processes with continuous time.

Consider a sequence of separable processes with (strictly) stationary increments $\{X^{(r)}(t);\ -\infty < t < \infty\}$, $r = 1, 2, \ldots$, and a process of the same type $\{X(t);\ -\infty < t < \infty\}$. We denote, in a similar way to earlier,

$$Y(t) = \sup_{v \leqslant t} (X(t) - X(v))$$

$$Y^{(r)}(t) = \sup_{v \leqslant t} (X^{(r)}(t) - X^{(r)}(v))$$

and put

$$Y(t, t-u) = \sup_{t-u \leqslant v \leqslant t} (X(t) - X(v))$$

$$Y^{(r)}(t, t-u) = \sup_{t-u \leqslant v \leqslant t} (X^{(r)}(t) - X^{(r)}(v))$$

so that $Y(t) = Y(t, -\infty)$.

THEOREM 3. *Let the random variables*

$$\xi_k = X(k) - X(k-1), \quad \xi_k^{(r)} = X^{(r)}(k) - X^{(r)}(k-1)$$

satisfy the conditions of theorem 1. *And, in addition,*

(IV) *For any finite u the distribution of* $Y^{(r)}(0, -u)$ *weakly converges, as* $r \to \infty$, *to the distribution of* $Y(0, -u)$, *so that* $\mathbf{E}Y^{(r)}(0, -u) \to \mathbf{E}Y(0, -u) < \infty$.

Then the finite-dimensional distributions of $Y^{(r)}(t)$ *weakly converge as* $r \to \infty$ *to the distribution of* $Y(t)$.

Proof. Repeating the arguments of the proof of theorem 1, we see (see (4.6) and condition (IV) of theorem 3) that it is sufficient for us to show that a relation similar to (4.5) holds. Namely, it must be shown that

$$\limsup_{r \to \infty} \mathbf{P}\left(\sup_{v \geqslant N} (X^{(r)}(0) - X^{(r)}(-v)) \geqslant 0\right) \to 0 \tag{4.9}$$

as $N \to \infty$. We already know that this is satisfied if the variable $v = k$ in (4.9) takes only integer values (see (4.5)).

In precisely the same way we could also establish that the probability of

$$B_{r,N} = \left\{ \sup_{k \geqslant N} \left(X^{(r)}(0) - X^{(r)}(-k) - \frac{ak}{2} \right) \geqslant 0 \right\} \quad (a = \mathbf{E}\xi_k < 0)$$

also converges to 0 as $r \to \infty$, $N \to \infty$. We consider now the estimation of the probability (4.9)

$$\mathbf{P}\left(\bigcup_{k=N}^{\infty} \left\{ \sup_{v \in [k,k+1]} (X^{(r)}(0) - X^{(r)}(-v)) \geqslant 0 \right\} \right)$$

$$\leqslant \mathbf{P}(B_{r,N}) + \mathbf{P}\left(\bigcup_{k=N}^{\infty} \{ Y^{(r)}(-k, -k-1) \geqslant -\tfrac{1}{2}ak \} \right)$$

$$\leqslant \mathbf{P}(B_{r,N}) + \sum_{k=N}^{\infty} \mathbf{P}(Y^{(r)}(0, -1) \geqslant -\tfrac{1}{2}ak). \tag{4.10}$$

But from condition (IV) it follows that the series on the right of the last inequality converges uniformly in r. This proves (4.9) and with it the result of the theorem. □

We note that condition (IV) is essential. It is not difficult to construct examples where the absence of (IV) makes the theorem false. However, in a great many special cases it can be avoided.

We consider the important special case when $X^{(r)}(t)$ is a difference

$$X^{(r)}(t) = X_+^{(r)} - X_-^{(r)}$$

where $X_\pm^{(r)}$ are independent nondecreasing processes with stationary increments whose finite-dimensional distributions converge to the distributions of processes $X_\pm(t)$ which are, obviously, of the same type as the $X_\pm^{(r)}(t)$. We denote

$$\xi_{k\pm}^{(r)} = X_\pm^{(r)}(k) - X_\pm^{(r)}(k-1), \quad \xi_{k\pm} = X_\pm(k) - X_\pm(k-1).$$

THEOREM 4. *Let the following conditions hold.*

(I) *The sequence* $\{\xi_k = \xi_{k+} - \xi_{k-}\}$ *is ergodic,* $\mathbf{E}\xi_1 < 0$.

(II) *The finite-dimensional distributions of* $X_\pm^{(r)}(t)$ *weakly converge as* $r \to \infty$ *to the distributions of* $X_\pm(t)$.

(III) $\mathbf{E}\xi_{1+}^{(r)} \to \mathbf{E}\xi_{1+}$.

Then the finite-dimensional distributions of $Y^{(r)}(t)$ *weakly converge as* $r \to \infty$ *to the distribution of* $Y(t)$.

Proof. We will again follow the scheme of proof of theorem 1. However, in this case, besides (4.9) we must also establish (see (4.6)) the convergence of the distributions of $Y^{(r)}(0, -u)$ for any fixed u. We first prove (4.9).

By theorem 2 ξ_k satisfy the conditions of theorem 1. This will mean that in (4.10)

$$\limsup_{r\to\infty} \mathbf{P}(B_{r,N}) \to 0$$

as $N \to \infty$,

$$\mathbf{P}(Y^{(r)}(0, -1) \geqslant -\tfrac{1}{2}ak) \leqslant \mathbf{P}(\xi^{(r)}_{0+} \geqslant -\tfrac{1}{2}ak).$$

Hence, and from conditions (II) and (III), (4.9) follows.

We now establish the convergence of the distribution of $Y^{(r)}(0, -u)$. Since the trajectories of $X^{(r)}_{\pm}(t)$ are *monotonic*, then the convergence of the finite-dimensional distributions will imply the convergence of the distributions of functionals of $X^{(r)}_{\pm}(t)$, which are continuous in the Skorokhod topology M_z (see Skorokhod, 1956), or what is the same, functionals which are continuous in the metric ρ_F, defined as follows (see Borovkov, 1972). Consider, for simplicity, the segment $[0, 1]$ and define the graph $\mathscr{X}$ of the function $x(t)$, $t \in [0, 1]$. That is, the simply-connected set in the (t, x)-plane such that its cross-section at t coincides with the segment $[\liminf_{u\to t} x(u), \limsup_{u\to t} x(u)]$. At each point of $\mathscr{X}$ construct an open sphere of radius ε. The domain in the plane, obtained as the intersection of the strip $0 \leqslant t \leqslant 1$ with the union of all the spheres, is denoted $G_\varepsilon(\mathscr{X})$. We will say that the distance $\rho_F(x, y) < \varepsilon$ if the graph $\mathscr{Y}$ of $y(t)$ belongs to $G_\varepsilon(\mathscr{X})$, and the graph $\mathscr{X}$ belongs to $G_\varepsilon(\mathscr{Y})$. In other words, if a and b are points of the plane and $R(a, b)$ is the Euclidean distance, then

$$\rho_F(x, y) = \max\left[\sup_{a\in\mathscr{X}} \inf_{b\in\mathscr{Y}} R(a, b), \sup_{a\in\mathscr{Y}} \inf_{b\in\mathscr{X}} R(a, b)\right].$$

Obviously, the functional $f(x) = \sup_{0\leqslant t\leqslant 1} x(t)$ is continuous in the metric ρ_F.

We return to the processes $X^{(r)}_{\pm}(t)$ on $[-u, 0]$ (for simplicity let $X^{(r)}_{\pm}(0) = 0$). If we use the method of Skorokhod (1956, 1961) and condition (II), then for $X^{(r)}_{+}$ and X_{+}, by the monotonicity of $X^{(r)}_{+}(t)$, we can construct stochastically equivalent processes $\tilde{X}^{(r)}_{+}$ and $\tilde{X}_{+}$, on the same probability space, and such that $\rho_F(\tilde{X}^{(r)}_{+}, \tilde{X}_{+}) \to 0$ almost surely. Similar constructions hold for $X^{(r)}_{-}$, X_{-}. Then, enlarging the probability space if necessary (this is possible by the independence of $X^{(r)}_{\pm}$), we obtain that with probability 1

$$\rho_F(\tilde{X}^{(r)}_{+} - \tilde{X}^{(r)}_{-}, \tilde{X}_{+} - \tilde{X}_{-}) \to 0$$

as $r \to \infty$. Since the distributions of $\tilde{X}^{(r)}_{+} - \tilde{X}^{(r)}_{-}$ and $X^{(r)}_{+} - X^{(r)}_{-}$ coincide, this obviously implies weak convergence of the distributions of $Y^{(r)}(0, -u)$. The theorem is proved. □

Evidently the result of the theorem still holds even for dependent components $X^{(r)}_{\pm}(t)$, when $(X^{(r)}_{+}(t), X^{(r)}_{-}(t))$ form an arbitrary process with stationary nonnegative increments.

4.2. Stability theorems for single-channel systems with waiting and systems with autonomous service

4.2.1. As is well known, a broad class of single-channel systems with waiting are described by an equation of the form

$$w_{n+1} = \max(0, w_n + \xi_n), \quad n \geqslant 1. \tag{4.11}$$

Consider, for example, systems $\langle G, G, G, 1\rangle$ (for the notation see Borovkov, 1976a). We recall that these are systems with waiting, where the requests arrive in groups of size v_j^e at intervals of time τ_j^e; the requests are served singly, the service time of the jth request being τ_j^s. The initial condition w_1 and the controlling sequence $\{\tau_j^e, v_j^e, \tau_j^s; j \geqslant 1\}$ are given. Then, if we put

$$\xi_k = S_k - \tau_k^e$$

where S_k is the sum of the service times of requests in the kth group, the waiting time w_{k+1} for the first request in the $(k+1)$st group is given by (4.11). It is well known (see, for example, Borovkov, 1976a) that the solutions of (4.11) are given by the formula

$$w_{n+1} = \max(X_n + w_1, X_n - X_1, \ldots, X_n - X_n), \quad X_n = \sum_{j=1}^{n} \xi_j. \tag{4.12}$$

We obtain a similar equality for systems $\langle E, G, G, G\rangle$ and $\langle G, G, E, G\rangle$, if w_k denotes the length of the queue at suitable moments of time (see Borovkov, 1976a).

Suppose now that we are considering a sequence of service systems controlled by stationary sequences $\xi^{(r)} = \{\xi_k^{(r)};\ k \geqslant 1\}$, $r = 1, 2, \ldots$. The sequences $\xi^{(r)} = \{\xi_k^{(r)}\}$ may be regarded as defined 'on the whole axis', that is, for $-\infty < k < \infty$. Assume that the finite-dimensional distributions of $\xi^{(r)}$ converge to the distributions of a stationary sequence $\xi = \{\xi_k\}$. The question arises, whether here the distributions of the sequences

$$w_n^{(r)} = \{w_{n,k}^{(r)};\ k \geqslant 0\} = \{w_{n+k}^{(r)};\ k \geqslant 0\}$$

where the $w_n^{(r)}$ are defined by (4.11) with initial condition $w_1^{(r)}$ and sequence $\{\xi_k^{(r)}\}$, approach the distributions of

$$w_n = \{w_{n,k};\ k \geqslant 0\} = \{w_{n+k};\ k \geqslant 0\}.$$

Also, whether the stationary (limit over n) distributions of these sequences approach each other.

We consider an example. The length of the communications τ_j, transmitted over communication channels, are often regarded as exponentially distributed ($\tau_j = \tau_j^s$ is the service time of a request in the corresponding system receiving this communication). However, this assumption may not be satisfied in the neighbourhood of 0 because of the presence of a compulsory auxiliary group of words (so that for small ε $\mathbf{P}(\tau_j^3 > \varepsilon) = 1$ and not $e^{-\alpha x}$). Nevertheless the assumption, as a rule, is justified.

The question, obviously, is this: when will these 'idealized' systems be close to the real systems in the sense of the characteristics considered?

We adopt the basic notation of the previous section

$$Y_{n,k} = \sum_{i=n-k}^{n-1} \xi_i, \quad Y_{n,0} = 0, \quad Y(n) = \sup_{k \geq 0} Y_{n,k}$$

and put

$$X_k = \sum_{i=1}^{k} \xi_i, \quad X_0 = 0.$$

An upper index (r) in this notation, as before, will mean that in the corresponding definition ξ_j is replaced by $\xi_j^{(r)}$.

In addition, we denote

$$w^k = Y(k+1), \quad \mathbf{w} = (w^0, w^1, w^2, \ldots) = (Y(1), Y(2), \ldots).$$

Then, as shown in Borovkov (1976a, p. 10), for any measurable B

$$|\mathbf{P}(\mathbf{w}^n \in B) - \mathbf{P}(\mathbf{w} \in B)| \leq \mathbf{P}\left(\min_{0 \leq j \leq n} X_j > -\max(w_1, Y(1)) \right) \tag{4.13}$$

from which it should be clear that the approach of the stationary and prestationary distributions is essentially guaranteed by theorem 1 of section 4.1. In more general form the answer to the above question is contained in the following theorem.

THEOREM 1. *Let the conditions of theorem 1 of section 4.1 be satisfied relative to the sequences $\xi^{(r)}$ and ξ (convergence of finite-dimensional distributions, ergodicity of ξ, $\mathbf{E}\xi_1 < 0$ and the convergence of the expectations of the positive parts of $\xi_1^{(r)}$ and ξ_1). In addition, let*

$$\mathbf{P}(w_1^{(r)} > c) \to 0 \tag{4.14}$$

as $c \to \infty$, uniformly in r. Then the finite-dimensional distributions of $\mathbf{w}_n^{(r)}$ weakly converge as $r \to \infty$, $n \to \infty$ to the distributions of $\mathbf{w}$.

From the remarks made at the beginning of the section it follows that the same kind of conditions should be satisfied for the stability of the distribution of the queue length in the systems $\langle E, G, G, G \rangle$ and $\langle G, G, E, G \rangle$.

Proof. For an arbitrary cylinder set B which is a set of continuity for the distribution of $\mathbf{w}$, the difference

$$|\mathbf{P}(\mathbf{w}_n^{(r)} \in B) - \mathbf{P}(\mathbf{w} \in B)|$$

is estimated by the sum of the differences

$$|\mathbf{P}(\mathbf{w}_n^{(r)} \in B) - \mathbf{P}(\mathbf{w}^{(r)} \in B)|, \quad |\mathbf{P}(\mathbf{w}^{(r)} \in B) - \mathbf{P}(\mathbf{w} \in B)|.$$

It is clear that convergence to 0 of the second difference is guaranteed by theorem 1 of section 4.1. By (4.13) the first difference, for given $\varepsilon > 0$, does not exceed

$$\varepsilon + \mathbf{P}\left(\min_{0 \leqslant j \leqslant n} X_j^{(r)} > -L \right) \tag{4.15}$$

where L is chosen so that for all large enough r

$$\mathbf{P}(\max(w_1^{(r)}, Y^{(r)}(1)) > L) \leqslant \varepsilon.$$

By the conditions of the theorem and the convergence of the distributions of $Y^{(r)}(1)$ such an L always exists. Since (4.15) is monotonically decreasing as n increases, then it is sufficient for us to show that there exist n and r_0 such that

$$\mathbf{P}(X_n^{(r)} > -L) < \varepsilon$$

for all $r \geqslant r_0$. But this immediately follows from the ergodicity of ξ (we must take n so that $\mathbf{P}(X_n > -L - 1) < \varepsilon/2$) and the weak convergence of the distributions of $X_n^{(r)}$ and X_n as $r \to \infty$. □

From the proof of the theorem it is clear that theorem 1 of section 4.1 is none other than a stability theorem for the stationary distribution of the waiting time, that is, for the vectors $\mathbf{w}^{(r)}$ and $\mathbf{w}$.

From theorem 2 of section 4.1 and theorem 1 follows immediately

THEOREM 2. *Let the sequence* $\{\xi_j^{(r)} = S_j^{(r)} - \tau_j^{(r)e}\}$, *where* $S_1 = \tau_1^s + \cdots + \tau_{\nu_1^e}^s$, $S_2 = \tau_{\nu_1^e+1}^s + \cdots + \tau_{\nu_1^e+\nu_2^e}^s, \ldots$ *etc., controlling a* $\langle G, G, G, 1 \rangle$ *system, be stationary for each* r *and, in addition, let the system* $\{\xi_j\}$ *be ergodic*

$$\mathbf{E}\xi_1 = \mathbf{E}(\tau_1^s + \cdots + \tau_{\nu_1^e}^s - \tau_1^e) < 0.$$

Then, if the finite-dimensional distributions of $\{\xi_j^{(r)}\}$ *weakly converge to the corresponding limits and*

$$\mathbf{E}S_1^{(r)} \to \mathbf{E}S_1 \tag{4.16}$$

then if (4.14) *is satisfied the result of theorem* 1 *still holds.*

Theorem 4 of section 4.1 and the results of Borovkov (1976a, section 6) allow us to obtain similar results for the virtual waiting time $w(t)$ (the waiting time of a request arriving at time t. For a precise definition see Borovkov, 1976a).

To conclude this section we give some numerical examples, which illustrate the speed of approach of the one-dimensional distributions of the waiting time under the conditions of theorem 2 and the importance of condition (4.16).

Let τ_j^e and τ_j^s be exponentially distributed

$$\mathbf{P}(\tau_j^e > x) = \mathrm{e}^{-x}, \quad \mathbf{P}(\tau_j^s > x) = \mathrm{e}^{-2x}.$$

We consider three methods of constructing the sequences $\{\tau_j^{(r)e}, \tau_j^{(r)s}\}$:

1. $\tau_j^{(r)e} = \tau_j^e \quad \tau_j^{(r)s} = \tau_j^s + 2^{-r}, \quad r = 1, 2, \ldots$

2. $\tau_j^{(r)e} = \tau_j^e \quad \tau_j^{(r)s} = \begin{cases} \tau_j^s & \text{with probability } 1-2^{-r} \\ \tau_j^s + 2^{r-3} & \text{with probability } 2^{-r}. \end{cases}$

3. $\tau_j^{(r)s} = \tau_j^s \quad \tau_j^{(r)e} = \begin{cases} \tau_j^e & \text{with probability } 1-2^{-r} \\ \tau_j^e + 2^{r-3} & \text{with probability } 2^{-r} \end{cases}$

where the choices with probabilities 2^{-r} and $1 - 2^{-r}$ occur independently for each j.

Obviously condition (4.16) is not satisfied in the second case since $\mathbf{E}\tau_j^{(r)s} = \mathbf{E}\tau_j^s + \frac{1}{8}$.

By Borovkov (1976a, section 10) the stationary distribution of w^k and the stationary distribution of the virtual waiting time coincide in our case. Therefore in the search for the distribution of w^k (or, what is the same, Y) we can use the Khinchin formula (see Borovkov, 1976a, section 8, corollary 3). By this formula

$$\mathbf{E}\,\mathrm{e}^{\mathrm{i}\lambda Y} = \frac{1 - \alpha \mathbf{E} S_1}{1 - \alpha(f(\lambda) - 1)/\mathrm{i}\lambda}$$

where under our conditions $S_1 = \tau_1^s$, $\mathbf{E}S_1 = \frac{1}{2}$, $\alpha = (\mathbf{E}\tau_j^e)^{-1} = 1$,

$$f(\lambda) = \mathbf{E}\,\mathrm{e}^{\mathrm{i}\lambda S_1} = 2\int_0^\infty \mathrm{e}^{\mathrm{i}\lambda x - 2x}\,\mathrm{d}x = \frac{2}{2 - \mathrm{i}\lambda}.$$

Therefore

$$\mathbf{E}\,\mathrm{e}^{\mathrm{i}\lambda Y} = \frac{2 - \mathrm{i}\lambda}{2(1 - \mathrm{i}\lambda)}, \quad \mathbf{P}(Y = 0) = \tfrac{1}{2}, \quad \mathbf{P}(Y > x) = \tfrac{1}{2}\mathrm{e}^{-x}$$

for $x \geqslant 0$. These formulae also follow directly from the results of chapters II and IV of Borovkov (1976a).

To obtain the empirical distribution of $Y^{(r)}$ for each of the five values $r = 2, 3, \ldots, 6$ and in each of the three cases 1000 trials were made. To estimate $\sup_x |\mathbf{P}(Y < x) - \mathbf{P}(Y^{(r)} < x)|$, points $x_0 = 0$, $x_1, \ldots, x_{199}$, $x_{200} = \infty$ were chosen so that

$$\mathbf{P}(Y \in (x_{j-1}, x_j)) = \tfrac{1}{400}, \quad j = i, \ldots, 200.$$

For the above values of r the following sample values A_r^* were obtained for the variable

$$A_r = \sup_{1 \leqslant i \leqslant 200} |\mathbf{P}(Y < x_i) - \mathbf{P}(Y^{(r)} < x_i)|.$$

In the first case

r	2	3	4	5	6
A_r^*	0.326	0.145	0.082	0.026	0.014

In the second case

r	2	3	4	5	6
A_r^*	0.146	0.167	0.185	0.200	0.178

In the third case

r	2	3	4	5	6
A_r^*	0.084	0.074	0.058	0.050	0.030

Table 4.1 Table for the graphs of $F^{(r)*}(x_{4k})$ in Fig. 1–4

	Second method		Third method			Second method		Third method	
k	$r=3$	$r=6$	$r=3$	$r=6$	k	$r=3$	$r=6$	$r=3$	$r=6$
0	0.374	0.401	0.565	0.527	100	0.587	0.616	0.819	0.763
4	388	413	569	533	104	597	630	824	773
8	398	424	578	546	108	606	640	832	783
12	407	432	597	554	112	615	649	837	789
16	410	439	608	567	116	624	660	841	802
20	416	445	613	576	120	633	664	849	807
24	422	453	626	584	124	649	668	862	817
28	433	461	636	592	128	663	678	875	829
32	440	472	644	605	132	678	685	883	840
36	450	484	655	620	136	694	692	891	850
40	460	493	663	624	140	701	701	894	858
44	471	504	673	633	144	715	707	899	869
48	476	515	689	638	148	729	711	906	875
52	481	522	704	651	152	737	726	910	884
56	488	532	710	663	156	754	729	919	895
60	501	541	721	668	160	763	733	933	906
64	508	545	730	681	164	780	739	939	914
68	522	552	741	692	168	793	744	948	922
72	529	557	751	701	172	814	758	956	935
76	533	566	761	712	176	827	768	967	946
80	545	572	771	720	180	849	777	975	957
84	553	585	778	729	184	869	782	984	962
88	563	593	786	738	188	888	792	989	969
92	570	603	794	745	192	911	806	996	980
96	578	612	807	756	196	945	830	999	987

Values $r > 6$ were not considered because in the chosen method of conducting trials, in the first and third cases for $r > 6$, A_r^* becomes an unsatisfactory estimate for A_r. While in the first and third cases, in accordance with theorem 2, A_r converges to 0, in the second case A_r stays close to 0.2 as r grows even though $\mathbf{P}(\xi_j^{(r)} \neq \xi_j) = 2^{-r} \to 0$ as $r \to \infty$.

In figures 4.1–4.4 we give the graphs obtained in these experiments for the empirical distribution functions $F^{(r)*}(x)$, corresponding to the functions

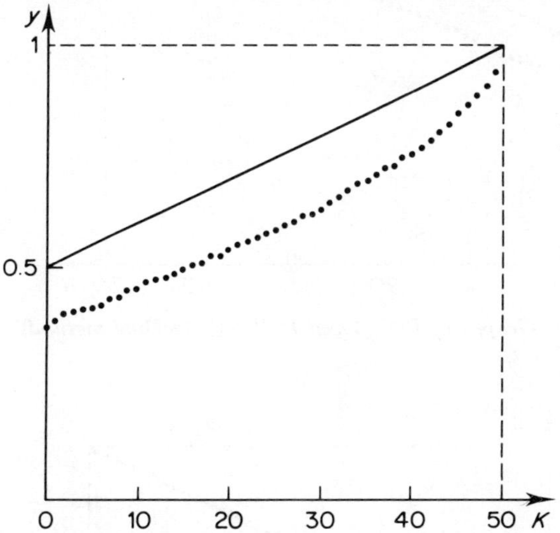

***Figure 1** Graphs of $F(x_{4k})$ and $F^{(3)*}(x_{4k})$ (second method)*

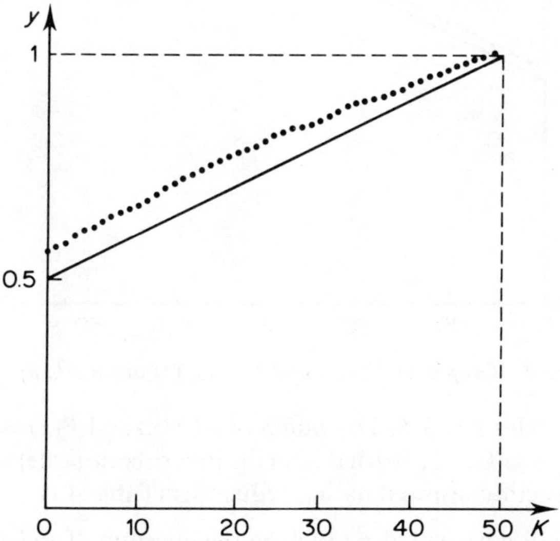

***Figure 2** Graphs of $F(x_{4k})$ and $F^{(3)}(x_{4k})$ (third method)*

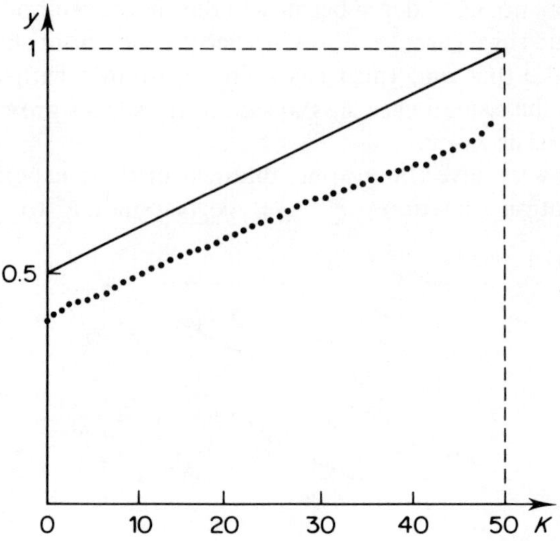

Figure 3 *Graphs of* $F(x_{4k})$ *and* $F^{(6)*}(x_{4k})$ *(second method)*

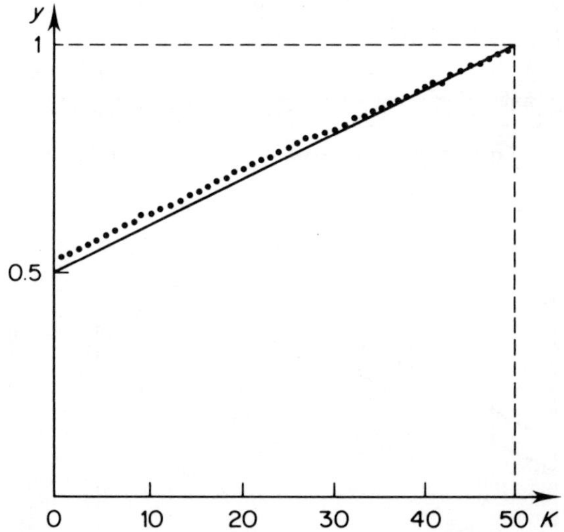

Figure 4 *Graphs of* $F(x_{4k})$ *and* $F^{(6)}(x_{4k})$ *(third method)*

$F^{(r)}(x) = \mathbf{P}(Y^{(r)} < x)$ for $r = 3, 6$. The values of $F^{(r)*}(x)$ and $F(x) = \mathbf{P}(Y < x)$ are given at points x_{4k}, $k = 0, \ldots, 50$ (that is, in the logarithmic scale) for the last two methods of constructing approximating sequences (Table 4.1).

4.2.2. In this section we will also touch on the question of *stability of systems with autonomous service* (see Borovkov, 1976a, chapter 8). A system $\langle G, G, G, G\rangle_A$

with autonomous service and controlling sequence

$$\{\tau_j^e, \nu_j^e, \tau_j^s, \nu_j^s; j \geqslant 1\} \tag{4.17}$$

differs from the usual system in that the service of requests begins at times 0, τ_1^s, $\tau_1^s + \tau_2^s, \ldots$ independently of the input stream and the presence of a queue. Service takes place in groups of size $\nu_1^s, \nu_2^s, \ldots$.

We introduce the processes $\{e(t)\}$ and $\{s(t)\}$. $e(t)$ describes the input stream: $e(t)$ is equal to the number of requests which have joined the system up to time t. $s(t)$ is defined similarly as the sum

$$s(t) = \nu_1^s + \cdots + \nu_{\eta(t)-1}^s$$

where $\eta(t)$ is the first passage time through the level t in a random walk with jumps $\tau_1^s, \tau_2^s, \ldots$. Roughly speaking $s(t)$ is the number of requests which would have been accepted into service up to time t for an infinite queue.

We denote $X(t) = e(t) - s(t)$ and by $q(t)$ the length of the queue at time t, not counting the requests already in service (here it is natural to separate busy channels from 'pure' queues, since several requests may be found in each channel). The processes $e(t)$, $s(t)$, and $q(t)$ are step processes; we will assume that they are continuous from the right. For example, requests whose service begins at time t are regarded as leaving the queue at time t (related to them are the requests which join the system at time t).

The distribution of $\{q(t)\}$ is completely determined, if in the probability space in which (4.17) (or, equivalently, $\{e(t), s(t)\}$) is given, there is another random variable $q(0)$ which defines the queue at time $t = 0$. Then, as shown in Borovkov (1976a),

$$q(t) = q(0) + X(t) - \text{üinf} \Rightarrow \inf_{u \varepsilon [0,t]} (0, q(t) + X(u))$$

$$= \sup_{u \varepsilon [0,t]} [q(0) + X(t), X(t) - X(u)]. \tag{4.18}$$

The equality (4.18) can also be obtained as the natural solution of the equation

$$\mathrm{d}q(t) = \mathrm{d}X(t) - \min (0, q(t-0) + \mathrm{d}X(t)).$$

(4.18) is completely analogous to the relation (4.12) for discrete time. From (4.18) it is not difficult to obtain that when $X(t)$ is a process with stationary increments such that

$$a = \mathbf{E}(X(1) - X(0)) > 0, \quad \frac{X(t)}{t} \xrightarrow[\text{a.s.}]{} a \quad \text{as } t \to \infty$$

then the distributions of the processes $\{q_t(u) = q(t+u); u \geqslant 0\}$ converge to the distributions of

$$Y(u) = \sup_{v \leqslant u} (X(u) - X(v)).$$

To obtain now a stability theorem for the stationary queue for systems with autonomous service it remains for us to use theorems 3 and 4 of section 4.1.

4.3. Some estimates for the speed of convergence

4.3.1. The stability theorems in sections 4.1 and 4.2 establish the fact of the approach of the stationary distributions of the waiting time under the approach of the distributions of the controlling sequences. These results are qualitative. The question arises, how quickly does this approach occur if the speed of approach (in some sense or other) of the distributions of the controlling sequences is known?

We consider here the simplest system $\langle G_I, 1, G_I, 1\rangle$. For these systems the random variables $\xi_j = \tau_j^s - \tau_j^e$ are independent. Here we will discuss for the beginning only the *one-dimensional stationary* distributions of the waiting time w_k, which coincide with the distributions of the supremum

$$Y = Y(0) = \sup_{k\geqslant 0} Y_{0,k}, \quad Y_{0,k} = \sum_{j=-k}^{-1} \xi_j, \quad Y_{0,0} = 0.$$

The upper index (r), as before, will denote the correspondence with the controlling sequences $\{\tau_j^{(r)e}, \tau_j^{(r)s}\}$.

For *prestationary* distributions of $w_n^{(r)}$ and w_n *under zero initial conditions* $w_1^{(r)} = w_1 = 0$ the estimate of their closeness may be better so that it is possible to obtain estimates which are *uniform in n*. For *nonzero* initial conditions the order of these estimates is preserved; however, the uniformity with respect to $w_1^{(r)}$ and w_1, certainly, is absent. Appropriate explanations will be made at the end of subsection 4.3.2.

There are some remarks on the possibility of generalizing these results to more complex systems and dependent ξ_j at the end of this section and also at the end of section 4.7. Estimates for various types of multichannel systems, close in order to the estimates of this section, are obtained by the renewal method in Akhmarov (1979a, b, c, d).

We mention here two ways to obtain estimates of the speed of convergence. The first is connected with the 'smoothness' of convergence of the distributions of $\xi_j^{(r)}$ and ξ_j (for example, with the requirement of convergence of distributions in variation). The second is connected with boundedness of the moments $\mathbf{E}(\xi_j^{(r)+})^\alpha$ of order $\alpha > 1$ (under weak convergence of the distributions of $\xi_j^{(r)}$ and ξ_j; $x^+ = \max(0, x)$). The first way is based on factorization identities and can be generalized to the case of dependent ξ_j really only in the case of Markov chains (see Borovkov, 1980). The second way is more universal (cf. Zolotarev, 1975a); however, the estimates obtained essentially depend on the number α which (as shown by the results of the first direction) is not always essential. The question of how much one can eliminate the shortcomings of the estimates mentioned above, for the present, remains largely opn (for some results on best estimates see subsection 4.3.3).

We now give results relating to the first direction. Denote by W ($W^{(r)}$) the distribution function of Y ($Y^{(r)}$) and by $U(F_1, F_2)$ and $V(F_1, F_2)$ respectively the distances between F_1 and F_2 in the uniform metric

$$U(F_1, F_2) = \sup_x |F_1(x) - F_2(x)|$$

and in variation $V(F_1, F_2) = \mathrm{Var}\,(F_1 - F_2) = \int |\mathrm{d}(F_1 - F_2)|$. Obviously $U(F_1, F_2) \leqslant V(F_1, F_2)$.

THEOREM 1. (*Borovkov, 1976a, p. 120*) *Let* $F(x) = \mathbf{P}(\xi_1 < x)$ *and* $F^{(r)}(x) = \mathbf{P}(\xi_1^{(r)} < x)$ *have absolutely continuous components,* $-\infty < \mathbf{E}\xi_1 < 0$. *Then*

$$V(W^{(r)}, W) \leqslant c(V(F^{(r)}, F) + V(Q^{(r)}, Q))$$

where c depends only on the distribution F; $Q^{(r)}$ *and Q are the 'tails' of the distributions* $F^{(r)}$ *and F:*

$$Q^{(r)}(x) = \int_{-\infty}^{x} F^{(r)}(t)\,\mathrm{d}t, \quad Q(x) = \int_{-\infty}^{x} F(t)\,\mathrm{d}t$$

so that $V(Q^{(r)}, Q) = \int |F^{(r)} - F|\,\mathrm{d}x$.

It is possible to also give the following estimate, which contains the distance in variation between the densities $\phi^{(r)} = \mathrm{d}F^{(r)}/\mathrm{d}x$ and $\phi = \mathrm{d}F/\mathrm{d}x$ (if they exist)

THEOREM 1A. (*Borovkov, 1976a, p. 121*) *Let the distributions F and* $F^{(r)}$ *have densities* ϕ *and* $\phi^{(r)}$ *so that*

$$V(\phi^{(r)}, \phi) = \int |\mathrm{d}(\phi^{(r)} - \phi)| < \infty.$$

Then

$$V(W^{(r)}, W) < c[V(Q^{(r)}, Q) + \sqrt{V(Q^{(r)}, Q)(V(\phi^{(r)}, \phi) + V(F^{(r)}, F))}].$$

These results contain no restrictions on the moments of $\xi_j^{(r)}$ of order greater than 1, but require smallness of $V(Q^{(r)}, Q)$ and the existence or smallness of the distances $V(F^{(r)}, F)$ and $V(\phi^{(r)}, \phi)$. From theorem 1 it follows that at least for smooth (one-cornered) distributions ξ_j the deviation of ξ_j by ε (we can put, for example, $\xi_j^{(r)} = \xi_j + \varepsilon$) implies the deviation of the same order ε in the distributions of $Y^{(r)}$ and Y. The subsequent discussions show that *to obtain an estimate of such order in the general case is not successful* (at least, under the conditions of subsection 4.3.3).

4.3.2. Let $L(F_1, F_2)$ denote the Levy distance between F_1 and F_2

$$L(F_1, F_2) = \inf\{\varepsilon: F_2(x - \varepsilon) - \varepsilon \leqslant F_1(x) \leqslant F_2(x + \varepsilon) + \varepsilon\}.$$

A function $G(x)$ is called a *power function* if $G(x) \sim x^{-\alpha}h(x)$ where $\alpha \geqslant 0$ and $h(x)$ is a slowly varying function. A function $G(x)$ is called *exponential* if $G(x) \leqslant c\,e^{-\beta x}$ for $x > 0$ and some $\beta > 0$.

THEOREM 2. *Let the following conditions hold.*

(1) *There is a power or exponential function G such that for sufficiently large x and all r*

$$1 - F(x) \leqslant G(x), \quad 1 - F^{(r)}(x) \leqslant G(x).$$

(2) $\varepsilon = \varepsilon_r = L(F^{(r)}, F) \to 0$ *as* $r \to \infty$.

(3) $a = \mathbf{E}\xi_1 < 0$.

Then, if G is a power function, for any $\gamma > 0$, *and* $b > 0$, *and all sufficiently small r*

$$L(W^{(r)}, W) \leqslant \varepsilon G^{-1}(\varepsilon b/d)(1 + b)(1 + \gamma)$$

where $G^{-1}(y)$ *is the solution of*

$$G(x) = y, \quad d = \frac{2W(+0)}{(\alpha - 1)|a|^{\alpha}}.$$

If G is exponential, then for any $\gamma > 0$ *and all sufficiently large r*

$$L(W^{(r)}, W) \leqslant \frac{\varepsilon \log \varepsilon}{\log m}(1 + \gamma) \tag{4.19}$$

where $m = \min_{\mu} \mathbf{E}\, e^{\mu\xi_1} < 1$.

If, instead of condition (2) *of the theorem, we require*

(2a) $\varepsilon = U(F^{(r)}, F) \to 0$

then both inequalities in the theorem will hold for $U(W^{(r)}, W)$.

From this theorem immediately follows

COROLLARY. *Let* $\xi_1^+ = \max(0, \xi_1)$. *Then if* $\mathbf{E}(\xi^{(r)+})^{\alpha} < c$ *for all r, and conditions* (2) *and* (3) *are satisfied, then*

$$L(W^{(r)}, W) \leqslant c_1 \varepsilon^{1-(1/\alpha)}.$$

But if $\mathbf{E}\, e^{\beta\xi_1^{(r)}} < c$ *for some* $\beta > 0$ *and for all r then* (4.19) *is valid. The same inequalities are valid for* $U(W^{(r)}, W)$ *under condition* (2a).

For the proof of the theorem we need the following simple lemma. In the following F_{ξ} will denote the distribution function of ξ.

LEMMA 1. *Let ξ_1 and η_1 be independent and let ξ_2 and η_2 be independent. Then*

$$L(F_{(\xi_1+\eta_1)^+}, F_{(\xi_2+\eta_2)^+}) \leqslant L(F_{\xi_1}, F_{\xi_2}) + L(F_{\eta_1}, F_{\eta_2}).$$

The same result also holds for the metric U.

The proof follows immediately from the relations

$$L(F_{\xi_1+\eta_1}, F_{\xi_2+\eta_2}) \leqslant L(F_{\xi_1}, F_{\xi_2}) + L(F_{\eta_1}, F_{\eta_2})$$
$$L(F_{\xi_1^+}, F_{\xi_2^+}) \leqslant L(F_{\xi_1}, F_{\xi_2})$$

which can be verified directly. □

The inequality for U is established in the same obvious way.

The following result follows from this lemma. We denote

$$\bar{Y}_{0,k} = \max_{0\leqslant j\leqslant k} Y_{0,j}, \quad W_k(x) = \mathbf{P}(\bar{Y}_{0,k} < x).$$

COROLLARY. $L(W_k^{(r)}, W_k) \leqslant kL(F^{(r)}, F)$.

This follows from lemma 1 and the fact that $\bar{Y}_{0,k+1}$ and $\bar{Y}_{0,k}$ are connected by the relation

$$\bar{Y}_{0,k+1} \underset{p}{=} \bar{Y}_{1,k+1} = (\bar{Y}_{0,k} + \xi_0)^+. \quad \square$$

Proof of theorem 2. We have

$$W^{(r)}(x) = \mathbf{P}(Y^{(r)} < x) \leqslant \mathbf{P}(\theta^{(r)} > N) + \mathbf{P}(\bar{Y}_{0,N}^{(r)} < x, \theta^{(r)} \leqslant N)$$

where θ is the smallest k for which $\bar{Y}_{0,k} = Y$. Hence, and from lemma 1, we obtain

$$\begin{aligned} W^{(r)}(x) &\leqslant \mathbf{P}(\theta^{(r)} > N) + \mathbf{P}(\bar{Y}_{0,N}^{(r)} < x) \\ &\leqslant \mathbf{P}(\theta^{(r)} > N) + \mathbf{P}(\bar{Y}_{0,N} < x + \varepsilon N) + \varepsilon N \\ &\leqslant \mathbf{P}(\theta^{(r)} > N) + \mathbf{P}(\theta > N) + \mathbf{P}(Y < x + \varepsilon N) + \varepsilon N \\ &\leqslant T_{r,N} + W(x + \varepsilon N) + \varepsilon N \end{aligned} \tag{4.20}$$

where $T_{r,N} = \mathbf{P}(\theta^{(r)} > N) + \mathbf{P}(\theta > N)$.

We now turn to the estimation of the numbers $T_{r,N}$. By the first condition of the theorem for any $\delta > 0$, $\delta < -a$ we may choose number M and a distribution function $\Phi(x)$, such that

(1) $1 - F^{(r)}(x) \leqslant 1 - \Phi(x), \quad 1 - F(x) \leqslant 1 - \Phi(x)$

(2) $1 - \Phi(x) = G(x) \quad \text{for } x \geqslant M$

(3) $\displaystyle\int x \, d\Phi(x) < a + \delta < 0.$

We denote by ξ_j^* independent random variables with distribution functions Φ. Then, with the obvious notation, we obtain

$$\begin{aligned} \mathbf{P}(Y_{0,k} > 0) &\leqslant \mathbf{P}(Y_{0,k}^* > 0) \\ \mathbf{P}(Y_{0,k}^{(r)} > 0) &\leqslant \mathbf{P}(Y_{0,k}^* > 0). \end{aligned} \tag{4.21}$$

Further, it is known (Borovkov, 1976a, p. 121) that the following identity holds:

$$\sum_{k=0}^{\infty} z^k \mathbf{P}(\theta = k) = \mathbf{P}(\theta = 0) \exp\left\{\sum_{k=1}^{\infty} \frac{z^k}{k} \mathbf{P}(Y_{0,k} > 0)\right\}.$$

By (4.21) the coefficients $\mathbf{P}(\theta = k)/\mathbf{P}(\theta = 0)$ will be majorized by coefficients of the series

$$g(z) = \exp\left\{\sum_{k=1}^{\infty} \frac{z^k}{k} \mathbf{P}(Y_{0,k}^* > 0)\right\}.$$

To estimate $\mathbf{P}(Y_{0,k}^* > 0)$ we use the following result on the probabilities of large deviations (Nagev, (1981)).

Let $G(x)$ be a power function, $\alpha > 1, \mathbf{E}|\xi_1^*| < \infty$. Then for $x > ck, c > 0, k \to \infty$

$$\mathbf{P}(Y_{0,k}^* - a^*k > ck) \sim \beta k G(ck) = kP(\xi_1^* > ck). \tag{4.22}$$

(We note that in fact it is enough for use to have an inequality for the left-hand side of (4.22) (more precisely for $\mathbf{P}(Y_{0,k}^* > 0)$.) Up to a constant the required inequality

$$\mathbf{P}(Y_{0,k}^* - a^*k > ck) < \beta k G(ck), \quad \beta > 1$$

can be obtained from Dao Kha Fuk and Nagaev (1971); (4.22) also allows us to suppose that for large k the coefficient β can be chosen arbitrarily close to 1. Under the additional condition $\mathbf{E}(\xi_1^*)^2 < \infty$ (4.22) is contained in Nagaev (1969).

For $c = -a^* > 0$ (4.22) implies

$$\mathbf{P}(Y_{0,k}^* > 0) \sim kG(-a^*k).$$

Using theorems on the behaviour of the coefficients of power series (Borovkov, 1976a, p. 258; see also Essen, 1973, and Rogozin, 1973) we find

$$g(z) = \sum_{k=0}^{\infty} z^k g_k, \quad g_k \sim \frac{\mathbf{P}(Y_{0,k}^* > 0)}{k} \sim G(-a^*k)$$

$$\mathbf{P}(\theta = k) \leqslant \mathbf{P}(\theta = 0) G(-a^*k)(1 + \gamma)$$

$$\mathbf{P}(\theta^{(r)} = k) \leqslant \mathbf{P}(\theta^{(r)} = 0) G(-a^*k)(1 + \gamma)$$

for any $\gamma > 0$ and all sufficiently large k uniformly with respect to r.

We may return now to the inequalities (4.20) in which, as we have explained,

$$T_{r,N} \leqslant 2(\alpha - 1)^{-1} \mathbf{P}(\theta = 0) NG(-aN)(1 + \gamma) = dNG(N)(1 + \gamma)$$

for all sufficiently large N and r and for any preassigned $\gamma > 0$ (the number a^* can be made arbitrarily close to a. In this equality $d = 2\mathbf{P}(\theta = 0)/(\alpha - 1)|a|^\alpha$, since $\limsup_{r\to\infty} \mathbf{P}(\theta^{(r)} = 0) \leqslant \mathbf{P}(\theta = 0)$). Putting N equal to the solution $G^{-1}(\varepsilon b/d)$ of the equation $dG(N) = \varepsilon b$, we obtain

$$W^{(r)}(x) \leqslant W\left(x + \varepsilon G^{-1}\left(\frac{\varepsilon b}{d}\right)\right) + \varepsilon G^{-1}\left(\frac{\varepsilon b}{d}\right)(1 + b)(1 + \gamma). \qquad (4.23)$$

For simplicity we assume here that ε takes values for which $G^{-1}(\varepsilon b/d)$ is an integer.

The converse inequality is established in a similar way which, together with (4.23), proves the first assertion of the theorem.

In order to pass to arbitrary sequences $\varepsilon \to 0$ it is necessary in the condition of the theorem to take ε to be the smallest number larger than $L(F^{(r)}, F)$ for which $G^{-1}(\varepsilon b/d)$ is an integer. This can add to the estimate of $L(W^{(r)}, W)$ only a multiplier $1 + o(1)$, which is contained in $1 + \gamma$.

Now let $G(x)$ be an exponential function. In this case, together with (4.22), we will have by a Chebyshev-type inequality

$$\mathbf{P}(Y^*_{0,k} > 0) \leqslant (m^*)^k$$

where $m^* = \inf_\mu f^*(\mu) < 1$, $f^*(\mu) = \mathbf{E}\, e^{\mu\xi_1^*}$. Using the above theorems again we obtain

$$\mathbf{P}(\theta = k) \leqslant \frac{c(m^*)^k}{k}, \quad \mathbf{P}(\theta^{(r)} = k) \leqslant \frac{c(m^*)^k}{k}.$$

For $\mathbf{P}(\theta \geqslant k)$ and $\mathbf{P}(\theta^{(r)} \geqslant k)$ there will obviously be inequalities of the same type. But m^* can be made arbitrarily close to $m = \inf_\mu f(\mu)$, $f(\mu) = \mathbf{E}\, e^{\mu\xi}$. Therefore for any $\gamma > 0$ and sufficiently large k we may write

$$\mathbf{P}(\theta \geqslant \mathrm{k}) \leqslant \tfrac{1}{2}(m + \gamma)^k, \quad \mathbf{P}(\theta^{(r)} \geqslant k) \leqslant \tfrac{1}{2}(m + \gamma)^k.$$

Since, in our case, $T_{r,N}$ in (4.20) does not exceed $(m + \gamma)^N$, then for $N = \log \varepsilon/\log(m + \gamma)$ (assuming again that N is an integer)

$$W^{(r)}(x) \leqslant W\left(x + \frac{\varepsilon \log \varepsilon}{\log(m + \gamma)}\right) + \frac{\varepsilon \log \varepsilon}{\log(m + \gamma)} + \varepsilon.$$

Together with the converse inequality this gives us the second assertion of the theorem for the metric L. For the metric U the proof is the same. □

We now consider an estimate of closeness of *prestationary* distributions of waiting time $w_n^{(r)}$ and w_n. As we know, the sequence of waiting times is defined by the initial value w_1 and the recurrence relation $w_{n+1} = \max(0, w_n + \xi_n)$, $n \geqslant 1$. Put $W_n(x) = \mathbf{P}(w_{n+1} < x)$, $W(x) = \mathbf{P}(Y(0) < x)$.

If $w_1 = w_1^{(r)} = 0$ then to obtain estimates for $L(W_n^{(r)}, W_n)$ the same as the estimates for $L(W^{(r)}, W)$ obtained in theorem 2, we must use, with no essential changes, the calculations of theorem 2: for $n \geqslant N$

$$\begin{aligned}W_n^{(r)}(x) &= \mathbf{P}(\bar{Y}_{0,n}^{(r)} < x) \leqslant \mathbf{P}(\theta^{(r)} > N) + \mathbf{P}(\bar{Y}_{0,N}^{(r)} < x)\\ &\leqslant \mathbf{P}(\theta^{(r)} > N) + \varepsilon N + \mathbf{P}(\bar{Y}_{0,n} < x + \varepsilon N)\\ &\leqslant \mathbf{P}(\theta^{(r)} > N) + \mathbf{P}(\theta > N) + \varepsilon N + \mathbf{P}(\bar{Y}_{0,n} < x + \varepsilon N)\\ &= T_{r,N} + \varepsilon N + W_n(x + \varepsilon N).\end{aligned}$$

Repeating all the remaining arguments in theorem 2, we obtain for $n \geqslant N$

$$L(W_n^{(r)}, W_n) \leqslant \varepsilon N + T_{r,N}.$$

Since for $n \geqslant N$

$$L(W_n^{(r)}, W_n) \leqslant \varepsilon n \leqslant \varepsilon N$$

then for $N = G^{-1}(\varepsilon b/d)$ we obtain a uniform estimate

$$\sup_n L(W_n^{(r)}, W_n) \leqslant (1 + b)\varepsilon N(1 + \gamma) = \varepsilon G^{-1}\left(\frac{\varepsilon b}{d}\right)(1 + b)(1 + \gamma).$$

The right-hand side of this estimate is the same as for the estimate of $L(W^{(r)}, W)$ which was established in theorem 2.

But if $w_1 = w_1^{(r)} \neq 0$ then the order of the estimate remains the same but the *uniformity* with respect to w_1, obviously, is impossible to obtain.

For simplicity suppose $w_1 = w_1^{(r)}$ is not random. Then, since

$$w_{n+1} \underset{d}{=} \max(\bar{Y}_{0,n}, Y_{0,n} + w_1)$$

we find as before that for $n \leqslant N$

$$L(W_n^{(r)}, W_n) \leqslant \varepsilon n \leqslant \varepsilon N.$$

For $n \geqslant N$ we have:

$$\begin{aligned}W_n^{(r)}(x) &\leqslant \mathbf{P}(\theta^{(r)} > N) + \mathbf{P}(\max(\bar{Y}_{0,N}^{(r)}, Y_{0,n}^{(r)} + w_1^{(r)}) < x; \theta^{(r)} \leqslant N)\\ &\leqslant \mathbf{P}(\theta^{(r)} > N) + \mathbf{P}(\bar{Y}_{0,N}^{(r)} < x)\\ &\leqslant \mathbf{P}(\theta^{(r)} > N) + \varepsilon N + \mathbf{P}(\bar{Y}_{0,N} < x + \varepsilon N)\\ &\leqslant \mathbf{P}(\theta^{(r)} > N) + \varepsilon N + \mathbf{P}(\max(\bar{Y}_{0,N}, Y_{0,n} + w_1) < x + \varepsilon N)\\ &\quad + \mathbf{P}(Y_{0,n} + w_1 \geqslant x + \varepsilon N)\\ &\leqslant T_{r,N} + \varepsilon N + \mathbf{P}(Y_{0,n} \geqslant -w_1) + W_n(x + \varepsilon N).\end{aligned}$$

The converse inequality for $W_n^{(r)}(x)$ is written similarly. In these inequalities for any $\gamma > 0$

$$\begin{aligned}\mathbf{P}(Y_{0,n} \geqslant -w_1) &\leqslant \mathbf{P}(Y_{0,n}^* \geqslant -w_1) \sim nG(-a^*n - w_1)\\ &\leqslant |a|^{-\alpha} nG(n)(1 + \gamma) \leqslant |a|^{-\alpha} NG(N)(1 + \gamma)\end{aligned}$$

for all sufficiently large N.

Since $T_{r,N} \leqslant dNG(N)$, then for sufficiently large N we obtain

$$\sup L(W_n^{(r)}, W_n) \leqslant \varepsilon N + NG(N)(d + |a|^{-\alpha})(1 + \gamma).$$

Putting $N = G^{-1}(\varepsilon b/(d + |a|^{-\alpha}))$, we obtain as before

$$\sup L(W_n^{(r)}, W_n) \leqslant \varepsilon G^{-1}\left(\frac{\varepsilon b}{d + |a|^{-\alpha}}\right)(1 + b)(1 + \gamma)$$

for any $b > 0$, $\gamma > 0$ and all sufficiently large r.

This result is easily extended to random w_1, $w_1^{(r)}$, not depending on $\{\xi_j\}$, $\{\xi_j^{(r)}\}$ and such that

$$L(W_0^{(r)}, W_0) < c\varepsilon G^{-1}(\varepsilon), \quad \mathbf{P}(w_1 > x) < cxG(x)$$

$$\mathbf{P}(w_1^{(r)} > x) < cxG(x).$$

The remarks made above about prestationary distributions also hold for the other results (except theorem 2) of this section.

4.3.3. We will now suppose that ξ_j *and* $\xi_j^{(r)}$ *are given on the same probability space and are close in the sense of the distance*

$$\rho(\xi, \eta) = \inf\{\delta: \mathbf{P}(|\xi - \eta| > \delta) < \delta\}.$$

Then Y and $Y^{(r)}$ will also be given on the same probability space. We obtain an estimate* of $\rho(Y^{(r)}, Y)$ by $\varepsilon = \varepsilon_r = \rho(\xi_1^{(r)}, \xi_1)$.

We note that giving ξ_j and $\xi_j^{(r)}$ on the same probability space and their closeness in the metric ρ may be acceptable in the general case, if it is known that the distance $\rho(F^{(r)}, F)$, between $F^{(r)}$ and F, is small, where by definition

$$\rho(\Phi, F) = \inf\{\delta: \text{mes}\,[x: |\Phi^{-1}(x) - F^{-1}(x)| > \delta] < \delta\}$$

where F^{-1} and Φ^{-1} are the functions inverse to F and Φ, defined by left continuity and, at 0, by right continuity, so that $F^{-1}(0) = \inf\{x: F(x) > 0\}$. In fact, if we put

$$\xi_j = F^{-1}(\omega_j), \quad \xi_j^{(r)} = (F^{(r)})^{-1}(\omega_j)$$

where ω_j are independent and uniformly distributed on $[0, 1]$, then we obtain† $\rho(\xi_j^{(r)}, \xi_j) = \rho(F^{(r)}, F)$.

In addition, if it is known that the distributions $\mathbf{P}^{(r)}$ and $\mathbf{P}$ of $\xi_j^{(r)}$ and ξ_j are close in the Levy–Prokhorov distance $\lambda(\mathbf{P}^{(r)}, \mathbf{P})$ (for the definition, see below), then by Strassen's theorem (Dudley, 1968; Geza Schay, 1974; Strassen, 1965) there exist $\tilde{\xi}_j^{(r)}$ and $\tilde{\xi}_j$ given on the same probability space such that $\rho(\tilde{\xi}_j^{(r)}, \tilde{\xi}_j) \leqslant \lambda(\mathbf{P}^{(r)}, \mathbf{P})$.

The following result, analogous to theorem 2, holds.

* It is not difficult to see that $\rho(Y^{(r)}, Y) < \delta$ implies $L(W^{(r)}, W) < \delta$.

† Thus $\rho(\xi_j^{(r)}, \xi_j) \leqslant \rho(F^{(r)}, F)$. The reverse inequality does not hold. For example, let ξ be uniformly distributed on $[0, N]$ and $\xi^{(r)} = \xi$, if $\xi \geqslant 1$; $\xi^{(r)} = \xi + N$ if $\xi < 1$. Then, obviously, $\xi^{(r)}$ is uniformly distributed on $[1, N + 1]$, $\rho(F^{(r)}, F) = 1$ and $\rho(\xi^{(r)}, \xi) = 1/N$.

THEOREM 3. *Let* $\{\xi_j\}$ *and* $\{\xi_j^{(r)}\}$ *be given on the same probability space. Suppose that conditions* (1), (3) *of theorem* 2 *are satisfied and as* $r \to \infty$

$$\varepsilon = \varepsilon_r = \rho(\xi_1^{(r)}, \xi_1) \to 0.$$

Then, if G is a power function, for all $b > 0, \gamma > 0$ *and for all sufficiently large r*

$$\rho(Y^{(r)}, Y) \leqslant \varepsilon G^{-1}\left(\frac{\varepsilon b}{d}\right)(1+b)(1+\gamma).$$

If G is an exponential function then

$$\rho(Y(r), Y) \leqslant \frac{\varepsilon \log \varepsilon}{\log m}(1+\gamma).$$

Here all notation has the same meaning as in theorem 2.

However, in this case, in addition to the result of theorem 3, it is possible to say that the estimates obtained are best possible.*

THEOREM 4. *There exist distributions for* ξ_j *satisfying conditions* (1), (3) *of theorem* 2, *such that if* $\xi_j^{(r)} = \xi_j + \varepsilon$, *then for a power function G, any* $\gamma > 0$, *and all sufficiently large r* ($\varepsilon = \varepsilon_r \to 0$ *as* $r \to \infty$) *we have*

$$\rho(Y^{(r)}, Y) \geqslant \varepsilon G^{-1}\left(\frac{2\varepsilon}{d}\right)(1-\gamma).$$

Here d has the same meaning as in theorems 2, 3. *For an exponential function G*

$$\rho(Y^{(r)}, Y) \geqslant \frac{\varepsilon \log \varepsilon}{\log m}(1-\gamma).$$

We note that these estimates may hold simultaneously with the estimate

$$V(W^{(r)}, W) < c_1\varepsilon.$$

This follows from theorem 1, the conditions of which are easily satisfied for $V(F^{(r)}, F) \sim c_2\varepsilon$, $V(Q^{(r)}, Q) \sim c_3\varepsilon$.

Proof of theorem 3. This is similar to the proof of theorem 2. We denote $\rho(\xi_1^{(r)}, \xi_1) = \varepsilon$,

$$B_N = \bigcup_{j=1}^{N} \{|Y_{0,j}^{(r)} - Y_{0,j}| > j\varepsilon\}$$

* Some nearby estimates of speed of convergence for the distributions of $w^{(r)k}$ and w^k (or $Y^{(r)}$ and Y), and also for the distributions of prestationary waiting times, are given in Zolotarev (1975b, 1977).

and note first that $B_N \subset \bigcup_{j=1}^{N} \{|\xi_{-j}^{(r)} - \xi_{-j}| > \varepsilon\}$,

$$\mathbf{P}(B_N) \leqslant \sum_{j=1}^{N} \mathbf{P}(|\xi_{-j}^{(r)} - \xi_{-j}| > \varepsilon) \leqslant N\varepsilon.$$

If we now denote

$$A_N = \{|Y^{(r)} - Y| > N\varepsilon\},$$

then we obtain

$$\mathbf{P}(A_N) \leqslant \mathbf{P}(A_N \bar{B}_N;\, \theta^{(r)} \leqslant N, \theta \leqslant N) + \mathbf{P}(B_N) + \mathbf{P}(\theta^{(r)} > N) + \mathbf{P}(\theta > N).$$

Since the intersection $A_N \bar{B}_N \{\theta^{(r)} \leqslant N, \theta \leqslant N\}$ is empty then we find

$$\mathbf{P}(A_N) \leqslant N\varepsilon + \mathbf{P}(\theta^{(r)} > N) + \mathbf{P}(\theta < N).$$

The choice of N on the right-hand side in this inequality occurs in precisely the same way as in the preceding theorem. As a result for $\rho(Y^{(r)}, Y)$ we obtain the same estimate as for $L(W^{(r)}, W)$ in theorem 2. The theorem is proved. □

Proof of theorem 4. Choose a smooth distribution ξ_1 so that $1 - F(x) = G(x)$ for $x > 0$ and put $\xi_j^{(r)} = \xi_1 + \varepsilon$, $\varepsilon = \varepsilon_r > 0$. Then $\rho(\xi_j^{(r)}, \xi_j) = \varepsilon$,

$$Y_{0,k}^{(r)} = Y_{0,k} + k\varepsilon, \quad Y^{(r)} \geqslant Y + \theta\varepsilon$$

and, consequently,

$$\rho(Y^{(r)}, Y) \geqslant \rho(0, \theta\varepsilon).$$

In the proof of theorem 2 we have established ($\xi_j^* = \xi_j$ in our case) that for a power function G

$$\mathbf{P}(\theta = k) \sim \mathbf{P}(\theta = 0) G(-ak)$$

as $k \to \infty$, and, consequently, as $N \to \infty$

$$\mathbf{P}(\theta > N) \sim \frac{\mathbf{P}(\theta = 0) N G(-aN)}{\alpha - 1} \sim d_1 N G(N)$$

$$d_1 = \frac{\mathbf{P}(\theta = 0)}{|a|^\alpha (\alpha - 1)} = \frac{d}{2}.$$

But

$$\rho(0, \theta\varepsilon) = \inf\{\delta\colon \mathbf{P}(\theta\varepsilon > \delta) < \delta\}.$$

Solving the equation

$$\mathbf{P}(\theta\varepsilon > \delta) \sim d_1\left(\frac{\varepsilon}{\delta}\right) G\left(\frac{\varepsilon}{\delta}\right) = \delta$$

relative to δ, we obtain $\delta = G^{-1}(\varepsilon/d_1)$,

$$\liminf_{r\to\infty} \frac{\rho(Y^{(r)}, Y)}{\varepsilon G^{-1}(\varepsilon/d_1)} \geqslant 1.$$

For an exponential function G we have

$$\mathbf{P}(Y_{0,k} > 0) \sim \frac{b_1 m^k}{\sqrt{k}}, \quad m = \inf_{\mu} \mathbf{E}\, e^{\mu\xi_1} < 1$$

$$\mathbf{P}(\theta = k) \sim \frac{b_2 m^k}{k^{3/2}}, \quad \mathbf{P}(\theta \geqslant k) \sim \frac{b_3 m^k}{k^{3/2}}.$$

Solving the equation

$$\mathbf{P}(\theta\varepsilon > \delta) \sim b_3 m^{\delta/\varepsilon}\left(\frac{\delta}{\varepsilon}\right)^{-3/2} = \delta = \left(\frac{\delta}{\varepsilon}\right)\varepsilon$$

(or, what is the same thing, the equation $b_3 m^N N^{-5/2} = \varepsilon$), we obtain

$$\frac{\delta}{\varepsilon} = \frac{\log \varepsilon}{\log m} + o\left(\frac{\log \varepsilon}{\log m}\right)$$

$$\rho(Y^{(r)}, Y) \geqslant \varepsilon\left[\frac{\log \varepsilon}{\log m} + o\left(\frac{\log \varepsilon}{\log m}\right)\right]. \quad \square$$

4.3.4. We note now that there is the possibility of obtaining estimates of the speed of convergence for *dependent* ξ_j and for *multidimensional distributions* $\{w^k\}$. If we keep in view the method of obtaining estimates, given above, then it is easy to see that these possibilities are determined by the following two factors:

(1) The presence of estimates for the probabilities $\mathbf{P}(\theta > N)$ and $\mathbf{P}(\theta^{(r)} > N)$.

(2) The possibility of describing the differences $\mathbf{P}((Y_{0,0}^{(r)}, \ldots, Y_{0,N}^{(r)}) \in B) - \mathbf{P}((Y_{0,0}, \ldots, Y_{0,N}) \in B)$ for sets B of a special form (cf. the proof of theorem 2).

Estimates for $\mathbf{P}(\theta > N)$, although rough, can be obtained using, for example, the relation

$$\{\theta > N\} \subset \bigcup_{k=N+1}^{\infty} \{Y_{0,k} > 0\}$$

from which it follows that

$$\mathbf{P}(\theta > N) \leqslant \sum_{k=N+1}^{\infty} \mathbf{P}(Y_{0,k} > 0).$$

The right-hand side here can be estimated using the condition of weak dependence of $\{\xi_j\}$ and the finiteness of the moments of ξ_j of sufficiently high order.

For example, if for $\zeta_j = \xi_j - a$

$$|\mathbf{E}\zeta_{j-b}\zeta_j\zeta_k\zeta_{k+d}| \leqslant m_4 \min[\phi(b), \phi(d)], \quad j \leqslant k$$

where

$$\phi = \sum n\phi(n) < \infty$$

then

$$\sum_{k=N+1}^{\infty} \mathbf{P}(Y_{0,k} > 0) \leqslant \frac{3m_4(8\phi + 1)}{Na^4}$$

(see Borovkov, 1976a, p. 48).

But if ξ_j is a function on the states of a homogeneous Markov chain, then on the basis of factorization identities much more precise estimates for $\mathbf{P}(\theta > N)$ are possible, similar to the estimates obtained for independent ξ_j.

We now consider the second of the above-mentioned factors. Here, evidently, it is convenient to use the Levy–Prokhorov distance $\lambda(\cdot, \cdot)$ between distributions. If $\mathbf{P}$ and $\mathbf{Q}$ are two distributions, say on R^k, then by definition

$$\lambda(\mathbf{P}, \mathbf{Q}) = \inf\{\varepsilon: \mathbf{P}(B) < \mathbf{Q}(B^\varepsilon) + \varepsilon, \mathbf{Q}(A) < \mathbf{P}(A^\varepsilon) + \varepsilon\}.$$

Here the relation under the inf sign must be satisfied for all closed sets A and B from R^k, A^ε denotes the ε-neighbourhood of A. Obviously we could also look for uniform closeness of $\mathbf{P}$ and $\mathbf{Q}$ on a given class of sets, if this class includes the sets that we need.

The introduction of the metric λ allows the characterization of closeness of finite-dimensional distributions of sequences $\{\xi_j^{(r)}\}$ and $\{\xi_j\}$. However, in our case, to use λ in many-dimensional spaces is evidently inappropriate. For us it is suitable to use a somewhat different characteristic of closeness of distributions. To some extent this may explain the following assertion. Let $\mathbf{P}_N$ be the distribution of the N-dimensional vector $(\xi_{j-1}, \ldots, \xi_{j-N})$ in R^N.

LEMMA 2. *Let* $\{\xi_j\}$ *and* $\{\xi_j^{(r)}\}$ *be sequences of independent variables,* $\lambda(\mathbf{P}_1^{(r)}, \mathbf{P}_1) < \varepsilon$. *Then for any closed* $B \subset R^N$

$$\mathbf{P}_N^{(r)}(B) \leqslant \mathbf{P}_N(B^{\varepsilon\sqrt{N}}) + \varepsilon N$$

$$\mathbf{P}_N(B) \leqslant \mathbf{P}_N^{(r)}(B^{\varepsilon\sqrt{N}}) + \varepsilon N.$$

*Proof.** It is sufficient to prove that if for independent $\zeta \in R^k$ and $\eta \in R^l$

$$\mathbf{P}(\zeta \in A) \leqslant \alpha_1 + \mathbf{P}(\zeta^{(r)} \in A^\alpha)$$

$$\mathbf{P}(\eta \in B) \leqslant \beta_1 + \mathbf{P}(\eta^{(r)} \in B^\beta)$$

for all closed $A \subset R^k$, $B \subset R^l$, then for each closed $C \subset R^{k+l}$

$$\mathbf{P}((\zeta, \eta) \in C) \leqslant \alpha_1 + \beta_1 + \mathbf{P}((\zeta^{(r)}, \eta^{(r)}) \in C^{\sqrt{\alpha^2 + \beta^2}}).$$

* The proof given below was suggested by V. V. Yurinskii.

We denote by $C_{.y}$ and $C_{x.}$ the sections of the set C

$$C_{.y} = \{x: (x, y) \in C\}, \quad C_{x.} = \{y: (x, y) \in C\}$$

and put

$$C^{\cdot\beta} = \bigcup_x \{z = (x, y): y \in (C_{x.})^\beta\}$$

$$C^{\alpha\cdot} = \bigcup_x \{z = (x, y): x \in (C_{.y})^\alpha\}.$$

Then

$$\begin{aligned}\mathbf{P}((\zeta, \eta) \in C) &= \int_{R^k} \mathbf{P}(\eta \in C_{x.})\mathbf{P}(\zeta \in \mathrm{d}x)\\ &\leqslant \int_{R^k} [\beta_1 + \mathbf{P}(\eta^{(r)} \in (C_{x.})^\beta]\mathbf{P}(\zeta \in \mathrm{d}x)\\ &\leqslant \beta_1 + \mathbf{P}((\zeta, \eta^{(r)}) \in C^{\cdot\beta})\\ &= \beta_1 + \int_{R^l} \mathbf{P}(\zeta \in (C^{\cdot\beta})_{.y})\mathbf{P}(\eta^{(r)} \in \mathrm{d}y)\\ &\leqslant \alpha_1 + \beta_1 + \int_{R^l} \mathbf{P}(\zeta^{(r)} \in [(C^{\cdot\beta})_{.y}]^\alpha \mathbf{P}(\eta^{(r)} \in \mathrm{d}y)\\ &= \alpha_1 + \beta_1 + \mathbf{P}((\zeta^{(r)}, \eta^{(r)}) \in (C^{\cdot\beta})^{\alpha\cdot}).\end{aligned} \tag{4.24}$$

But $(C^{\cdot\beta})^{\alpha\cdot} \subset C^{\sqrt{\alpha^2+\beta^2}}$. In fact, let $(x, y) \in (C^{\cdot\beta})^{\alpha\cdot}$. Then there is a point $(x_0, y) \in C^{\cdot\beta}$ such that $|x_0 - x| < \alpha$. This in turn means that there is a point $(x_0, y_0) \in C$ with $|y_0 - y| < \beta$. Thus the distance between (x, y) and (x_0, y_0) does not exceed

$$\sqrt{(x_0 - x)^2 + (y_0 - y)^2} \leqslant \sqrt{\alpha^2 + \beta^2}.$$

The lemma is proved. □

On the basis of this lemma and for reasons which will become clear later, it is natural to characterize closeness of distributions in R^N by the numbers

$$\begin{aligned}\varepsilon(N) &= \lambda^*(\mathbf{P}_N^{(r)}, \mathbf{P}_N)\\ &= \inf\{\varepsilon: \mathbf{P}_N^{(r)}(B) < \mathbf{P}_N(B^{\varepsilon\sqrt{N}}) + \varepsilon N, \mathbf{P}_N(A) < \mathbf{P}_N^{(r)}(A^{\varepsilon\sqrt{N}}) + \varepsilon N\}.\end{aligned}$$

Then, say, for a set $B \subset R^1$ of the form $B = [b_1, b_2]$ we will have ($\bar{Y}_{0,N} = \max_{0 \leqslant j \leqslant N} Y_{0,j}$)

$$\mathbf{P}(Y^{(r)} \in B) \leqslant \mathbf{P}(\theta^{(r)} > N) + \mathbf{P}(\bar{Y}_{0,N}^{(r)} \in B). \tag{4.25}$$

Corresponding to the event $\{\bar{Y}_{0,N} \in B\}$ there is a domain A in R^N contained between two similar polygonal surfaces, such that $\{\bar{Y}_{0,N} \in B\} = \{\xi_{-1}, \ldots, \xi_{-N} \in A\}$ (recall that $\{\bar{Y}_{0,N} \leqslant b\} = \{\xi_{-1} \leqslant b,\ \xi_{-1} + \xi_{-2} \leqslant b,\ \ldots,$

$\xi_{-1} + \cdots + \xi_{-N} \leqslant b\}$). Using the condition of closeness of $\mathbf{P}_N^{(r)}$ and $\mathbf{P}_N$, we may write

$$\mathbf{P}(\bar{Y}_{0,N}^{(r)} \in B) \leqslant \mathbf{P}((\xi_{-1}, \ldots, \xi_{-N}) \in A^{\varepsilon(N)\sqrt{N}}) + \varepsilon(N)N.$$

Since

$$\{\bar{Y}_{0,N} < b\}^{\alpha} = \{\xi_{-1} < b + \alpha, \xi_{-1} + \xi_{-2} < b + \alpha\sqrt{2}, \ldots, \xi_{-1} + \cdots + \xi_{-N} < b + \alpha\sqrt{N}\}$$
$$\subset \{\bar{Y}_{0,N} < b + \alpha\sqrt{N}\}$$

then it is easy to obtain

$$\mathbf{P}(\bar{Y}_{0,N}^{(r)} \in B) \leqslant \mathbf{P}(\bar{Y}_{0,N} \in B^{\alpha N}) + N\varepsilon(N).$$

Together with (4.25) this gives us

$$\mathbf{P}(Y^{(r)} \in B) \leqslant T_N + N\varepsilon(N) + \mathbf{P}(Y \in B^{N\varepsilon(N)})$$

where T_N is such that

$$\mathbf{P}(\theta^{(r)} > N) + \mathbf{P}(\theta > N) \leqslant T_N \to 0 \quad \text{as } N \to \infty.$$

This relation is preserved even for arbitrary closed B. Taking $\varepsilon(N) = \varepsilon\psi(N)$, $\varepsilon = \varepsilon_r \to 0$ as $r \to \infty$ ($\psi(N) = 1$ for independent ξ_j), and solving for N the equation

$$T_N = N\psi(N)\varepsilon \tag{4.26}$$

we obtain

$$\lambda(\mathbf{Q}^{(r)}, \mathbf{Q}) < N_\varepsilon \varepsilon \psi(N_\varepsilon)$$

where N_ε is the solution of (4.26) and $\mathbf{Q}$ is the distribution of Y. Passing on to consideration of the multidimensional case, when $\mathbf{Q}$ is the distribution of $(Y(m_1), \ldots, Y(m_k))$, no essential changes in the above arguments are needed.

4.3.5. In conclusion we indicate one further class of estimates which, like the estimates of theorem 1, are best possible relative to order of smallness. As a preliminary we note that one of the aims of obtaining different kinds of estimates between the distributions $Y^{(r)}$ and Y (or $w_n^{(r)}$ and w_n in the prestationary case) is to estimate the closeness of $\mathbf{E}f(Y^{(r)})$ and $\mathbf{E}f(Y)$ for various kinds of smooth functions f. We assume that

(1) conditions (1), (3) of theorem 2 are satisfied; here

$$\int_1^\infty xG(x)\,dx < \infty$$

(2) the random variables $\xi_1^{(r)}$ and ξ_1 can be given on the same probability space so that $|\xi_1^{(r)} - \xi_1| \leqslant \varepsilon$ a.s.

Under these conditions for any f satisfying a Lipschitz condition

$$|f(x + \Delta) - f(x)| < c\Delta \tag{4.27}$$

we have

$$|\mathbf{E}f(Y^{(r)}) - \mathbf{E}f(Y)| < c_1\varepsilon. \tag{4.28}$$

The proof follows immediately from

$$|\mathbf{E}f(Y^{(r)}) - \mathbf{E}F(Y)| \leqslant \sum_{n=0}^{\infty} \mathbf{E}[|f(Y^{(r)}) - f(Y)|;\ \max(\theta^{(r)}, \theta) = n]$$

$$\leqslant \sum_{n=0}^{\infty} c\varepsilon n \mathbf{P}(\max(\theta^{(r)}, \theta) = n)$$

$$= c\varepsilon \mathbf{E} \max(\theta^{(r)}, \theta)$$

where, according to the proof of theorem 2,

$$\mathbf{P}(\max(\theta^{(r)}, \theta) > N) \leqslant \mathbf{P}(\theta > N) + \mathbf{P}(\theta^{(r)} > N)$$

$$= T_{r,N} \leqslant dNG(N)(1 + \gamma).$$

By condition (1) this means that $\mathbf{E} \max(\theta^{(r)}, \theta) < c_2$, where c_2 does not depend on r and, consequently, (4.28) is proved.

We may also note that (4.28) implies $L(W^{(r)}, W) \leqslant c_3\sqrt{\varepsilon}$.

4.4. Ergodicity and stability theorems for random walks in a strip and their application to single-channel systems with constraints

In this section the question is of ergodicity and stability theorems for random walks in a strip and the application of the results obtained to systems of type $\langle G, 1, G, 1\rangle$ with restricted waiting time or restricted sojourn time in the system, and to systems of type $\langle E, G, G, G\rangle$, $\langle G, G, E, G\rangle$ with restricted queues.

4.4.1. In systems with restricted waiting time requests are refused if their waiting time (first type of system) or waiting time of the preceding requests (second type of system) exceeds some given level N. In this case the relations describing the waiting time of the $(n+1)$st request will be (cf. (4.11)):

$$w_{n+1} = \max(0, w_n + y_n), \quad n \geqslant 1$$

where, for systems of the first type

$$y_n = \begin{cases} \xi_n & \text{if } w_n + \xi_n \leqslant N \\ -\tau_n^e & \text{otherwise} \end{cases} \tag{4.29}$$

for systems of the second type

$$y_n = \begin{cases} \xi_n & \text{if } w_n \leqslant N \\ -\tau_n^e & \text{otherwise.} \end{cases} \tag{4.30}$$

Here the notations $w_n, \tau_n^e, \tau_n^s, \xi_n = \tau_n^s - \tau_n^e$ have the same meaning as in sections 4.2 and 4.3.

Instead of (4.30) and (4.29) it is possible to consider the controls

$$y_n = \begin{cases} \xi_n & \text{if } w_n + \xi_n + \tau^s_{n+1} \leqslant N \\ -\tau^e_n & \text{otherwise} \end{cases} \tag{4.31}$$

or

$$y_n = \begin{cases} \xi_n & \text{if } w_n + \tau^s_n \leqslant N \\ -\tau^e_n & \text{otherwise} \end{cases} \tag{4.32}$$

meaning that the $(n + 1)$st request leaves the system if its sojourn time in the system (or the sojourn time of the previous request in the case (4.32)) exceeds the level N. Sometimes (see Afanas'eva, 1965; Afanas'eva and Martynov, 1969) 'continuous' control in a system with restricted sojourn time is discussed, for example, of the form

$$w_{n+1} = \begin{cases} \max(0, w_n + \xi_n) & \text{if } w_n + \tau^s_n \leqslant N \\ \max(0, w_n - \tau^e_n) & \text{if } w_n > N \\ \max(0, N - \tau^e_n) & \text{if } w_n \leqslant N < w_n + \tau^s_n \end{cases} \tag{4.33}$$

(if $w_n > N$ then the nth request is refused; if $w_n \leqslant N < w_n + \tau^s_n$ then the nth request waits for a time w_n and is served only for the 'remaining' time $N - w_n < \tau^s_n$, after which it is removed from the system).

It must be noted that in all of these problems w_n is in some sense the virtual waiting time, since it is possible that the nth request is refused.

Concerning the sequence (4.33) we observe here that it is majorized (under the same initial conditions) by the sequence determined by (4.30). This allows us to reduce the study of (4.33) to that of (4.30). Equation (4.33) itself will, in some sense, be simpler than (4.29)–(4.32) since the function $f(x, \tau^e_n, \tau^s_n)$ in the equality $w_{n+1} = f(w_n, \tau^e_n, \tau^s_n)$ for relation (4.33) is continuous and monotone with respect to x. This makes it possible in the case (4.33) to apply a method proposed by Loynes (1962) (which has been done in Afanas'eva, 1965, Afanas'eva and Martynov, 1969, and Krupin, 1976). The approach, considered below, is more universal and makes it possible, alongside the ergodic theorems, to also obtain stability theorems.

In what follows we will discuss a more general formulation of the problem when the bounding level N depends on n (one for each request) and is a random variable. In this case (4.29)–(4.32) cease to differ and take the simplified and more general form

$$w_{n+1} = \max(0, w_n + y_n), \quad n \geqslant 1, \; w_1 \leqslant N_1$$
$$y_n = \begin{cases} \xi_n & \text{if } w_n \leqslant N_n \\ -\tau^e_n & \text{otherwise} \end{cases} \tag{4.34}$$

where $\{N_n, \tau^e_n, \tau^s_n\}$, $\xi_n = \tau^s_n - \tau^e_n$ is the given controlling random sequence.

In this section we consider, one more type of equation, no less important from the point of view of applications, which takes the form

$$w_{n+1} = \min\left[N_{n+1}, \max\left(M_{n+1}, w_n + \xi_n\right)\right] \tag{4.35}$$

where $\{\xi_n, M_{n+1}, N_{n+1}\}$ is a given control sequence. By such equations is described, in particular, the queue length for systems $\langle E, G, G, G\rangle$ and $\langle G, G, E, G\rangle$ in the case when this length at the nth step is bounded by a random level K_n (cf. Borovkov, 1976a). For example, for systems $\langle E, G, G, G\rangle$ the queue length q_n of the queue before the arrival epoch of the nth group of requests, bounded by the level K_n, satisfies the equation

$$q_{n+1} = \max\left[0, \min\left(K_n, q_n + v_n^e\right) - v_n\right]$$

where v_n^e is the number of requests in the nth group, v_n is the number of requests served in the gap between the arrivals of the nth and $(n+1)$st groups (see Borovkov, 1976a, section 2, formula (7)). This equation reduces to the form (4.35) if we put $w_n = -q_n$, $\xi_n = v_n - v_n^e$, $M_n = -K_n + v_{n-1}$, $N_n = 0$.

In what follows we will suppose, with no loss of generality, that $M_n = 0$ in equation (4.35). The general case reduces to this by means of the introduction of new variables $w_n^* = w_n - M_n$, $\xi_n^* = \xi_n - M_{n+1} + M_n$, $N_n^* = N_n - M_n$.

Thus we will consider (4.34) and (4.35) for $M_n \equiv 0$. The methods of investigation of these equations are very close, but all the same have some differences. We consider firstly (4.35).

This equation for fixed $Nn = N$ was investigated in Borovkov (1972b) and Kingman (1962), where conditions of ergodicity were clarified. The case when $Nn = N$ and ξ_k are independent was studied in Lindley (1959), Pakes (1978) and Phatarfod *et al.* (1971).

4.4.2. First of all we clarify the conditions under which the solution $\{w_{n+k}; k \geqslant 0\}$ of (4.35) converges, as $n \to \infty$, to a stationary sequence.

Let, as before (see sections 4.2 and 4.3)

$$Y_{n,k} = \sum_{i=n-k}^{n-1} \xi_i, \quad \bar{Y}_{n,k} = \max_{0 \leqslant j \leqslant k} Y_{n,k}, \quad X_n = \sum_{j=1}^{n} \xi_j.$$

Further we denote

$$Z_{n,k} = N_n - N_{n-k} - Y_{n,k}, \quad \bar{Z}_{n,k} = \max_{0 \leqslant j \leqslant k} Z_{n,j}$$

and introduce the random variables

$$v_1(n) = \min\left\{k > 1:\ \min_{k-1 \geqslant j \geqslant 1}\left(N_{n-k+1} + X_{n-k+j} - X_{n-k}\right) < 0\right\}$$

$$v_2(n) = \min\left\{k > 1:\ \min_{k-1 \geqslant j \geqslant 1}\left(X_{n-k+j} - X_{n-k} - N_{n-k+j+1}\right) > 0\right\}$$

$$v_3(n) = \min\left\{k > 1: N_{n-k+1} < 0\right\}$$

$$v(n) = \min\left\{v_1(n), v_2(n), v_3(n)\right\}.$$

In order that these differences make sense even for negative n (this is needed later), the difference

$$X_{n-k+j} - X_{n-k} = \xi_{n-k+1} + \cdots + \xi_{n-k+j}$$

in the formulae for $v_1(n)$ and $v_2(n)$ must be written in the form

$$Y_{n,k-1} - Y_{n,k-j-1}.$$

The random variables $v_1(n)$, $v_2(n)$ may not be defined on the whole space of elementary outcomes. Where they are not defined we put them equal to ∞.

THEOREM 1. *On the set* $\{v(n) \leqslant n\}$ *the solution of* (4.35) *satisfies*

$$w_n = \begin{cases} \bar{Y}_{n,v(n)} & \text{if } v_1(n) < v_2(n) \\ N_n - \bar{Z}_{n,v(n)} & \text{if } v_2(n) < v_1(n). \end{cases}$$

The event $\{v_1(n) = v_2(n)\}$ *is possible only in the case* $v_1(n) = v_2(n) \leqslant v(n) = v_3(n)$. *Here for* w_n *we will have both the above equality and the equalities*

$$w_{n-v(n)+1} = N_{n-v(n)+1} < 0, \quad w_n = N_{n-v(n)+1} + X_{n+1} - X_{n-v(n)}.$$

The result of this theorem gives an effective algorithm for the construction of solutions on the set $\{v(n) \leqslant n\}$. On this set w_n does not depend on the initial condition w_1 but only on the random number of random variables $\xi_{n-1}, \ldots, \xi_{n-v(n)}$.

Proof. Suppose first that $v_3 > v$. The value $k = v_1(n)$ defines the least, relative to k, time interval $[n-k+1, n-1]$ on which the trajectory

$$N_{n-k+1} + \xi_{n-k+1} + \xi_{n-k+2} + \cdots + \xi_j, \quad j = n-k, \ldots, n-1$$

starting for $j = n-k$ at a point $N_{n-k+1} > 0$, reaches the domain of negative values at time $j = n-1$. $v_2(n)$ has a similar meaning. It defines the least, relative to k, time interval $[n-k+1, n-1]$, on which the trajectory $\sum_{i=n-k+1}^{j} \xi_i, j = n-k, \ldots, n-1$, starting for $j = n-k$ at 0 upcrosses the variable boundary $N_{j+1} > 0, j = n-j, \ldots, n-1$. From these remarks it is not difficult to see (this will be clear also from what follows) that the event $v(n) > v_1(n) = v_2(n) \leqslant n$ is impossible.

If $v_1(n) = k$, then there is a maximal index $\theta_1 < v_1(n) = k$ such that

$$N_{n-k+1} + X_{n-\theta_1} - X_{n-k} < 0.$$

Since $w_{n-k+1} \leqslant N_{n-k+1}$, then by (4.35) this means that

$$w_{n-\theta_1+1} = 0. \tag{4.36}$$

(a) Consider the event $\{v_1(n) < v_3(n), \ v_1(n) < v_2(n)\}$. For elementary outcomes of this event the trajectory w_{n-j} for $\theta_1 \geqslant j \geqslant 0$ never touches the upper boundary $N_{n-j}, j = \theta_1, \theta_1 - 1, \ldots, 0$ (see (4.36) and the definition of $v_2(n)$ on the set $\nu_2(n) > k$, from which it follows that on this set

$$\max_{\theta_1 - 1 \geqslant j \geqslant 1} (X_{n-\theta_1+j} - X_{n-\theta_1} - N_{n-\theta_1+j+1}) \leqslant 0).$$

Consequently,

$$\begin{aligned} w_n &= \max_{\theta_1 \geqslant j \geqslant 1} (X_{n-1} - X_{n-j}) = \max_{0 \leqslant k \leqslant \theta_1} Y_{n,k} \\ &= \bar{Y}_{n,\theta_1} = \bar{Y}_{n,\nu_1(n)}. \end{aligned} \tag{4.37}$$

The last equality uses the fact that the increment $Y_{n-\theta_1,l}$ for $\theta_1 - l \geqslant \nu_1$ is negative.

(b) Now consider the set $\{\nu_2(n) < \nu_3(n), \nu_2(n) < \nu_1(n)\}$. Introduce the sequence $\Delta_n = N_n - w_n$, which defines the distance of w_n from the upper bound. It is easy to see that Δ_n satisfies an equation similar to (4.35):

$$\Delta_{n+1} = \min\,[N_{n+1}, \max\,(0, \Delta_n + \eta_n)]$$

where $\eta_n = N_{n+1} - N_n - \xi_n$. If we denote

$$Z_n = \sum_{k=1}^{n} \eta_k = -X_n + N_{n+1} - N_1$$

and consider the inequality in the definition of $\nu_1(n)$, then we obtain

$$\min_{k-1 \geqslant j \geqslant 1} (N_{n-k+1} + Z_{n-k+j} - Z_{n-k}) = -\max_{k-1 \geqslant j \geqslant 1} (X_{n-k+j} - X_{n-k} - N_{n-k+j+1})$$

and, consequently, $\nu_2(n)$ coincides with the $\nu_1(n)$ defined by the new sequence $\{\eta_k\}$ and conversely. Hence and from (4.37) it follows immediately that on $\{\nu_1(n) > \nu_2(n)\}$

$$\Delta_n = \bar{Z}_{n,\nu_2(n)}$$

where $Z_{n,k} = \sum_{j=n-k}^{n-1} \eta_j = N_n - N_{n-k} - Y_{n,k}$.

It remains for us to clarify what happens on the occurrence of an event different from (a) or (b). For this it is convenient to carry out several other classifications of all possible outcomes. We will establish here that in all cases w_n is defined by the formulae obtained in cases (a) and (b). Consider the trajectory $\{w_{n-j}, j = 0, 1, \ldots\}$ 'in reverse direction' from time n down to 1. If this trajectory touches the lower boundary: $w_{n-j} = 0$ (the upper boundary: $w_{n-j} = N_{n-j}$) at j and if at this point the action of the boundary is essential, that is, $w_{n-j-1} + \xi_{n-j-1} < 0$ ($w_{n-j-1} + \xi_{n-j-1} > N_{n-j} > 0$, see (4.35)), then we will say that the boundary was touched essentially and denote this by the symbol $w_{n-j} \otimes 0$ ($w_{n-j} \otimes N_{n-j}$). We denote

$$\theta_{(1)} = \min\,\{k \geqslant 1: w_{n-k+1} \otimes N_{n-k+1};\ w_{n-j+1} \otimes 0 \ \textit{for at least one}\ j \leqslant k\}$$

$$\theta_{(2)} = \min\,\{k \geqslant 1: w_{n-k+1} \otimes 0;\ w_{n-j+1} \otimes N_{n-k+1} \ \textit{for at least one}\ j \leqslant k\}.$$

The random variable $\theta_{(1)}$ ($\theta_{(2)}$) is defined on trajectories which first essentially touch the lower (upper) boundary and later essentially touch the upper (lower) boundary. If the contact of lower and upper boundaries takes place simultaneously at a point k (this is possible only if $N_{n-k+1} < 0$), then we put $\theta_{(1)} = \theta_{(2)} = k$. On those trajectories for which $\theta_{(1)}$ ($\theta_{(2)}$) is not defined we put $\theta_{(1)} = \infty$ ($\theta_{(2)} = \infty$). There are in all six possibilities

(1) $\theta_{(1)} < \theta_{(2)} = \infty, \quad N_{n-\theta_{(1)}+1} \geqslant 0$

(2) $\theta_{(1)} < \theta_{(2)} = \infty, \quad N_{n-\theta_{(1)}+1} < 0.$

Possibilities (3) and (4) differ from (1) and (2) by the transposition of the indices (1) and (2)

(5) $\theta_{(1)} = \theta_{(2)} < \infty$

(6) $\theta_{(1)} = \theta_{(2)} = \infty.$

The first possibility means that $\nu_1(n) \leqslant \theta_{(1)} < \nu_2(n), \theta_{(1)} < \nu_3(n)$. If we suppose, for example, that $\nu_2(n) \leqslant \theta_{(1)}$, then we would have obtained on the strength of (a) that $w_{n-j+1} \circledS N_{n-j}$ for $j < k$, which contradicts $\theta_{(1)} = k$. Obviously the first possibility is contained in (a).

Similarly we can verify that the second possibility means that $\nu_1(n) \leqslant \theta_{(1)} = \nu_3(n) < \nu_2(n)$, $w_{n-\nu_3(n)+1} = N_{n-\nu_3(n)+1} < 0$, so that the trajectory $\{w_{n-j};\ j = 0, \ldots, \theta_{(1)} - 2\}$ (and hence the trajectory $\{w_{n-j};\ j = 0, \ldots, \nu(n) - 1\}$) never essentially touches the upper boundary and we may again use the discussion of (a) by which $w_n = \bar{Y}_{n,\nu(n)}$.

The third and fourth possibilities are discussed in precisely the same way.

The fifth possibility means $\nu(n) = \nu_3(n) = \theta_{(1)} = \theta_{(2)}, \nu_1(n) > \nu(n), \nu_2(n) > \nu(n)$, $w_{n-\nu(n)+1} = N_{n-\nu(n)+1} < 0$. In this case the trajectory $\{w_{n-j}, j \leqslant \nu(n) - 2\}$ does not touch essentially either the lower or the upper boundary and the discussion of both parts (a) and (b) serves for the computation of $w_n = N_{n-\nu(n)+1} + X_{n-1} - X_{n-\nu(n)}$.

In the sixth possibility $\max(\nu_1(n), \nu_2(n)) = \infty$ and only one of the boundaries can be essentially touched (for example, the lower boundary if $\nu_1(n) < \nu_2(n) = \infty$). Here again it is possible to use part (a) if $\nu_1(n) < \nu_2(n)$ and part (b) if $\nu_2(n) < \nu_1(n)$.

The theorem is proved. □

Remark. If the statement of theorem 1 is changed somewhat then its proof can be simplified. Suppose that either $\mathbf{P}(\min_k N_k < 0) = 0$ or $\mathbf{P}(\min_k N_k < 0) = 1$. In the first case $\nu_3(n) = \infty$ with probability 1, the statement of the basic result concerning w_n is unchanged and the discussions in (a) and (b) above are sufficient for its proof. In the second case it is possible, in general, not to consider the random variables $\nu_1(n)$ and $\nu_2(n)$ and to restrict oneself to the assertion that on the set $\{\nu(n) \equiv \nu_3(n) \leqslant n\}$ the value w_n is defined by the value $w_{n-\nu+1} = N_{n-\nu+1} < 0$ and ν-multiple use of the recurrence relation (4.35). It is just such an assertion which is required by us later for the study of the ergodic properties of the sequence w_n.

We assume now that the sequence $\{N_k, \xi_k\}$ is *stationary*. We will take it as given on the whole axis, that is, for $-\infty < k < \infty$. Then the following result follows from theorem 1. We denote

$$\mathbf{w}_n = (w_n, w_{n+1}, \ldots).$$

THEOREM 2. *If* $\{N_k, \xi_k\}$ *is stationary and* $\mathbf{P}(\nu(0) < \infty) = 1$, *then as* $n \to \infty$ *there is a limit stationary distribution for the sequence* $\mathbf{w}_n$, *coinciding with the distribution of* $\mathbf{w} = (w^0, w^1, \ldots)$, *where*

$$w^k = \begin{cases} \bar{Y}_{k,\nu(k)} & \text{if } \nu_1(k) < \nu_2(k) \\ N_k - \bar{Z}_{k,\nu(k)} & \text{if } \nu_1(k) > \nu_2(k). \end{cases} \tag{4.38}$$

In addition, for any measurable B

$$|\mathbf{P}(\mathbf{w}_n \in B) - \mathbf{P}(\mathbf{w} \in B)| \leqslant \mathbf{P}(\nu(0) > n). \tag{4.39}$$

The condition $\mathbf{P}(\nu(0) < \infty) = 1$ is satisfied, obviously, if $\{\xi_k\}$ is ergodic, $\mathbf{E}\xi_k \neq 0$. If $\mathbf{E}\xi_k = 0$, then this condition will also be satisfied provided that ξ_k does not take the form $\xi_k = \zeta_k - \zeta_{k-1}$, where $\{\zeta_k\}$ is also ergodic (cf. Borovkov, 1972b, 1976a).

If ξ_k are independent then this condition is satisfied everywhere provided $\xi_k \not\equiv 0$.

Proof of theorem 2. Relation (4.39), from which the result of the theorem virtually follows, is obtained if we note that $A_n = \{\nu(n) \leqslant n\} \subset \{\nu(n+k) \leqslant n + k\}$, $k \geqslant 0$, and that, consequently, by theorem 1, on this set

$$w_{n+k} = w^{n+k}, \quad k = 0, 1, \ldots.$$

Since the sequence $\{N_k, \xi_k\}$ is stationary, then

$$\mathbf{P}(\nu(n) \geqslant n) = \mathbf{P}(\nu(0) \leqslant n)$$

$$\begin{aligned} \mathbf{P}(\mathbf{w}_n \in B) &= \mathbf{P}(\mathbf{w}_n \in B; A_n) + \mathbf{P}(\mathbf{w}_n \in B; \bar{A}_n) \\ &= \mathbf{P}((w^n, w^{n+1}, \ldots) \in B; A_n) + \mathbf{P}(\mathbf{w}_n \in B; \bar{A}_n) \\ &= \mathbf{P}(\mathbf{w} \in B) - \mathbf{P}((w^n, w^{n+1}, \ldots) \in B; \bar{A}_n) + \mathbf{P}(\mathbf{w}_n \in B; \bar{A}_n). \end{aligned}$$

The theorem is proved. □

4.4.3. *We now consider* (4.34). Here it is possible to suppose for simplicity that $N_k > 0$ for all k (in problems leading to (4.34) N_k are the upper boundaries of the waiting times and cannot be negative). We will follow subsection 4.4.2 but here we require somewhat different characteristics of the trajectories. We observe as a preliminary that if we denote by μ the value

$$\mu = \min\{j \geqslant 1 : w_{n-j} \leqslant N_{n-j}\}$$

then

$$w_n \leqslant N_{n-\mu} + \xi_{n-\mu} - \tau^e_{n-\mu+1} - \cdots - \tau^e_{n-1}$$

and, consequently ($w_1 \leqslant N_1$),

$$w_n \leqslant \max_{n>j\geqslant 1} \{N_{n-j} + \xi_{n-j} - \tau^e_{n-j+1} - \cdots - \tau^e_{n-1}\}.$$

For simplicity we will suppose that $\{N_k, \tau^e_k, \tau^s_k\}$ is stationary and defined on 'the whole axis' (that is, for $-\infty < k < \infty$). Then, if we denote

$$\Delta_k = N_{k-1} - N_k + \xi_{k-1} - \xi_k - \tau^e_k \tag{4.40}$$

we obtain

$$w_n \leqslant V_n \equiv N_{n-1} + \xi_{n-1} + v_{n-1} \tag{4.41}$$

where

$$v_{n-1} = \sup_{k\geqslant 0}\left(\sum_{j=n-k}^{n-1} \Delta_j\right).$$

The sequences $\{v_n\}$, $\{V_n\}$, obviously, are also stationary and are proper if $\mathbf{E}\Delta_k < 0$ and $\{\Delta_k\}$ is metrically transitive. As before we denote

$$\nu(n) = \min\left\{k > 1 \colon \min_{1\leqslant j\leqslant k-1} (V_{n-k+1} + X_{n-k+j} - X_{n-k}) < 0\right\}. \tag{4.42}$$

Since $w_n \leqslant V_n$, then by the former arguments we obtain the following.

THEOREM 1A. *On the set* $\{\nu(n) \leqslant n\}$ *the random variable* w_n, *defined by* (4.34), *can be defined as the value* w^*_ν *obtained by using* (4.34) $\nu = \nu(n)$ *times* (*with all variables asterisked*) *with the zero initial condition* $w^*_0 = 0$ *and with the random variables* $\tau^{e*}_k = \tau^e_{n-\nu+k}$, $\tau^{s*}_k = \tau^s_{n-\nu+k}$, $\xi^*_k = \tau^{s*}_k - \tau^{e*}_k$, $N^*_k = N_{n-\nu+k}$, $k = 0, \ldots, \nu - 1$.

Remark 1. The essential distinction of (4.34) from (4.35) is that for this equation the presence on $[n-k, n-1]$ of 'overshoot' of the trajectory $\{w_k\}$ 'upwards' (that is, the event $\{\nu_2(n) < k\}$) does not introduce 'renewal' in the trajectory, that is, does not remove the dependence on the whole prehistory. This is connected with the fact that in (4.34) the upper boundary for the random walk is not so 'rigid' as the lower, or as the upper for (4.35). The 'rënewal' method will be discussed in a more general form in section 4.6. The results of this section serve as a nice illustration of it.

Remark 2. From the proof of theorem 1 it can be seen that any number from $[0, V_{n-\nu}]$ can be taken as the initial value w^*_0 in theorem 1A—w_n does not depend on this number.

Remark 3. In contrast to (4.35) the definition of $\nu(n)$ in the condition of theorem 1A requires, in general, a knowledge of the entire sequence $\{N_k, \tau^e_k, \tau^s_k\}$ (see (4.41)). An exception, for example, is the case

$$N_k + \xi_k \leqslant N \tag{4.43}$$

where N is fixed (this is the case, for example, for (4.29)). In this case $w_k \leqslant N$ and $\nu(n)$, as with $\nu(n)$ in theorem 1, is 'Markovian' (is a stopping time), that is, the event $\{\nu(n) \leqslant k\}$ belongs to the σ-algebra generated by

$$(N_{n-1}, \tau_{n-1}^e, \tau_{n-1}^s), \ldots, (N_{n-k-1}, \tau_{n-k-1}^e, \tau_{n-k-1}^s).$$

We may now state the analogue of theorem 2 for the case of a stationary sequence $\{N_k, \tau_k^e, \tau_k^s\}$.

THEOREM 2A. *If* $\{N_k, \tau_k^e, \tau_k^s\}$ *is stationary and* $\mathbf{P}(\nu(0) < \infty) = 1$, *then, as* $n \to \infty$, *for* (4.34) *there is a limit stationary distribution for* $\mathbf{w}_n$ *which coincides with the distribuion of* $\mathbf{w} = (w^0, w^1, \ldots)$. *An algorithm for obtaining* w^k *is described in theorem* 1A. *Here for any measurable* B

$$|\mathbf{P}(\mathbf{w}_n \in B) - \mathbf{P}(\mathbf{w} \in B)| \leqslant \mathbf{P}(\nu(0) > n).$$

The proof of this theorem is no different from the proof of theorem 2.

Theorems 1A, 2A, as theorems 1, 2, contain the construction of an effective algorithm by which it is possible to construct the stationary distributions of the solution w^k of (4.34) and (4.35) using only a finite (but random) number of elements $(N_{k-1}, \tau_{k-1}^e, \tau_{k-1}^s) \ldots (N_{k-\nu}, \tau_{k-\nu}^e, \tau_{k-\nu}^s)$ of the sequence $\{N_k, \tau_k^e, \tau_k^s\}$.

The finiteness of $\nu(0)$ with probability 1 in theorem 2A is obviously a stronger finiteness condition than in theorem 2 and is essential. For example, if $\xi_j > 0$ with probability 1 then $\mathbf{P}(\nu(0) < \infty) = 1$ under the conditions of theorem 2 and $\mathbf{P}(\nu(0) < \infty) = 0$ under the conditions of theorem 2A and here the conclusion of theorem 2A will be false since, for example, for integer and even ξ_j, τ_j^e the value (parity) of w_n in (4.34) will depend for all n on the initial condition w_1.

The condition $\mathbf{P}(\nu(0) < \infty) = 1$ in theorem 2A will be satisfied if $\{V_k\}$ is a proper sequence, $\{\xi_k\}$ is ergodic, and $\mathbf{E}\xi_k < 0$. If $\mathbf{E}\xi_k = 0$, then the condition will also be satisfied except in the case mentioned in the remark to theorem 2. If $\xi_k, \xi_{k+1}, \ldots$ are independent and do not depend on V_k, $\mathbf{P}(\xi_k < 0) > 0$, then the condition $\mathbf{P}(\nu(0) < \infty) = 1$ will always be satisfied.

In order that the sequence $\{V_k\}$ be proper, it is sufficient that the sequence

$$-\Delta_k = N_k - N_{k-1} + \xi_k - \xi_{k-1} + \tau_k^e$$

be metrically transitive and

$$\mathbf{E}(N_k - N_{k-1} + \xi_k - \xi_{k-1})$$

exists.

4.4.4. We now pass on to *stability theorems*. With the controlling sequence $S = \{N_k, \tau_k^e, \tau_k^s\}$ let there be given a series of controlling sequences $S^{(r)} = \{N_k^{(r)}, \tau_k^{(r)e}, \tau_k^{(r)s}\}$. Again we consider (4.35) first and its corresponding stationary (see (4.38)) and prestationary sequences $\mathbf{w}$ and $\mathbf{w}_n$. Alongside these sequences we introduce the sequences $\mathbf{w}^{(r)}$ and $\mathbf{w}_n^{(r)}$ corresponding to $S^{(r)}$. The index (r) in the other notations introduced above will also mean relative to $S^{(r)}$.

THEOREM 3. *Let the sequences $S^{(r)}$, S be stationary and satisfy the following conditions:*

(1) *The finite-dimensional distributions of $S^{(r)}$ weakly converge as $r \to \infty$ to the distributions of S.*
(2) *For S we have $\mathbf{P}(\nu(0) < \infty) = 1$, where $\nu(n)$ is defined in theorem 1.*

Then the finite-dimensional distributions of $\mathbf{w}_n^{(r)}$, $\mathbf{w}^{(r)}$, as defined by (4.35), converge as $n \to \infty$, $r \to \infty$ to the finite-dimensional distributions of $\mathbf{w}$.

Remark. From comparison with theorem 1 of section 4.2 it is clear that there are no additional conditions on $S^{(r)}$, apart from the convergence of the finite-dimensional distributions. This is because $\nu(0)$ is 'Markovian' and 'does not depend on the future', in contrast to the epoch of attainment of $\sup_k Y_{n,k}$, present in sections 4.1 and 4.2.

Proof. As in theorem 1 of section 4.2, it is necessary for us to show convergence to zero, as $n \to \infty$, $r \to \infty$, of the differences

$$\begin{aligned} &|\mathbf{P}(\mathbf{w}_n^{(r)} \in B) - \mathbf{P}(\mathbf{w}^{(r)} \in B)| \\ &|\mathbf{P}(\mathbf{w}^{(r)} \in B) - \mathbf{P}(\mathbf{w} \in B)| \end{aligned} \tag{4.44}$$

for any cylinder set B which is a continuity set for the distribution of $\mathbf{w}$.

We first show

$$\mathbf{P}(\nu^{(r)}(0) > n) \to 0 \tag{4.45}$$

uniformly in r as $n \to \infty$. We denote (see the definition of $\nu_1(n)$, $\nu_2(n)$)

$$A_1(n) = \min_{1<k\leqslant n} \min_{1\leqslant j\leqslant k-1} (N_{-k+1} + Y_{0,k-1} - Y_{0,k-j-1})$$

$$A_2(n) = \min_{1\leqslant k\leqslant n} \min_{1\leqslant j\leqslant k-1} (N_{-k+j+1} - Y_{0,k-1} + Y_{0,k-j-1})$$

$$A(n) = \min\,[A_1(n), A_2(n)].$$

Finiteness of $\nu(0)$ with probability 1 means that given $\varepsilon > 0$ there is $n = n(\varepsilon)$ such that

$$\mathbf{P}(A(n) < 0) > 1 - \frac{\varepsilon}{3}.$$

Using the properties of probability and weak convergence we can find an arbitrarily small $\alpha > 0$ for which $-\alpha$ is a point of continuity of the distribution of $A(n)$ and

$$\mathbf{P}(A(n) < -\alpha) > 1 - \frac{\varepsilon}{2}.$$

But

$$\{v^{(r)}(0) > n\} = \{A^{(r)}(n) \geqslant 0\} \subset \{A^{(r)}(n) \geqslant -\alpha\}.$$

Hence it follows that

$$\limsup_{r\to\infty} \mathbf{P}(v^{(r)}(0) > n) \leqslant \frac{\varepsilon}{2}$$

and this means that $\mathbf{P}(v^{(r)}(0) > n) \leqslant \varepsilon$ for all sufficiently large r and n. (4.45) is proved. Since, by theorem 2, the first difference in (4.44) does not exceed $\mathbf{P}(v^{(r)}(0) > n)$, then the convergence of this difference to 0 is also proved.

We now consider the second difference in (4.44). Theorem 2 means that on $\{v^{(r)}(0) < \infty\}$ the vector $(w^{(r)0}, \ldots, w^{(r)l})$, for each l, is a function of $v^{(r)}(0) + l$ elements of the controlling sequence. This function is given by the right-hand side of (4.35) and is continuous. From here and from (4.45) it is easy to derive the result of the theorem ((4.45) means essentially that $(w^{(r)0}, \ldots, w^{(r)l})$ is a continuous function of a finite number of elements of the controlling sequence). □

Since the plan of the proof of theorem 3 is the same as for theorem 1 of section 4.1, then here clearly, as in section 4.3, an estimate of speed may be obtained, but under simpler assumptions. Since, for weak dependence between the ξ_k, the decrease of $\mathbf{P}(v(0) > n)$ will be exponential, then the estimate will be of order $\varepsilon \log 1/\varepsilon$, where ε approximates the closeness of the distributions of $\xi^{(r)}$ and ξ (see Akhmarov, 1979d, for details).

It is quite clear that passage to (4.34) brings no essential changes to the argument of theorem 3. It is only necessary to replace $A(n)$ by $A_1(n)$ and N_k by V_k in the appropriate places. However, in contrast to (4.35), the function defined by the right-hand side of (4.34) is discontinuous at one point $w_n = N_n$. We avoid the difficulties connected with this by the additional requirement that for the limit distribution the probability of hitting this point is equal to zero. Thus we may state the following.

THEOREM 3A. *Let $S^{(r)}$ and S be stationary and satisfy the following conditions:*

(1) *The finite-dimensional distributions of $\{V_k^{(r)}, \tau_k^{(r)e}, \tau_k^{(r)s}\}$ weakly converge as $r \to \infty$ to the distributions of $\{V_k, \tau_k^e, \tau_k^s\}$* (for the definition of V_k see (4.40), (4.41)).

(2) $\mathbf{P}(v(0) < \infty) = 1$.

(3) *For each k the distribution of $w_k - N_k$, where w_k is defined by* (4.34) *and the initial condition $w_1 = 0$, is continuous at* 0.

Then, for equation (4.34), *the finite-dimensional distributions of $\mathbf{w}_n^{(r)}, \mathbf{w}^{(r)}$ converge to the finite-dimensional distributions of $\mathbf{w}$ as $n \to \infty$, $r \to \infty$.*

The required convergence of the distributions of $\{V_k^{(r)}, \tau_k^{(r)e}, \tau_k^{(r)s}\}$ in condition 1 will be a corollary of the convergence of the distributions of $\{N_k^{(r)}, \tau_k^{(r)e}, \tau_k^{(r)s}\}$ and the strong convergence of the moments $\mathbf{E}|\xi_k^{(r)}| \to \mathbf{E}|\xi_k|$, $\mathbf{E}N_k^{(r)} \to \mathbf{E}N_k$ (see section

4.1 and (4.40) and (4.41), defining V_k). In particular, for (4.30) the convergence of the distributions of $\{\tau_k^{(r)e}, \tau_k^{(r)s}\}$ and the convergence $\mathbf{E}\tau_k^{(r)s} \to \mathbf{E}\tau_k^s$ is enough.

4.5. Stability theorems for systems with an infinite number of service channels

We recall the description of the service systems of the title of this section. On the basic probability space let there be given a controlling sequence

$$\{\tau_j^e, \nu_j^e, \boldsymbol{\tau}_j^s; j \geqslant 1\} \tag{4.46}$$

where τ_j^e, τ_j^s and the coordinates of the ν_j^e-dimensional vector $\boldsymbol{\tau}_j^s = (\tau_{j,1}^s, \ldots, \tau_{j,\nu_j^e}^s)$ are nonnegative and in addition ν_j^e is integer-valued. τ_j^e and ν_j^e characterize the input stream of requests: requests join the systems at intervals of time $\tau_1^e, \tau_2^e, \ldots$ in groups of size $\nu_1^e, \nu_2^e, \ldots$ where in each group there is an order which the requests follow. The vector $\boldsymbol{\tau}_j^s$ describes the service times of the requests which arrive in the jth group. The service channels are numbered and the number of them is unbounded. If a request is in service in some channel then this channel is regarded as busy throughout the service time of the given request. For definiteness it can be assumed that in the first place the channels of lowest numbers are occupied. Under these assumptions the assignment of the sequence (4.46) will completely determine the evolution of the system if we specify a vector of initial conditions $(q_1; \rho_1, \ldots, \rho_{q_1})$ where q_1 is the number of busy channels at the initial epoch and $\rho_1, \ldots, \rho_{q_1}$ are the service times of the 'initial' requests in these busy channels.

When the distributions of the sequences $\{\tau_j^e\}$, $\{\nu_j^e\}$, $\{\boldsymbol{\tau}_j^s\}$ are arbitrary the notation $\langle G, G, G/\infty, 1\rangle$ was adopted for this system in Borovkov (1976a). If $\{\tau_j^e\}$ is a sequence of identically distributed random variables, not depending on the remaining sequences, then the initial symbol G is replaced by G_I. If, in addition, $\mathbf{P}(\tau_j^e > x) = \mathrm{e}^{-ax}$, then we write $\langle E, G, G/\infty, 1\rangle$. Similar changes are made to the second and third symbols with the introduction of the corresponding assumptions regarding $\{\nu_j^e\}$ and $\{\boldsymbol{\tau}_j^s\}$. If $\mathbf{P}(\nu_j^e = 1) = 1$, then the notation takes the form $\langle G, 1, G/\infty, 1\rangle$. The basic object of study is the sequence q_n which denotes the number of busy lines at the arrival epoch of the nth group of requests.

In all that follows we will assume that the controlling sequence is strictly stationary. In this case, without loss of generality, we can take the sequence $\{\tau_j^e, \nu_j^e, \boldsymbol{\tau}_j^s\}$ to be given on the whole axis, that is, for $-\infty < j < \infty$.

We will study stability conditions (or the continuous dependence) of the distribution of the stationary sequence $\{q^k\}$ (the limit of $\{q_{n,k} = q_{n-k}; k \geqslant 0\}$, see theorem 1) under small variations of the distribution of the controlling sequences.

As an illustration of the problem considered under the conditions of this section we consider the following example. As is well known (see, for example, Borovkov, 1976a), the distribution of q^k for a system $\langle E, 1, G_I/\infty, 1\rangle$ is Poisson.

Will the distribution of q^k for such systems be close to Poisson if

$$\mathbf{P}(\tau_j^e \geqslant x) = P(x, p, r) = (1-p)\,\mathrm{e}^{-\alpha x} + p\delta_r(x)$$

where p is small,

$$\delta_r(x) = \begin{cases} 1 & x \leqslant r \\ 0 & x > r? \end{cases}$$

A similar question can be put for systems 'close' to $\langle G_I, 1, E/m, 1\rangle$ (see Borovkov, 1976a, section 29), when

$$\mathbf{P}(\tau_j^s \geqslant x) = P(x, p, r).$$

In the first case the answer to the question on the stability of the stationary distribution of q^k for small p is affirmative; in the second case it depends on r (more precisely, on the product pr).

We first consider the simpler systems when $\nu_j^e \equiv 1, q_1 = 0$. For such systems the following ergodic theorem was obtained in Borovkov (1976a).

THEOREM A. *Let the sequence* $\{\tau_j^e, \tau_j^s;\ -\infty < j < \infty\}$ *be strictly stationary and, in addition let* $\{\tau_j^e\}$ *be metrically transitive. Then, if* $\mathbf{E}\tau_j^s < \infty$, *the distribution of* $\{q_{n+k};\ k \geqslant 0\}$ *monotonely converges as* $n \to \infty$ *to the distribution of the proper stationary sequence*

$$q^k = I(\tau_k^s > \tau_k^e) + I(\tau_{k-1}^s > \tau_{k-1}^e + \tau_k^s) + I(\tau_{k-2}^s > \tau_{k-2}^e + \tau_{k-1}^e + \tau_k^e) + \cdots \tag{4.47}$$

where $I(B)$ *is the indicator of the event* B.

The proof of this theorem is fairly brief and we will need it later. Therefore we reproduce it here. We have

$$\begin{aligned} q_n = {} & I(\tau_1^s > \tau_1^e + \cdots + \tau_{n-1}^e) + I(\tau_2^s > \tau_2^e + \cdots + \tau_{n-1}^e) \\ & + \cdots + I(\tau_{n-1}^s > \tau_{n-1}^e). \end{aligned} \tag{4.48}$$

Using stationarity we find*

$$\begin{aligned} q_{n+k} \underset{d}{=} q_{n,k} \equiv {} & I(\tau_k^s > \tau_k^e) + I(\tau_{k-1}^s > \tau_{k-1}^e + \tau_k^e) \\ & + \cdots + I(\tau_{-n+2}^s > \tau_{-n+2}^e + \cdots + \tau_k^e) \end{aligned}$$

so that $q_{n,k} \uparrow q^k$ as $n \to \infty$. It remains to show that q^0 is a proper random variable. In fact, for some $b > 0$

$$\mathbf{P}(q^0 > N) \leqslant \mathbf{P}\Bigg(\bigcup_{j=N}^{\infty} \{I(\tau_{-j}^s > \tau_{-j}^e + \cdots + \tau_0^e) = 1\}\Bigg)$$

* The relation $\xi =_d \eta$ means that the distributions of ξ and η coincide.

$$\leqslant \mathbf{P}\left(\bigcup_{j=N}^{\infty}\{\tau_0^e+\cdots+\tau_{-j}^e<b(j+1)\}\right)$$

$$+\mathbf{P}\left(\bigcup_{j=N}^{\infty}\{\tau_{-j}^s>b(j+1)\}\right). \qquad (4.49)$$

For a suitable $b>0$ the first term here converges to 0 as $N\to\infty$ by the strong law of large numbers. The second term is majorized by the sum $\sum_{j=N}^{\infty}\mathbf{P}(\tau_0^s>b(j+1))$ and also converges to 0, since $\mathbf{E}\tau_0^s<\infty$. The theorem is proved. □

Theorem A gives an explicit expression for the stationary distribution of the number of busy lines via the distribution of the elements of the controlling sequence on the basis of which it is possible to obtain, for example, for a $\langle G_I, 1, G_I/\infty, 1\rangle$ system, explicit formulae for the distribution of q^k (see, for example, Borovkov, 1976a, section 30).

To give a precise statement of the stability problem we must introduce the systems $\langle G, 1, G/\infty, 1\rangle^{(r)}$ controlled by stationary sequences $\{\tau_j^{(r)e}, \tau_j^{(r)s}\}$, depending on a parameter $r=1,2,\ldots$. In all the notations adopted earlier and relating to the system $\langle\cdot,\cdot,\cdot,\cdot\rangle^{(r)}$ we add an upper index (r).

Let the following conditions be satisfied.

(A) *There is a sequence* $\{\tau_j^e, \tau_j^s\}$ *satisfying the conditions of theorem A, such that the finite-dimensional distributions of* $\{\tau_j^{(r)e}, \tau_j^{(r)s}\}$, *also satisfying the conditions of theorem A, converge to the corresponding finite-dimensional distributions of* $\{\tau_j^e, \tau_j^s\}$.

(B) $\mathbf{E}\tau_1^{(r)s}\to\mathbf{E}\tau_1^s$ *as* $r\to\infty$.

(C) *The distribution of* $\tau_{-j}^s - X_{-j}^e$, *where* $X_{-j}^e=\sum_{k=-j}^{0}\tau_k^e$, *for all* $j\geqslant 0$, *is continuous at* 0. *In other words, for any integer* $j\geqslant 0$

$$\mathbf{P}(\tau_{-j}^s - X_{-j}^e=0)=0.$$

THEOREM 1. *Under conditions* (A), (B), *and* (C) *the finite-dimensional distributions of the stationary sequences* $\{q^{(r)k}: k\geqslant 0\}$ *converge as* $r\to\infty$ *to the corresponding distributions of* $\{q^k: k\geqslant 0\}$.

Remark. All of these conditions are essential. Failure of at least one of them immediately allows the construction of an example where convergence of the distributions of $q^{(r)k}$ fails.

Proof. We show first that in the inequality (4.49) for $q^{(r)0}$ the right-hand side converges to 0 uniformly in r as $N\to\infty$. In fact, for any fixed $b>0$ the last term in (4.49)

$$\mathbf{P}\left(\bigcup_{j=N}^{\infty}\{\tau_{-j}^{(r)s}>b(j+1)\}\right)$$

$$\leqslant \sum_{j=N+1} \mathbf{P}(\tau_0^{(r)s}>b_j)\leqslant \int_N^{\infty} \mathbf{P}(\tau_0^{(r)s}>bx)\,\mathrm{d}x=\frac{1}{b}\int_{bN}^{\infty} \mathbf{P}(\tau_0^{(r)s}>x)\,\mathrm{d}x.$$

But the uniform in r convergence to 0 as $N\to\infty$ of the last integral is necessary and sufficient for $\mathbf{E}\tau_0^{(r)s}\to\mathbf{E}\tau^s$ as $r\to\infty$ (condition (B)).

Now consider the first term on the right-hand side in (4.49)

$$\mathbf{P}\left(\bigcup_{j=N}^{\infty}\{\tau^{(r)e}+\cdots+\tau^{(r)e}<b(j+1)\}\right)\leqslant \mathbf{P}\left(\sup_{k\geqslant N} X^{(r)}>0\right)$$

where $X_k^{(r)}=\sum_{j=0}^{k}(b-\tau_{-j}^{(r)e})$. Here the required convergence follows from the following lemma, contained in the proof of theorem 1 of section 4.1.

LEMMA. *Let* $\{\xi_j\}$ *and* $\{\xi_j^{(r)}\}$, $r=1,2,\ldots$, *be strictly stationary such that the finite-dimensional distributions of* $\{\xi_j^{(r)}\}$ *converge as* $r\to\infty$ *to the corresponding distributions of* $\{\xi_j\}$. *Then, if* $\{\xi_j\}$ *is metrically transitive,* $\mathbf{E}\xi_j<0$ *and* $\mathbf{E}(\xi_j^{(r)})^+\to\mathbf{E}_j^+$ *then* $\mathbf{P}(\max_{k\geqslant N}\sum_{j=0}^{k}\xi^{(r)}>0)\to 0$ *uniformly in* r *as* $N\to\infty$.

We now consider for a noninteger y the difference

$$\mathbf{P}(q^{(r)0}\geqslant y)-\mathbf{P}(q^0\geqslant y)=\mathbf{P}\left(\sum_{j=0}^{N} I(\tau_{-j}^{(r)s}>X_{-j}^{(r)e})\geqslant y\right)$$

$$-\mathbf{P}\left(\sum_{j=0}^{N} I(\tau_{-j}^{s}>X_{-j}^{e})\geqslant y\right)+\varepsilon(r,N)+\varepsilon(N) \qquad (4.50)$$

where, by what has been proved, $\varepsilon(N)\to 0$.

$$\varepsilon(r,N)\leqslant \mathbf{P}\left(\sum_{j=N}^{\infty} I(\tau_{-j}^{(r)s}>X_{-j}^{(r)e})>0\right)\to 0$$

uniformly in r as $N\to\infty$. Let $R(x)$ be a function uniformly distributed on $[0,1]$. For some $\delta>0$ put

$$I_{j-}=R\left(\frac{\tau_{-j}^{s}-X_{-j}^{e}}{\delta}\right),\quad I_{j+}=R\left(1+\frac{\tau_{-j}^{s}-X_{-j}^{e}}{\delta}\right)$$

$$I_j=I(\tau_{-j}^{s}>X_{-j}^{e}).$$

Then, obviously, $I_{j-}\leqslant I_j\leqslant I_{j+}$. Since $R(x)$ is a continuous function then, with the natural notation, for $r\to\infty$

$$\mathbf{P}(I_{j\pm}^{(r)}<x)\Rightarrow \mathbf{P}(I_{j\pm}<x).$$

Therefore there are arbitrarily small $\varepsilon > 0$ such that

$$\limsup_{r\to\infty} \mathbf{P}\left(\sum_{j=0}^{N} I(\tau_{-j}^{(r)s} > X_{-j}^{(r)e}) \geqslant y\right)$$

$$\leqslant \limsup_{r\to\infty} \mathbf{P}\left(\sum_{j=0}^{N} I_{j+}^{(r)} \geqslant y\right) \leqslant \mathbf{P}\left(\sum_{j=0}^{N} I_{j+} \geqslant y - \varepsilon\right). \quad (4.51)$$

But by condition (C) the probability of the event

$$A_{n,\delta} = \bigcup_{j=0}^{N} \{I_{j+} \neq I_{j-}\} \quad (4.52)$$

equal to $\mathbf{P}(\bigcup_{j=0}^{N} \{|\tau_{-j}^{s} - X_{-j}^{e}| \leqslant \delta\})$, tends to 0 as

$$\mathbf{P}\left(\sum_{j=0}^{N} I_{j+} \geqslant y - \varepsilon\right) - \mathbf{P}\left(\sum_{j=0}^{N} I_{j} \geqslant y\right)$$

$$= \mathbf{P}\left(\sum_{j=0}^{N} I_{j+} \geqslant y - \varepsilon; \bar{A}_{n,\delta}\right) - \mathbf{P}\left(\sum_{j=0}^{N} I_{j} \geqslant y; \bar{A}_{n,\delta}\right) + \varepsilon_1(\delta, N)$$

$$= \mathbf{P}\left(\sum_{j=0}^{N} I_{j} \in [y - \varepsilon, y); \bar{A}_{n,\delta}\right) + \varepsilon_1(n, \delta)$$

where $|\varepsilon_1(\delta, N)| \leqslant \mathbf{P}(A_{N,\delta})$. If we choose ε so that the semi-interval $[y - \varepsilon, y)$ contains no integers, then the event $\{\sum_{j=0}^{N} I_j \in [y - \varepsilon, y)\}$ is empty, and by (4.50) and (4.51) we obtain

$$\limsup_{r\to\infty} (\mathbf{P}(q^{(r)0} \geqslant y) - \mathbf{P}(q^{0} \geqslant y))$$

$$\leqslant \limsup_{r\to\infty} \varepsilon(r, N) + \varepsilon(N) + \varepsilon_1(\delta, N). \quad (4.53)$$

From these properties of $\varepsilon(r, N)$, $\varepsilon_1(\delta, N)$, and $\varepsilon(N)$ and the independence of the left-hand side in (4.53) on δ and N, it follows that the estimated lim sup does not exceed 0. In just the same way via the inequality

$$\liminf_{r\to\infty} \mathbf{P}\left(\sum_{j=0}^{N} I(\tau_{-j}^{(r)s} > X_{-j}^{(r)e}) \geqslant y\right) \geqslant \mathbf{P}\left(\sum_{j=0}^{N} I_{j-} \geqslant y + \varepsilon\right)$$

we show that

$$\liminf_{r\to\infty} (\mathbf{P}(q^{(r)0} \geqslant y) - \mathbf{P}(q^{0} \geqslant y)) \geqslant 0$$

for any noninteger y. The result of the theorem on convergence of the one-dimensional distributions of $q^{(r)k}$ is thus proved.

The idea behind the proof was fairly simple: in the representation

$$q^{(r)k} = \sum_{j=0}^{\infty} I(\tau_{k-j}^{(r)s} > X_k^{(r)e} - X_{k-j-1}^{(r)e}) \tag{4.54}$$

the 'tail' $\sum_{j=N}^{\infty}$ turns out to be small uniformly in r for large N. The convergence of the distributions of the finite sums $\sum_{j=0}^{N}$ in (4.54) is determined under condition (C) by the convergence of the distributions of the controlling sequences. Therefore it is quite clear that if we use the same method we also obtain convergence of arbitrary finite-dimensional distributions of the sequences $\{q^{(r)k}: k \geqslant 0\}$ to the corresponding distributions of the sequence $\{q^k; k \geqslant 0\}$. The proof of the theorem is complete. □

We move on now to *stronger continuity theorems for systems* $\langle G, G, G/\infty, 1\rangle$. We consider the more general situation when v_j^e is arbitrary and $q_1 \neq 0$ and we consider a more precise characterization of the state of the system with the help of the processes

$$q_n(x) = \sum_{j=1}^{q_1} I(\rho_j > \tau_1^e + \cdots + \tau_{n-1}^e + x)$$
$$+ \sum_{j=1}^{v_1^e} I(\tau_{1,j}^s > \tau_1^e + \cdots + \tau_{n-1}^e + x)$$
$$+ \cdots + \sum_{j=1}^{v_{n-1}^e} I(\tau_{n-1,j}^s > \tau_{n-1}^e + x)$$

such that $q_n(0) = q_n$.

The process $\{q_n(x): x \geqslant 0\}$ is obviously stepped and nonincreasing. It can be uniquely described by the initial value $q_n(0) = q_n$ and the locations $x_1, \ldots, x_{q_n}$ of the points of unit jumps. The vector $(x_1, \ldots, x_{q_n})$ is a vector of the times during which requests which joined the system before the nth group are still being served. The variable $q_n(x)$ itself is the number of requests from those that arrived before the arrival epoch t_n of the nth group and which remain in the system a time x beyond the epoch t_n.

The corresponding stationary sequence of processes $q^k(x)$ is defined by

$$q^k(x) = \sum_{j=1}^{v_k^e} I(\tau_{k,j}^s > \tau_k^e + x) + \sum_{j=1}^{v_{k-1}^e} I(\tau_{k-1,j}^s > \tau_{k-1}^e + \tau_k^e + x) + \cdots.$$

As before we find that

$$\mathbf{P}\left(q^0(x) > \sum_{j=0}^{N} v_{-j}^e + q_1\right)$$
$$\leqslant \mathbf{P}\left(\bigcup_{j=N}^{\infty} \{\tau_0^e + \cdots + \tau_{-j}^e < b(j+1)\}\right) + \mathbf{P}\left(\bigcup_{j=N}^{\infty} \left\{\max_k \tau_{-j,k}^s > b(j+1)\right\}\right).$$

Since $\max_k \tau^s_{0,k} \leqslant [\tau^s_0] \equiv \sum_{j=1}^{\nu^e_0} \tau^s_{0,j}$ then, assuming that the initial conditions are proper (q_1 and ρ_1 are finite), we obtain the following result.

THEOREM B. *Let* $\{\tau^e_n, \nu^e_n, \tau^s_n;\ -\infty < n < \infty\}$ *be strictly stationary and, in addition, let* $\{\tau^e_j\}$ *be metrically transitive. Then if* $\mathbf{E}[\tau^s_0] < \infty$, *the distribution of the sequence* $\{q_{n+k}(x)\colon k \geqslant 0, x \geqslant 0\}$ *converges as* $n \to \infty$ *to the distribution of the proper, stationary in k, process* $\{q^k(x); k \geqslant 0, x \geqslant 0\}$.

More precisely: there are processes $\{\tilde{q}^k(x) = \tilde{q}^k_n(x); k \geqslant 0, x \geqslant 0\}, n = 1, 2, \ldots$ *distributed for all n the same as* $\{q_k(x); k \geqslant 0, x \geqslant 0\}$ *and such that as* $n \to \infty$

$$\mathbf{P}\left(\bigcup_{k=0}^{\infty} \bigcup_{x \geqslant 0} \{q_{n+k}(x) \neq q^k_n(x)\}\right) \to 0.$$

We appeal now to the *continuity theorem*. We will say that *the distribution of the processes* $\{q^{(r)0}(x); x \geqslant 0\}$ *weakly converges as* $r \to \infty$ *to the distribution of* $\{q^0(x); x \geqslant 0\}$, *if for some everywhere dense set S the distribution of* $(q^{(r)0}(x_1), \ldots, q^{(r)0}(x_m))$ *weakly converges to the distribution of* $(q^0(x_1), \ldots, q^0(x_m))$ *for* $x_1 \in S, \ldots, x_m \in S$.

We denote by S_0 the set of $x \geqslant 0$ which are points of continuity of the distributions of $\tau^s_{-j,k} - X^e_{-j}$ for all $j \geqslant 0$ and k.

THEOREM 2. *Let condition* (A) *and condition* ($\mathbf{B}_1$): $\mathbf{E}[\tau^{(r)s}] \to \mathbf{E}[\tau^s]$ *as* $r \to \infty$, *be satisfied.*

Then the distribution of the processes $\{q^{(r)0}(x); x \geqslant 0\}$ *weakly converges as* $r \to \infty$ *to the distribution of* $\{q^0(x); x \geqslant 0\}$. *Here the set of convergence S coincides with* S_0.

We note that theorem 1 in part, concerning one-dimensional distributions, is a corollary of theorem 2 and that the conditions of theorem 2 are weaker than those of theorem 1, which include condition (C). This becomes clear when we pass to the proof of the theorem and repeat the argument of theorem 1 under the new conditions and obtain the convergence of the distribution of $q^{(r)0}(x)$ for all $x \geqslant 0$ which are continuity points of the distribution of $\tau^s_{-j,k} - X^e_{-j}$ (the probability $\mathbf{P}(\bigcup_{j=0}^{N} \bigcap_{l=0}^{\nu^e_j} \{|\tau^s_{-j,k} - X^e_{-j} - x| \leqslant \delta\})$ must converge to 0 as $\delta \to 0$ (cf. (4.45)). Since this set of x is the halfline $x \geqslant 0$ with a not more than countable set removed then the result of theorem 2 is proved. □

In a completely similar way we obtain

THEOREM 2A. *Under the conditions of theorem 2 convergence of the distributions of 'random fields'* $\{q^{(r)k}(x); k \geqslant 0, x \geqslant 0\}$ *will hold. That is, for any* $k_1, \ldots, k_m$ *and* $x_1 \in S_0, \ldots, x_m \in S_0$ *the distributions of the vectors* $(q^{(r)k_1}(x_1), \ldots, q^{(r)k_m}(x_m))$ *will converge.*

We note that it is possible to obtain the convergence of the distributions of

$q^{(r)k}(x)$ even at random moments of time. We state here the following result for systems $\langle G, 1, G/\infty, 1\rangle$, which obviously follows from the preceding and will be needed in section 4.6.

THEOREM 3. *Let conditions* (A), (B), *and* (C) *of theorem* 1 *hold.*
Then we will have convergence as $r \to \infty$ *of the distributions of the vectors*

$$(q^{(r)k}(0), q^{(r)k}(\tau_{k+1}^{(r)e}), \ldots, q^{(r)k}(\tau_{k+1}^{(r)e} + \cdots + \tau_{k+N}^{(r)e})).$$

Here the joint distributions of these vectors for different k will also converge.

4.6. General ergodic theorems and stability theorems for sequences $\mathbf{w}_{n+1} = f(\mathbf{w}_n, \boldsymbol{\tau}_n)$

In this section we explain a certain approach to the proof of ergodic theorems and stability theorems which can be applied in the investigation of various types of service processes (see, for example, section 4.7). In essence it has already been used by us in sections 4.1, 4.2, and 4.4, and also in Borovkov (1972e, 1976a) for the proof of the ergodic theorem and the stability theorem for systems with refusals, and in Akhmarov (1979a, b, c, d) and Borovkov (1977a) to obtain estimates of the speed of convergence for different types of service systems (for ergodic theorems, see also Akhmarov and Leont'eva, 1976; Borovkov, 1976a; Leont'eva, 1977, 1978).

The point at issue is an approach which could be called the 'renewal method'. It is of independent interest since it may turn out to be useful for solving problems outside queuing theory. We require here a more general statement of the problem.

Let $\{\boldsymbol{\tau}_j; -\infty < j < \infty\}$ be a vector-valued, stationary, metrically transitive sequence (in what follows bold face symbols will denote vectors). Consider a vector-valued sequence $\mathbf{w}_n$ (the dimensions of $\mathbf{w}_n$ and $\boldsymbol{\tau}_j$, in general, are different) which is determined by an initial value $\mathbf{w}_1$ and the recurrence relations

$$\mathbf{w}_{n+1} = \mathbf{f}(\mathbf{w}_n, \boldsymbol{\tau}_n), \quad n \geqslant 1. \tag{4.55}$$

Later we will see that the basic types of service systems have state characteristics described by equations of the form (4.55): for example, multichannel systems with queues and refusals (see section 4.7).

The problem is the elucidation of conditions under which the sequence $\{\mathbf{w}_{n+k}; k \geqslant 0\}$ converges as $n \to \infty$ to some stationary sequence $\{\mathbf{w}^k; k \geqslant 0\}$, and stability conditions for this second sequence (or $\{\mathbf{w}_n\}$) under small changes of the controlling sequence $\{\boldsymbol{\tau}_j\}$.

If $\mathbf{w}_1 = 0$ and the function $\mathbf{f} \geqslant \mathbf{0}$ is left continuous and monotone in its first argument, then, as established by Loynes (1962) (see also Borovkov, 1976a), ergodicity always holds (the case $\mathbf{w}^k = \infty$ is not excluded). However, the above requirement on $\mathbf{w}_1$ and $\mathbf{f}$ is very restrictive and the method of proof in Loynes (1962) is not successful in obtaining a stability theorem.

Some approaches to the study of the existence and uniqueness of the stationary solution of equation (4.55) were considered in Lisek (1982).

Let $\mathfrak{F}_{n,l}$ be the σ-algebra generated by $\boldsymbol{\tau}_n, \ldots, \boldsymbol{\tau}_l$. We denote by T the one-to-one measure preserving shift transformation on sets of $\mathfrak{F} = \mathfrak{F}_{-\infty,\infty}$, so that

$$T\{\omega: \boldsymbol{\tau}_j \in \mathbf{B}_j; j = 1, \ldots, k\} = \{\omega: \boldsymbol{\tau}_{j+1} \in \mathbf{B}_j; j = 1, \ldots, k\}$$

for any choice of Borel sets $\mathbf{B}_j$ in the domain of $\boldsymbol{\tau}_j$. The corresponding transformation on $\mathfrak{F}$-measurable functions is denoted by U, so that

$$\boldsymbol{\tau}_{j+1}(\omega) = U\boldsymbol{\tau}_j(\omega).$$

The transformations T^{-1}, U^{-1} are defined similarly. If $\zeta(\omega)$ is measurable relative to $\mathfrak{F}$ then the sequence $\zeta_n = U^n\zeta$, $-\infty < n < \infty$, alongside $\{\boldsymbol{\tau}_j\}$, will be stationary and metrically transitive.

Regarding the initial condition $\mathbf{w}_1$, it can be assumed that it either does not depend on $\{\boldsymbol{\tau}_j\}$ or is measurable relative to $\mathfrak{F}_{-\infty,\infty}$ (the first easily reduces to the second by extending the corresponding σ-algebras). In what follows, to avoid minor hypotheses regarding the initial condition, we will take $\mathbf{w}_1$ to be fixed.*

Definition. *The events $A_n \in \mathfrak{F}_{-\infty,n+L}$, $n \geqslant n_0$ are called renewing on the intervals $[n, n + L]$ if the random variables $\mathbf{w}_{n+k} = \mathbf{w}_{n+k}(\omega)$, for $k > L$, $\omega \in A_n$ can be expressed in the form*

$$\mathbf{w}_{n+k} = \boldsymbol{\phi}(\boldsymbol{\tau}_n, \ldots, \boldsymbol{\tau}_{n+k-1}) \tag{4.56}$$

where $\boldsymbol{\phi}$ depends only on the number of arguments and is determined by the choice of sequence A_n.

In other words, the following two properties are satisfied:

1. $\mathbf{w}_{n+k}$ for $n > L$ is measurable on A_n relative to $\mathfrak{F}_{n,n+k-1}$ (that is, $\{\omega: \mathbf{w}_{n+k} \in \mathbf{B}\}A_n \in \mathfrak{F}_{n,n+k-1}A_n$ for any Borel set $\mathbf{B}$).
2. For $\omega \in A_n T^{-s} A_{n+s}$ and any $s > 0$, $k > L$ we have

$$\mathbf{w}_{n+k}(\omega) = U^{-s}\mathbf{w}_{n+k+s}(\omega).$$

A hit of $\mathbf{w}_n$ on a fixed point $\mathbf{x}$ is obviously a renewing event. In particular, the event $A_n = \{\mathbf{w}_n = \mathbf{0}\} \in \mathfrak{F}_{1,n-1}$ will be renewing, since for $\omega \in A_n$ we have $\mathbf{w}_{n+1} = \mathbf{f}(\mathbf{0}, \boldsymbol{\tau}_n)$, $\mathbf{w}_{n+2} = \mathbf{f}(\mathbf{f}(\mathbf{0}, \boldsymbol{\tau}_n), \boldsymbol{\tau}_{n+1})$, etc.

In the conditions of section 4.4 (that is, for functions $\mathbf{f}$ defined by (4.34) and (4.35)) we used renewing events of precisely this form. The event $\{\nu(n + L) \leqslant L\}$ in theorems 1A and 2A of section 4.4 means that in $[n, n + L]$ there is at least one hit on the point $\mathbf{0}$ so the event $\{\nu(n + L) \geqslant L\}$ is also renewing.

* If $\mathbf{w}_1$ does not depend on $\{\boldsymbol{\tau}_j\}$, then, in general, this assumption will not limit generality. We could, for example, consider the conditional distributions of $\mathbf{w}_n$ relative to $\mathbf{w}_1$.

In section 4.3 we gave examples of more complex renewing events for the functions $\mathbf{f}$ which turn up in the description of multichannel systems.

A renewing event A_n is called *stationary* if

$$A_n = T^n A_0, \quad A_0 \in \mathfrak{F}_{-\infty,L}$$

We now prove the following result.

THEOREM 1. *Let there exist a sequence of renewing events A_j such that*

$$\mathbf{P}\left(\bigcap_{l=l_0}^{\infty} \bigcup_{j=n_0}^{n} A_j T^{-l} A_{j+l}\right) \to 1 \tag{4.57}$$

as $n \to \infty$ for any $l_0 > 0$, $n_0 > 0$. Then there is a (proper) vector $\mathbf{w}^0$ such that $U^{-n}\mathbf{w}_n \to \mathbf{w}^0$ a.s. The stationary sequence $\{\mathbf{w}^k = U^k \mathbf{w}^0;\ k \geqslant 0\}$ satisfies the recurrence relation (4.55). *The distribution of the sequence $\{\mathbf{w}_{n+k}; k \geqslant 0\}$ converges as $n \to \infty$ to the distribution of $\{\mathbf{w}^k;\ k \geqslant 0\}$ in the following strong sense*

$$\mathbf{P}(U^{-n-k+1}\mathbf{w}_{n+k+s} = \mathbf{w}^{s+1} \text{ for all } k \geqslant 0, s \geqslant 0) \to 1. \tag{4.58}$$

Hence it follows, in particular, that

$$\mathbf{P}(\mathbf{w}_{n+s} = \mathbf{w}^{n+s} \text{ for all } s \geqslant 0) \to 1$$

$$\mathbf{P}(U^{-n-k}\mathbf{w}_{n+k} = \mathbf{w}^0 \text{ for all } k \geqslant 0) \to 1 \tag{4.58a}$$

as $n \to \infty$.

If A_n is stationary then $T^{-l}A_{j+l} = A_j$ and for (4.57) *it is necessary and sufficient that*

$$\mathbf{P}(A_0) > 0.$$

The following converse is also valid.

Let there be a vector $\mathbf{w}^0$ such that the convergence (4.58a) *holds. Then there are stationary renewing events $A_n = T^n A_0$, $\mathbf{P}(A_0) > 0$.*

The converse is due to S. G. Foss.

Remark 1. In the first part of the theorem the metric transitivity of $\{\tau_j\}$ is not required.

Remark 2. If $\mathbf{w}_n$ form a Markov chain (say in the case when the τ_j are independent), then the criterion of ergodicity, given in the theorem, comes close to the condition of existence of a positive recurrent state.

Proof. Consider the values $\mathbf{w}_n$ and $\mathbf{w}_{n+k}$, $n > 0$, $k > 0$, which are measurable with respect to $\mathfrak{F}_{1,n-1}$ and $\mathfrak{F}_{1,n+k-1}$ respectively. The variables $\mathbf{u}_n = U^{-n+1}\mathbf{w}_n$ and $\mathbf{u}_{n+k} = U^{-n-k+1}\mathbf{w}_{n+k}$ will be measurable with respect to $\mathfrak{F}_{-n+2,0}$, $\mathfrak{F}_{-n-k+2,0}$. From the definition of renewing events it follows that

$$(T^{-n+1}A_{n-l}) \cap (T^{-n-k+1}A_{n+k-l}) \subset \{U^{-n+1}\mathbf{w}_{n+s} = U^{-n-k+1}\mathbf{w}_{n+k+s}\} \quad (4.59)$$

for any $s \geqslant 0$ and $l \geqslant L$ (for 'shifted' trajectories in both cases renewal occurs in the time interval $[-l+1, -l+L+1]$).

Further, from (4.59) it follows that

$$B_{n,k} = \bigcup_{l=L+1}^{n-1} (T^{-n+1}A_{n-l})(T^{-n-k+1}A_{n+k-l})$$
$$\subset \{U^{-n+1}\mathbf{w}_{n+s} = U^{-n-k+1}\mathbf{w}_{n+k+s};\ s = 0, 1, \ldots\}$$

and

$$\bigcap_{k=l_0}^{\infty} B_{n,k} \subset \{U^{-n+1}\mathbf{w}_{n+s} = U^{-n-k+1}\mathbf{w}_{n+k+s};$$
$$k = l_0, l_0+1, \ldots;\ s = 0, 1, \ldots\}. \quad (4.60)$$

By (4.57) the probability $\mathbf{P}(\bigcap_{k=l_0}^{\infty} B_{n,k}) = \mathbf{P}(\bigcap_{k=l_0}^{\infty} T^{n-1}B_{n,k})$ converges to 1 as $n \to \infty$. By (4.60) this means that the sequence $U^{-n+1}\mathbf{w}_{n+s}$ is fundamental as $n \to \infty$ in the sense of almost sure convergence. Consequently, with probability 1,

$$\lim_{n\to\infty} U^{-n+1}\mathbf{w}_{n+s} = \mathbf{w}^{s+1} \quad (4.61)$$

exists. Here

$$\mathbf{w}^{s+1} = \lim_{n\to\infty} U^{-n+1}\mathbf{f}(\mathbf{w}_{n+s-1}, \tau_{n+s-1})$$
$$= \lim_{n\to\infty} \mathbf{f}(U^{-n+1}\mathbf{w}_{n+s-1}, \tau_s).$$

If we allow for the nature of the convergence (coincidence with the limit, beginning with some time), we obtain that for any function $\mathbf{f}$

$$\mathbf{w}^{s+1} = \mathbf{f}(\lim U^{-n+1}\mathbf{w}_{n+s-1}, \tau_s) = \mathbf{f}(\mathbf{w}^s, \tau_s). \quad (4.62)$$

Since

$$U\mathbf{w}^{s+1} = U \lim_{n\to\infty} U^{-n+1}\mathbf{w}_{n+s} = \lim U^{-n+2}\mathbf{w}_{n+s} = \mathbf{w}^{s+2}$$

then $\mathbf{w}^s = U^s\mathbf{w}^0$. The nature of the convergence (4.58) follows from (4.60) and (4.61).

We will prove the result of the theorem on stationary A_j. In this case the probability in (4.57) is equal to $\mathbf{P}(\bigcup_{j=1}^{n} A_j)$ and the necessity of the condition $\mathbf{P}(A_0) > 0$ is obvious. We prove its sufficiency. Put $B = \bigcup_{j=1}^{\infty} A_j$. Since $A_j = T^jA_0$, then $TB = \bigcup_{j=1}^{\infty} TA_j = \bigcup_{j=2}^{\infty} A_j \subset B$. In addition, since T preserves measure, then $\mathbf{P}(TB) = \mathbf{P}(B)$. Hence it follows that $TB = B$ up to a set of measure zero, and, consequently, B is invariant relative to T. By the metric transitivity of $\{\tau_j\}$ this means that $\mathbf{P}(B) = 1$ or 0. Since $\mathbf{P}(B) \geqslant \mathbf{P}(A_0) > 0$ then $\mathbf{P}(B) = 1$.

Consequently

$$\mathbf{P}\left(\bigcup_{j=1}^{n} A_j\right) \to 1$$

as $n \to \infty$ and (4.57) is true.

We pass on now to the converse assertion of the theorem. Put $B_n = \{\mathbf{w}_{n+s} = \mathbf{w}^{n+s}$ for all $s \geqslant 0\}$. Then the event

$$A_n^{(1)} = (T^n B_L) \cap B_{n+L} \subset \{U^n \mathbf{w}_L = \mathbf{w}^{n+L} = \mathbf{w}_{n+L}\}$$

will be renewing. In fact, $\mathbf{w}_L$ is uniquely defined by the values $\tau_1, \ldots, \tau_{L-1}$, that is, admits a representation $\mathbf{w}_L = \phi(\tau_1, \ldots, \tau_{L-1})$ (see (4.55)). Consequently, on $A_n^{(1)}$ the values $\mathbf{w}^{n+L}$ and $\mathbf{w}_{n+L}$ will equal $\phi(\tau_{n+1}, \ldots, \tau_{n+L-1})$. It could be verified that these renewing sets $A_n^{(1)}$ satisfy (4.57) for corresponding L, l_0, and n_0. However, the verification is more suitably carried out with stationary events. Put

$$C_n = \{U^{-n-k} \mathbf{w}_{n+k} = \mathbf{w}^0 \text{ for all } k \geqslant 0\}$$

$$b_n = \{\mathbf{w}_n = \mathbf{w}^n\}.$$

Then $T^{n+L} C_{n+L} \subset \{\mathbf{w}_{n+L} = \mathbf{w}^{n+L}\}$, so that putting

$$A_n^{(2)} = (T^n C_L) \cap (T^{n+L} C_{n+L})$$

we will again have the required inclusion

$$A_n^{(2)} \subset \{U^n \mathbf{w}_L = \mathbf{w}^{n+L} = \mathbf{w}_{n+L}\}$$

meaning that $A_n^{(2)}$ are renewing events. Since $C_n \subset \{U^{-n} \mathbf{w}_n = \mathbf{w}^0\} = T^{-n} b_n$, then, by (4.58a), $\mathbf{P}(C_n) \to 1$, $\mathbf{P}(b_n) \to 1$, $\mathbf{P}(T^{n+L} C_{n+L}) = \mathbf{P}(C_{n+L}) \to 1$ as $n \to \infty$. Consequently, there are L and n_0 such that

$$\mathbf{P}(B_L) > 0, \quad \mathbf{P}(A_{n_0}^{(2)}) = \mathbf{P}(B_L \cap T^L C_{n_0+L})$$
$$\geqslant \mathbf{P}(B_L) - [1 - \mathbf{P}(C_{n_0+L})] \geqslant \tfrac{1}{2}\mathbf{P}(B_L) > 0.$$

Here we have used the inequality $\mathbf{P}(AB) \geqslant \mathbf{P}(A) - \mathbf{P}(B)$. Now put

$$A_n = (T^n b_L) \cap (T^{n+L} C_{n_0+L}).$$

Since $C_{n_0} \subset C_n$ for $n \geqslant n_0$, then $A_n \subset A_n^{(2)}$ and, together with $A_n^{(2)}$, is a renewing event. In addition, A_n is stationary:

$$A_n = T^{n-n_0} A_{n_0}$$

$\mathbf{P}(A_n) = \mathbf{P}(A_{n_0}) = \mathbf{P}(A_{n_0}^{(2)}) > 0$. The theorem is proved. □

We remark once again that in this section we have assumed the initial condition $\mathbf{w}_1$ to be fixed. The role of initial conditions in the construction of renewing events is discussed in the concrete examples of the following section.

Within the area of general results there is interest in the question of conditions for the existence of renewing events independent of the initial condition $\mathbf{w}_1$, when $\mathbf{w}_1$ belongs to some given set K. Such renewing events (not depending on $\mathbf{w}_1$) have been constructed by us in section 4.4 (theorems 1, 2). These theorems, and also the results of section 4.7, show that the existence of such renewing events is connected in the first place with the compactness of K.

We pass on to *stability theorems*. We will consider conditions, connected with the renewal method, under which a small change in the distribution of the sequence $\{\tau_j\}$ leads to a small change in the stationary or prestationary sequence $\{\mathbf{w}_n; n \geqslant 1\}$. More precisely, let $\{\tau_j^{(r)}\}$, $r = 1, 2, \ldots$ be a family of controlling sequences depending on a parameter r and let $\{\mathbf{w}^{(r)k}; k \geqslant 0\}$, $\{\mathbf{w}_n^{(r)}; n \geqslant 1\}$, $r = 1, 2, \ldots$ be the corresponding stationary and prestationary sequences (in what follows an upper index (r) in various notations means relative to the controlling sequence $\{\tau_j^{(r)}\}$). We assume that $\{\tau_j^{(r)}\}$ satisfies the conditions of theorem 1 so that $\{\mathbf{w}_{n+k}^{(r)}; k \geqslant 0\}$ converges (in the sense of (4.58)) to $\mathbf{w}^{(r)k}$. We now suppose that condition (A) is satisfied.

(A) *The finite-dimensional distributions of* $\{\tau_j^{(r)}\}$ *converge as* $r \to \infty$ *to the distributions of* $\{\tau_j\}$.

Does this mean convergence of the distributions of $\{\mathbf{w}^{(r)k}\}$ and $\{\mathbf{w}^k\}$ or the distributions of $\{\mathbf{w}_{n+k}^{(r)}\}$ and $\{\mathbf{w}^k\}$ as $r \to \infty$, $n \to \infty$?

To simplify the exposition we limit ourselves here to the case when there are stationary renewing events A_n. From the proof of theorem 1 it is clear that the $\mathbf{w}^k$, even the $\mathbf{w}_k$, may be taken to be measurable with respect to $\mathfrak{F}_{-\infty,k-1}$ (see (4.61)), therefore in what follows we may regard $\{\mathbf{w}_n\}$ as both a stationary and a prestationary sequence. The stationarity of $\{\mathbf{w}_n\}$, or its absence, as a rule, will not influence the course of our arguments and the use of one notation for both cases shortens the text.

We will suppose further that condition (B) is satisfied.

(B) *The renewing events* $A_n^{(r)}$ *and* A_n *have the properties*

$$\mathbf{P}(A_0) > 0, \quad \liminf_{r\to\infty} \mathbf{P}\left(\bigcup_{j=0}^{k} A_j^{(r)}\right) \geqslant 1 - p(k) > 0$$

where $p(k) \to 0$ *as* $k \to \infty$.

This condition is always satisfied if we have

(B$_1$)

$$\mathbf{P}(A_0) > 0, \quad \liminf_{r\to\infty} \mathbf{P}\left(\bigcup_{j=0}^{k} A_j^{(r)}\right) \geqslant \mathbf{P}\left(\bigcup_{j=0}^{k} A_j\right)$$

for any $k \geqslant 0$.

The last condition is more transparent and, in its turn, is always satisfied if

(B$_2$) *The events* A_k, $\mathbf{P}(A_k) > 0$, *admit a representation as open sets*

$$A_k = \{\gamma_{k,1} > 0, \ldots, \gamma_{k,s} > 0\}$$

for some $s \geqslant 1$, *where* $\gamma_{k,j}$ *are measurable relative to* $\mathfrak{F}_{-\infty,k+L}$, $U\gamma_{k,j} = \gamma_{k+1,j}$ *and the distribution of* $(\gamma_{0,1}^{(r)}, \ldots, \gamma_{0,s}^{(r)})$ *weakly converges as* $r \to \infty$ *to the distribution of* $(\gamma_{0,1}, \ldots, \gamma_{0,s})$ *(here it is sufficient to require convergence in a neighbourhood of* 0).

We also need a condition of a third type connected with the properties of the function $\boldsymbol{\phi}$, which appears in the definition of renewing events.

(C) *The function* $\boldsymbol{\phi}(\mathbf{y}_1, \ldots, \mathbf{y}_k)$ *in* (4.56) *is continuous almost surely relative to the distribution of* $(\boldsymbol{\tau}_1, \ldots, \boldsymbol{\tau}_k)$.

Remark. To verify condition (C) there can arise the problem of finding a function $\boldsymbol{\phi}$ relative to $\mathbf{f}$ and the events A_k. Suppose that $A_n \subset B_n$, where B_n is a broader renewing event with the property

$$B_n \in \sigma(\mathbf{w}_n, \ldots, \mathbf{w}_{n+L}; \boldsymbol{\tau}_n, \ldots, \boldsymbol{\tau}_{n+L}) \subset \mathfrak{F}_{-\infty,n+L}.$$

In all the applications referred to in this chapter B_n is given in the form of certain relations

$$\boldsymbol{\psi}(\mathbf{w}_n, \ldots, \mathbf{w}_{n+L}; \boldsymbol{\tau}_n, \ldots, \boldsymbol{\tau}_{n+L}) \in \mathbf{G}$$

where the vector-valued function $\boldsymbol{\psi}$ and the set $\mathbf{G}$ do not depend on n. We denote by $\mathbf{a}$ any value of $\mathbf{w}_{n-1}$ from which 'it is possible to reach' B_n, that is,

$$\mathbf{P}(\boldsymbol{\psi}(\mathbf{w}'_n, \ldots, \mathbf{w}'_{n+L}; \boldsymbol{\tau}_n, \ldots, \boldsymbol{\tau}_{n+L}) \in \mathbf{G}) > 0$$

where

$$\mathbf{w}'_n = \mathbf{f}(\mathbf{a}, \boldsymbol{\tau}_{n-1})$$

$$\mathbf{w}'_{n+1} = \mathbf{f}_2(\mathbf{a}, \boldsymbol{\tau}_{n-1}, \boldsymbol{\tau}_n) = \mathbf{f}(\mathbf{f}(\mathbf{a}, \boldsymbol{\tau}_{n-1}), \boldsymbol{\tau}_n), \ldots .$$

Then, obviously,

$$\mathbf{w}'_{n+k} = \mathbf{f}_{k+1}(\mathbf{a}, \boldsymbol{\tau}_{n-1}, \ldots, \boldsymbol{\tau}_{n+k-1})$$

where

$$\mathbf{f}_{k+1}(\mathbf{a}, \mathbf{y}_1, \ldots, \mathbf{y}_{k+1}) = \mathbf{f}(\mathbf{f}_k(\mathbf{a}, \mathbf{y}_1, \ldots, \mathbf{y}_k), \mathbf{y}_{k+1})$$

and we must put

$$\boldsymbol{\phi}(\mathbf{y}_1, \ldots, \mathbf{y}_k) = \mathbf{f}_{k+1}(\mathbf{a}, \mathbf{y}_0, \mathbf{y}_1, \ldots, \mathbf{y}_k).$$

Here the right-hand side in the renewing event does not depend on $\mathbf{y}_0$ and $\mathbf{a}$ (if $\mathbf{a}$ is changed into another point from which it is possible to reach B_n).

In a quite similar way take as $\mathbf{a}$ a value of $\mathbf{w}_{n-s}$, for any $s > 1$, from which it is possible to reach B_n. Then we would obtain

$$\boldsymbol{\phi}(\mathbf{y}_1, \ldots, \mathbf{y}_k) = \mathbf{f}_{k+s}(\mathbf{a}, \mathbf{y}_{-s+1}, \ldots, \mathbf{y}_k)$$

where the right-hand side depends only on the last k arguments.

THEOREM 2. *Let conditions* (A), (B), *and* (C) *be satisfied. Then the finite-dimensional distributions of* $\{\mathbf{w}_{n+k}^{(r)}; k \geqslant 0\}$ *weakly converge as* $n \to \infty$, $r \to \infty$ *to the finite-dimensional distributions of* $\{\mathbf{w}^k; k \geqslant 0\}$.

Remark. Obviously, if $\{\mathbf{w}_n^{(r)}\}$, $\{\mathbf{w}_n\}$ are stationary then the condition $n \to \infty$ becomes superfluous. For the nonstationary case it will in fact be proved that the distributions, *simultaneously for all sequences* $\{\mathbf{w}_{n-k}^{(r)}; k \geqslant 0\}$, $n \geqslant n_0$, converge as $n_0 \to \infty$, $r \to \infty$ to the distribution of $\{\mathbf{w}^k; k \geqslant 0\}$.

The question of closeness of the distributions of $\mathbf{w}_n$ and $\mathbf{w}_n^{(r)}$ for finite n does not usually give rise to difficulties and is guaranteed by condition (A) and the almost sure continuity of $\mathbf{f}_{n-1}(\mathbf{w}_1, \mathbf{y}_1, \ldots, \mathbf{y}_{n-1})$ relative to the distribution of $(\boldsymbol{\tau}_1, \ldots, \boldsymbol{\tau}_{n-1})$.

Proof. We note first of all that for sufficiently large r the sequences $\{\mathbf{w}_n^{(r)}\}$ also satisfy the conditions of theorem 1. In fact, from condition (B) it follows that

$$\liminf_{r\to\infty} \mathbf{P}(A_0^{(r)}) > 0 \tag{4.63}$$

and, consequently, from some point $\mathbf{P}(A_0^{(r)}) > 0$.

We consider a random variable v, equal to the least $k > L$ for which A_{n-k} occurs. In other words, $v(\omega) = k$ for $\omega \in B_k = \bar{A}_{n-L-1}\bar{A}_{n-L-2}\cdots\bar{A}_{n-k+1}A_{n+k}$. Since $\mathbf{P}(\bigcup_{k>L} B_k) = \mathbf{P}(\bigcup_{k<n-L} A_k) = 1$, then given $\varepsilon > 0$ we can find N such that

$$\mathbf{P}(v > N) < \varepsilon, \quad p(N) < \varepsilon.$$

Now let $\mathbf{x}$ be a point of continuity of the distribution of $\boldsymbol{\phi}(\boldsymbol{\tau}_1, \ldots, \boldsymbol{\tau}_{n+L})$. Then, for any $n > N + L$

$$\limsup_{r\to\infty} \mathbf{P}(\mathbf{w}_n^{(r)} < \mathbf{x}) \leqslant \limsup_{r\to\infty} \mathbf{P}(\mathbf{w}_n^{(r)} < \mathbf{x}, v^{(r)} \leqslant N) + \varepsilon. \tag{4.64}$$

Using the renewal condition we obtain

$$\mathbf{P}(\mathbf{w}_n^{(r)} < \mathbf{x};\, v^{(r)} \leqslant N) = \mathbf{P}\left(\mathbf{w}_n^{(r)} < \mathbf{x};\, \bigcup_{j=1}^{N} A_{n-L-j}^{(r)}\right)$$
$$= \mathbf{P}\left(\boldsymbol{\phi}(\boldsymbol{\tau}_{n-L-N}^{(r)}, \ldots, \boldsymbol{\tau}_{n-1}^{(r)}) < \mathbf{x};\, \bigcup_{j=1}^{N} A_{n-N-j}^{(r)}\right).$$

Therefore, extending (4.64), we find by conditions (A) and (C)

$$\limsup_{r\to\infty} \mathbf{P}(\mathbf{w}_n^{(r)} < \mathbf{x}) \leqslant \mathbf{P}(\boldsymbol{\phi}(\boldsymbol{\tau}_1, \ldots, \boldsymbol{\tau}_{n+L}) < \mathbf{x}) + \varepsilon.$$

In a similar way we find

$$\mathbf{P}(\mathbf{w}^n < \mathbf{x}) \geqslant \mathbf{P}\left(\mathbf{w}^n < \mathbf{x};\, \bigcup_{j=1}^{N} A_{n-L-j}\right)$$
$$= \mathbf{P}\left(\boldsymbol{\phi}(\boldsymbol{\tau}_{n-L-N}, \ldots, \boldsymbol{\tau}_{n-1}) < \mathbf{x};\, \bigcup_{j=1}^{N} A_{n-L-j}\right)$$
$$\geqslant \mathbf{P}(\boldsymbol{\phi}(\boldsymbol{\tau}_1, \ldots, \boldsymbol{\tau}_{N+L}) < \mathbf{x}) - \varepsilon.$$

Therefore, for $n > N + L$, $N = N(\varepsilon)$

$$\limsup_{r\to\infty} P(\mathbf{w}_n^{(r)} < \mathbf{x}) \leqslant \mathbf{P}(\mathbf{w}^n < \mathbf{x}) + 2\varepsilon.$$

The converse inequality for the lim inf is established similarly.

The convergence of arbitrary finite-dimensional distributions is proved in precisely the same way. The theorem is proved. □

4.7. Ergodic theorems and stability theorems for multichannel systems with refusals and with queues

The systems discussed in this section differ from the $\langle G, G, G/\infty, 1\rangle$ systems, described in section 4.5, as follows. The number m of service channels is finite: $m < \infty$. If at the arrival epoch of the next group of requests the number of free channels is less than the number of arriving requests then the 'surplus' (in order of arrival) requests form a queue for systems with waiting, or are refused and leave the discussion for systems with refusals. For systems with waiting the requests in the queue proceed to service as the channels are freed.

For simplicity here we will only consider the systems $\langle G, 1, G/m, 1\rangle$ and $\langle G, 1, G/m, 1\rangle_R$, in the notation of Borovkov (1976a), for which $v_j^e \equiv v_j^s \equiv 1$. The passage to the systems $\langle G, G, G/m, 1\rangle$ and $\langle G, G, G/m, 1\rangle_R$ (that is, to the case $v_j^e \not\equiv 1$) introduces only technical problems without changing the essence of the discussion (cf. section 4.5 and Borovkov, 1976a, chapters 5, 7).

One of the fundamental characteristics of these systems is the vector of waiting times

$$\mathbf{w}_n = (w_{n,1}, \ldots, w_{n,m})$$

where $w_{n,i}$ is the time from the arrival epoch of the nth request up to the time when the ith channel is free of requests which arrived prior to the nth (that is, of the first $(n-1)$ requests). Clearly the usual characteristic for systems with refusals, q_n, the number of busy (free) channels at the arrival epoch of the nth request, is equal to the number of nonzero (zero) coordinates of w_n.

Let $x^+ = \max(0, x)$ and for a vector $\mathbf{x}$ put

$$\mathbf{x}^+ = (x_1^+, \ldots, x_m^+).$$

$\mathbf{R}(\mathbf{x})$ will denote the vector obtained from $\mathbf{x}$ by ordering the coordinates so that the first coordinate of $\mathbf{R}(\mathbf{x})$ is $\min(x_1, \ldots, x_m)$. In addition, let

$$\mathbf{e} = (1, 0, \ldots, 0), \quad \mathbf{i} = (1, 1, \ldots, 1)$$

and let $I(A)$ be the indicator of the set A. Then it is not difficult to see that for systems with refusals the following recurrence relation holds for $\mathbf{w}_n$:

$$\mathbf{w}_{n+1} = \mathbf{R}(\mathbf{w}_n + \tau_n^e \mathbf{e} I(w_{n,1} = 0) - \tau_n^e \mathbf{i})^+ \tag{4.65}$$

where τ_n^e, τ_n^s are elements of the controlling sequence (see (4.46)) (the operations $\mathbf{R}$ and $(\cdot)^+$ commute).

For systems with queues (waiting) instead of (4.65) we will have the well-known equality

$$\mathbf{w}_{n+1} = \mathbf{R}(\mathbf{w}_n + \tau_n^s \mathbf{e} - \tau_n^e \mathbf{i})^+. \tag{4.66}$$

Relation (4.66), as also (4.65), is obviously an equality of type (4.55), studied in the previous section for $\boldsymbol{\tau}_n = (\tau_n^e, \tau_n^s)$.

For these multichannel systems with *arbitrary* stationary controlling sequences (τ_j^e, τ_j^s) *and arbitrary initial conditions* we do not have at present such complete ergodicity and stability theorems as were given in sections 4.2, 4.3, and 4.5. The renewing events which we use force us to place certain conditions on the systems which, although not strongly restricting, are all the same not apparently connected with essential matters. We must note that for simpler systems, for example, single-channel, this is not the case. The results of sections 4.2, 4.3, and, in part, section 4.4 contain no superfluous conditions, although they are completely based on the renewal method. It is just this that allows us to reduce the dependence of the characteristics under discussion on the controlling sequence essentially to dependence on a finite number of its elements.

It appears (see theorem 1 of section 4.6) that for multichannel systems $\langle G, 1, G/m, 1\rangle$ with queues and with refusals, under natural restrictions on $\{\tau_j^e, \tau_j^s\}$, renewing events always exist. In every case for systems $\langle G_I, 1, G_I, 1\rangle$ this is so (see theorem 5 and Akhmarov and Leont'eva, 1976). In this case we have succeeded in finding renewing events whose existence is not connected with the additional inessential conditions and, consequently, does not decrease generality. For multichannel systems the situation is somewhat different. The simplest renewing events which we use in a series of cases force us to impose additional restrictions (guaranteeing the occurrence of these events with positive probability) which are not always connected with the essential matters. According to which multichannel systems are considered this additional condition, roughly speaking consists of the requirement of unboundedness of the random variables τ_j^e (see theorem 3). This condition, as already noted in the introduction to this chapter, is apparently superfluous.

We first consider the application of the *ergodic* theorem 1 of section 4.6 to the systems $\langle G, 1, G/m, 1\rangle$ and $\langle G, 1, G/m, 1\rangle_R$. Here in the role of $\mathbf{f}$ are the functions on the right in (4.66) and (4.65). Since no conditions are put on $\mathbf{f}$ in theorem 1 of section 4.6, then its influence will only be felt in the construction of renewing sets.

We will suppose, of course, that the sequence $\boldsymbol{\tau}_j = (\tau_j^e, \tau_j^s)$ is *metrically transitive and does not depend on* $\mathbf{w}_1$.

We will begin with some examples which will be essential later. We consider the sequence $\mathbf{w}_n$ and put

$$A_n = \left\{ w_{n,1} = 0,\ w_{n,j} \leqslant \sum_{k=0}^{j-2} \tau_{n+k}^e;\ j = 2, \ldots, m \right\}. \tag{4.67}$$

We can establish directly from the recurrence formulae (4.65) and (4.66) that A_n is a renewing event for $L = m - 2$, since, for $k > n + m - 2$, $\mathbf{w}_{n+k}$ on A_n are functions only of (τ_j^e, τ_j^s) for $j \geqslant n$.

In (4.67) we have the simplest form of renewal. Other renewals are possible, for example (Akhmarov and Leont'eva, 1976)

$$\left\{ w_{n+j,1} = 0;\ j = 0, \ldots, L;\ w_{n,m} \leqslant \sum_{k=0}^{j-2} \tau_{n+k}^e \right\}. \tag{4.68}$$

It is not difficult here to see that $\mathbf{w}_{n+L+1}$ is a function only of (τ_j^e, τ_j^s) for $n \leqslant j \leqslant n+L$. We note that the event (4.67) includes (4.68).

The verification of a renewal condition of the form (4.67), (4.68) may present real difficulty. This is connected above all with the fact that the distribution of the sequence $\mathbf{w}_n$, as a rule, is not known to us. In many cases (particularly where the existence of a stationary sequence $\mathbf{w}^n$ has not been established) it is natural, say, to replace conditions of type (4.67) by

$$\left\{v_{n,1} = 0, v_{n,j} < \sum_{k=0}^{j-2} \tau_{n+k}^e; j = 2, \ldots, m\right\} \tag{4.69}$$

where $v_{n,j}$ is a known (in some way or other) sequence which majorizes $w_{n,j}$:

$$w_{n,j} \leqslant v_{n,j}.$$

Both for systems with waiting and for systems with refusals such a majorant may be constructed.

We consider first *systems with refusals* $\langle G, 1, G/m, 1\rangle_R$ controlled by a stationary metrically transitive sequence $\{\tau_j^e, \tau_j^s;\ -\infty < j < \infty\}$. The initial condition we will first take as zero, $\mathbf{w}_1 = 0$. We consider here two majorants.

(a) Let $\langle G, 1, G/\infty, 1\rangle$ be a system with an infinite number of channels, controlled by the sequence $\{\tau_j^e, \tau_j^s\}$ and let $q^k(x)$ denote its stationary (with respect to k) process

$$q^k(x) = \sum_{j=-\infty}^{k} I(\tau_j^s > \tau_j^e + \tau_{j-1}^e + \cdots + \tau_k^e + x) \tag{4.70}$$

defined in section 4.5 ($q^k(0)$ is the stationary number of busy channels, $q^k(x)$ indicates how many channels remain busy after time x, with no account taken of newly arriving requests). If $v_k(i)$ denotes the waiting time of the kth request in a system $\langle G, 1, G/\infty, 1\rangle$ until more than i channels are occupied by requests arriving before the kth, then the event $\{v_k(i) < x\}$, obviously, is equivalent to the event $\{q^{k-1}(x) \leqslant i\}$ ($v_k(i)$ coincides with the epoch of the jump of $q^{k-1}(x)$, when $q^{k-1}(x)$ changes value from $i+1$ to i).

But it is obvious that for our m-channel systems

$$w_{n,i} \leqslant v_n(m-i)$$

and (4.69) and (4.67) will be satisfied if

$$q^{n-1}(0) \leqslant m-1, \quad q^{n-1}\left(\sum_{k=0}^{j} \tau_{n+k}^e\right) \leqslant m-2-j, \quad j=0,\ldots,m-2. \tag{4.71}$$

Here (4.71) is equivalent to (4.69) with $v_{n,i} = v_n(m-i)$, and $q^{n-1}(x)$ is defined explicitly by (4.70).

(b) We now construct another majorant, based on an estimate of the last coordinate $w_{n,m}$ of $\mathbf{w}_n$.

LEMMA 1. *Under the zero initial condition* $\mathbf{w}_1 = \mathbf{0}$ *we have*

$$w_{n,m} \leqslant \zeta_{n-1}^+ + V_{n-1}$$

for all n, where

$$\zeta_n = \tau_n^s - \tau_n^e, \quad V_n = \sup_{s \leqslant n}\left(\sum_{j=s}^{n} \eta_j, 0\right)$$

$$\eta_n = \zeta_{n-1}^+ - \zeta_n^+ - \tau_n^e, \quad \mathbf{E}\eta_n < 0.$$

Proof. It is obvious that for $m \geqslant 2$

$$w_{1,m} = 0, \quad w_{2,m} = (\tau_1^s - \tau_1^e)^+$$

$$w_{3,m} = \max(w_{2,m} - \tau_2^e, \tau_2^s - \tau_2^e)^+$$

for $n > 2$

$$w_{n+1,m} = \max(w_{n,m} - \tau_n^e, (\tau_n^s - \tau_n^e)I(w_{n,1} = 0))^+$$

$$\leqslant \max(w_{n,m} - \tau_n^e, \tau_n^s - \tau_n^e)^+. \tag{4.72}$$

We denote

$$\zeta_n = \tau_n^s - \tau_n^e \tag{4.73}$$

and consider the sequence z_n: $z_1 = 0$,

$$z_{n+1} = \max(z_n - \tau_n^e, \zeta_n)^+ = \max(z_n - \tau_n^e, \zeta_n, 0)$$

$$= \max(z_n - \tau_n^e, \zeta_n^+).$$

Clearly $w_{n,m} \leqslant z_n$,

$$z_{n+1} - \zeta_n^+ = \max(z_n - \tau_n^e - \zeta_n^+, 0).$$

We denote $v_n = z_n - \zeta_{n-1}^+$, $v_1 = z_1 - \zeta_0^+ \leqslant 0$. Then for $n \geqslant 1$

$$v_{n+1} = \max(v_n + \eta_n, 0), \quad \text{where } \eta_n = \zeta_{n-1}^+ - \zeta_n^+ - \tau_n^e. \tag{4.74}$$

It is not difficult to see that $\mathbf{E}\eta_n < 0$. This follows at once from $\mathbf{E}\tau_j^s < \infty$ and $\eta_n \leqslant \tau_{n-1}^s - \zeta_n - \tau_n^e = \tau_{n-1}^s - \tau_n^s$. Since $v_n \leqslant \max_{1 \leqslant s \leqslant n-1}(\sum_{j=s}^{n-1} \eta_j, 0)$ (see section 4.1), then

$$w_{n,m} \leqslant z_n \leqslant \zeta_{n-1}^+ + V_{n-1}. \tag{4.75}$$

The lemma is proved. □

Thus if, for example,

$$\tau_{n-1}^e > \zeta_{n-2}^+ + V_{n-2}, \quad \sum_{j=0}^{m-2} \tau_{n+j}^e > \tau_n^s \tag{4.76}$$

then (4.67) is satisfied. It is obvious that the events A_n defined by (4.71) and (4.76) are *stationary* renewing events.

How does a renewing event change for *nonzero initial conditions*? Let $w_{1,m} > 0$.

The recurrence relation (4.72) obviously still holds. Since the solution of (4.74) is known explicitly:

$$v_n = \max\left(v_1 + \sum_{j=1}^{n-1} \eta_j, \max_{1<s\leqslant n} \sum_{j=n}^{n-1} \eta_j\right)\left(\sum_{j=n}^{n-1} \eta_j = 0\right)$$

where $v_1 = z_1 - \zeta_0^+ = w_{1,m} - \zeta_0^+$, then by the earlier arguments we obtain

$$w_{n,m} \leqslant z_n \leqslant \zeta_{n-1}^+ + \max\left(w_{1,m} - \zeta_0^+ + \sum_{j=1}^{n-1} \eta_j, \max_{1<s\leqslant n} \sum_{j=s}^{n-1} \eta_j\right)$$

$$= \zeta_{n-1}^+ + \max_{1<s\leqslant n} \sum_{j=s}^{n-1} \eta_j + \varepsilon_{n-1} \leqslant \zeta_{n-1}^+ + V_{n-1} + \varepsilon_{n-1} \tag{4.77}$$

where $\varepsilon_n = 0$ if and only if $\min_{1<s\leqslant n} (v_1 + \sum_{j=1}^{s} \eta_j) \leqslant 0$, so that $\varepsilon_n = 0$ implies $\varepsilon_{n+1} = 0$. Thus, for $w_{1,m} > 0$, the renewing event B_n, analogous to (4.76), takes the form

$$\begin{gathered} \tau_{n-1}^e > \zeta_{n-2}^+ + V_{n-2} + \varepsilon_{n-2} \\ \sum_{j=0}^{m-2} \tau_{n+j}^e > \tau_n^s \end{gathered} \tag{4.78}$$

where $\varepsilon_n = 0$ for all $n > N(\omega) = \min\{k: \varepsilon_k = 0\}$, and $\mathbf{P}(N(\omega) < \infty) = 1$ since $\mathbf{E}\eta_k < 0$.

We now consider the event, in our case $\bigcap_{l=l_0}^{\infty} \bigcup_{j=1}^{n} B_j T^{-l} B_{j+l}$, which is present in (4.57). We denote by A_n the stationary event (4.76). Then, obviously, $B_j \subset A_j$ and $T^{-l}B_{j+l}$ is the event

$$\begin{gathered} \tau_{j-1}^e > \zeta_{j-2}^+ + V_{j-2} + U^{-l}\varepsilon_{j+l-2} \\ \sum_{k=0}^{m-2} \tau_{j+k}^e > \tau_j^s. \end{gathered}$$

Let d_N denote the set

$$\left\{\min_{1\leqslant s\leqslant N}\left(w_{1,m} + \sum_{k=1}^{s} \eta_k\right) \leqslant 0\right\}$$

and $U_{N,l}$ the set

$$\{U^{-l}\varepsilon_{N+l} = 0, U^{-l}\varepsilon_{N+l+1} = 0, \ldots\}.$$

Then, from the properties of ε_n, we conclude:

$$T^{-l}d_{N+l} \subset U_{N,l}$$

$$\begin{aligned} D_N \equiv \bigcap_{l=0}^{\infty} T^{-l}d_{N+l} &\subset U_N \\ &\equiv \{U^{-l}\varepsilon_{N+l+s} = 0 \text{ for all } l \geqslant 0, s \geqslant 0\}. \end{aligned}$$

But

$$T^{-l}d_{N+l} = \left\{ \min_{-l+1 \leqslant s \leqslant N} \left(\sum_{k=-l+1}^{s} \eta_k \right) \leqslant -w_{1,m} \right\}$$

$$\left\{ \sum_{k=-l+1}^{0} \eta_k + \min_{1 \leqslant s \leqslant N} \sum_{k=1}^{s} \eta_k \leqslant -w_{1,m} \right\} \subset T^{-l}d_{N+l}$$

$$\left\{ \min_{1 \leqslant s \leqslant n} \sum_{k=1}^{s} \eta_k \leqslant -w_{1,m} - V_0 \right\} \subset D_N.$$

Since $w_{1,m}$ and V_0 are proper random variables then, by metric transitivity, $\mathbf{P}(D_N) \to 1$ as $N \to \infty$.

On the other hand, on D_N the inequalities defining A_j and $B_j T^{-l} B_{j+l}$, for $j \geqslant N + 2$, are the same. Consequently

$$B_j T^{-l} B_{j+l} D_N = A_j D_N \quad \text{for } j \geqslant N + 2.$$

This means

$$\mathbf{P}\left(\bigcap_{l=0}^{\infty} \bigcup_{j=1}^{\infty} B_j T^{-l} B_{j+l} \right) \geqslant \mathbf{P}\left(\bigcap_{l=0}^{\infty} \bigcup_{j=N+2}^{\infty} B_j T^{-l} B_{j+l} \right) \geqslant \mathbf{P}\left(\bigcup_{j=N+2}^{n} A_j D_N \right)$$

$$\geqslant -\mathbf{P}(\bar{D}_N) + \mathbf{P}\left(\bigcup_{j=N+2}^{n} A_j \right)$$

$$= -\mathbf{P}(\bar{D}_N) + \mathbf{P}\left(\bigcup_{j=2}^{n-N} A_j \right) \to 1$$

as $N = n/2 \to \infty$, if $\mathbf{P}(A_0) > 0$.

Thus, we have proved the following result:

If $\mathbf{P}(A_0) > 0$ *for the stationary renewing events defined in* (4.76) (*in this case the conditions of theorem* 1 *for null initial conditions will be satisfied*), *then the conditions of theorem* 1 *of section* 4.6 *will hold even for the nonstationary renewing events* B_n *defined in* (4.78) *for* $\mathbf{w}_1 \neq \mathbf{0}$.

We leave it to the reader to show that the same is true even with renewing events of the form (4.71).

We have therefore obtained the following corollary of theorem 1 of section 4.6. Let A_n be the stationary events defined by (4.76) or (4.71).

THEOREM 1. *If* $\mathbf{P}(A_0) > 0, \mathbf{E}\tau_j^s < \infty$, *then for the vector* $\mathbf{w}_n$ *of waiting times for a system with refusals* $\langle G, 1, G/m, 1 \rangle_R$ (4.58) *holds for any initial condition* $\mathbf{w}_1$.

Remark 1. We note that suitable stationary renewing events for any initial condition do not, as a rule, exist.

Remark 2. To verify $\mathbf{P}(A_0) > 0$ *for the event* A_0 defined in (4.76) we must keep in mind that always $\mathbf{P}(V_{n-2} = 0) > 0$. In Akhmarov and Leont'eva (1976) it is shown that if $\mathbf{P}(m\tau_0^e > \tau_0^s) > 0$ for $\langle G_I, 1, G_I/m, 1 \rangle_R$ systems then $\mathbf{P}(A_0) > 0$ (for A_0 in (4.71)) is always satisfied.

A result similar to theorem 1 can be obtained for *systems with queues.* We first consider $\langle G, 1, G/m, 1\rangle$ systems *with zero left-hand initial conditions.* For such systems a theorem on the monotone convergence of $\mathbf{w}_n$ to a stationary sequence $\mathbf{w}^k$ has been proved (see Loynes, 1962, and also Borovkov, 1976a). Therefore $\mathbf{w}_n$ will be majorized by $\mathbf{w}^n$ and for the stability theorems it may be useful, to some extent, to use stationary renewing events in the form (4.67) and (4.68), where the $w_{n,i}$ are replaced by the components w_i^n of $\mathbf{w}^n$.

In Loynes (1962) (see also Borovkov, 1976a) there is an implicit construction of yet another majorant, in some sense similar to (4.75).

Denote $[\mathbf{w}_n] = \sum_{i=1}^m w_{n,i}$.

LEMMA 2. *For* $\mathbf{w}_1 = \mathbf{0}$ *we have*

$$[\mathbf{w}_n] \leqslant W_n \equiv \sup_{k \leqslant n} \left[(m-1)^2 \tau_{k-1}^s + (m-1) v_{k-1} + \sum_{j=k}^{n-1} \xi_j \right] \tag{4.79}$$

where

$$\xi_k = \tau_k^s - m\tau_k^e, \quad v_k = \sup_{s \leqslant k} \left(\sum_{j=s}^{k} \zeta_j \right)^+$$

$$\zeta_j = -\tau_j^s + (m-1)(\tau_{j-1}^s - \tau_j^s).$$

*Proof.** We put $D_k \mathbf{x} = \sum_{i=1}^m (x_k - x_i)$ and show first that

$$D_m \mathbf{w}_n \leqslant v_{n-1} + (m-1)\tau_{n-1}^s. \tag{4.80}$$

It is easy to see that

$$D_m \mathbf{R}(\mathbf{w}_n + \tau_n^s \mathbf{e}) = D_m \mathbf{w}_n - \tau_n^s$$

if $w_{n,1} + \tau_n^s \leqslant w_{n,m}$.

But if $w_{n,1} + \tau_n^s > w_{n,m}$, then

$$\begin{aligned} D_m \mathbf{R}(\mathbf{w}_n + \tau_n^s \mathbf{e}) &= \sum_{i-2}^{m} (w_{n,1} + \tau_n^s - w_{n,i}) \\ &= D_m \mathbf{w}_n + (m-1)\tau_n^s - m(w_{n,m} - w_{n,1}) \\ &\leqslant (m-1)\tau_n^s. \end{aligned}$$

Hence, and from the recurrence formula for $\mathbf{w}_n$, we obtain

$$D_m \mathbf{w}_{n+1} \leqslant \max\left(D_m \mathbf{w}_n - \tau_n^s, (m-1)\tau_n^s\right). \tag{4.81}$$

* In Borovkov (1972a, p. 220) there is an inequality for $w_{n,m}$ which, side by side with W_n, allows the necessary majorant to be constructed by comparison of $\langle G, 1, G/m, 1\rangle$ with the so-called 'cyclic' service system in which requests numbered $n = km + l$, $0 < l \leqslant m$, are directed into the lth channel. As was observed independently by Stoyan (1973) and Neveu this inequality is not valid. It is valid only in the broader 'stochastic' sense (see Foss, 1980, 1981; Gittins, 1978; Vasicer, 1977; Whitt, 1982; Wolff).

Subtracting $(m-1)\tau_n^s$ from both sides and putting

$$u_n = D_m \mathbf{w}_n - (m-1)\tau_{n-1}^s$$

we obtain

$$u_{n+1} \leqslant \max(0, u_n + \zeta_n), \quad u_1 = -(m-1)\tau_{n-1}^s \leqslant 0.$$

Since $u_n \leqslant v_n$ (4.80) is proved.

We remark now that $\sum_i w_{n,i} \leqslant w_{n,1} + (m-1)w_{n,m}$ implies

$$D_1 \mathbf{w}_n \leqslant (m-1) D_m \mathbf{w}_n.$$

Further, let $w_{k,1} = 0$ be the last of the variables $w_{1,1} = 0, w_{2,1}, \ldots, w_{n,1}$ which is equal to zero. Then, obviously, $D_1 \mathbf{w}_k = [\mathbf{w}_k]$ and by the recurrence relation for $\mathbf{w}_n$ ($w_{j,1} > 0$ for $j > k$)

$$[\mathbf{w}_n] = D_1 \mathbf{w}_k + \sum_{j=k}^{n-1} \xi_j.$$

Hence it follows that

$$\begin{aligned}
[\mathbf{w}_n] &\leqslant \sup_{1 \leqslant k \leqslant n} \left(D_1 \mathbf{w}_k + \sum_{j=k}^{n-1} \xi_j \right) \\
&\leqslant \sup_{1 \leqslant k \leqslant n} \left[(m-1) D_m \mathbf{w}_k + \sum_{j=k}^{n-1} \xi_j \right] \\
&\leqslant \sup_{1 \leqslant k \leqslant n} \left[(w-1)^2 \tau_{k-1}^s + (m-1) v_{k-1} + \sum_{j=k}^{n-1} \xi_j \right].
\end{aligned}$$

Lemma 2 is proved. □

It is possible to construct other majorants for $[\mathbf{w}_n]$. For example, we can use the inequality $D_1 \mathbf{w}_n \leqslant m\Delta_n$, where $\Delta_n = w_{n,m} - w_{n,1}$, as is easily seen, satisfies

$$\Delta_{n+m} \leqslant \max[\tau_n^s, \ldots, \tau_{n+m-1}^s, \Delta_n - \min(\tau_n^s, \ldots, \tau_{n+m-1}^s)].$$

LEMMA 3. *If* $\mathbf{P}(\tau_k^s \geqslant \varepsilon) = 1$ *for some* $\varepsilon > 0$ *then the probability that the* sup *in* (4.79) *is attained for* $k < n - N$ *is estimated from above by the expression*

$$\mathbf{P}\left(\sup_{k \geqslant N} \sum_{j=1}^{K} (\xi_j + \varepsilon) > 0 \right) + \sum_{k=0}^{\infty} \mathbf{P}\left(\tau_0^s \geqslant \frac{\varepsilon(N+k)}{(m-1)^2} \right).$$

Proof. For brevity we denote $(m-1)^2 \tau_k^s + (m-1) v_k = d_k$ and we apply the shift transformation T^{-n} to both parts of (4.79). Then the event C_N, whose probability is to be estimated in lemma 3, includes the event

$$C_N' = \bigcup_{k \geqslant N} \{ d_{k-1} + \xi_{-1} + \cdots + \xi_{-k} \geqslant 0 \}. \tag{4.82}$$

Consider the event

$$E_N = \left\{ \sup_{k \geqslant N} \sum_{j=1}^{k} (\xi_{-j} + \varepsilon) \leqslant 0 \right\}.$$

It is easy to see that

$$C_N E_N \subset C'_N E_N \subset \bigcup_{k \geqslant 0} \{d_{-N-k-1} - \varepsilon(N+k) \geqslant 0\}. \tag{4.83}$$

Further, from the proof of lemma 2, it is clear that $z_k = (m-1)\tau_k^s + v_k$ satisfies (see the initial inequality of (4.81))

$$z_k = \max(z_{k-1} - \tau_{k-1}^s, (m-1)\tau_{k-1}^s).$$

Therefore

$$\{z_k \geqslant M\} = \bigcup_{j \geqslant 1} \left\{ (m-1)\tau_{k-j}^s - \sum_{i=1}^{j-1} \tau_{k-1}^s \geqslant M \right\}$$

$$\subset \bigcup_{j \geqslant 1} \{(m-1)\tau_{k-j}^s \geqslant M + (j-1)\varepsilon\}. \tag{4.84}$$

But $d_k = (m-1)z_1$. By (4.83) and (4.84), this means that

$$C_N E_N \subset \bigcup_{k \geqslant 0} \left\{ z_{-M-k-1} \geqslant \frac{\varepsilon(N+k)}{m-1} \right\}$$

$$= \bigcup_{k \geqslant 0} \bigcup_{r \geqslant 1} \left\{ (m-1)\tau_{N-k-j-1}^s \geqslant \frac{\varepsilon(N+k)}{m-1} + (j-1)\varepsilon \right\}$$

$$\subset \bigcup_{k \geqslant 0} \left\{ \tau_{-N-k-2}^s \geqslant \frac{\varepsilon(N+k)}{(m-1)^2} \right\}.$$

Since $\mathbf{P}(C_N) \leqslant \mathbf{P}(\bar{E}_N) + \mathbf{P}(C_N E_N)$, then the lemma is proved. □

It is clear from lemma 2 that the sequence W_n majorizing $[\mathbf{w}_n]$ is stationary. If $\mathbf{E}\tau_0^s < \infty$, $\mathbf{E}\xi_0 = \mathbf{E}(\tau_0^s - m\tau_0^e) < 0$, then, putting $\varepsilon \leqslant -\mathbf{E}\xi_0/3$, we obtain from lemma 3 that the sequence W_n is proper (finite with probability 1), provided $\mathbf{P}(\tau_0^s > \varepsilon) = 1$. But if the last condition is not satisfied then putting $\tau_j^{*s} = \max(\tau_j, \varepsilon)$ we obtain $\tau_j^s \leqslant \tau_j^{*s}$,

$$[\mathbf{w}_n] \leqslant [\mathbf{w}_n^*] \tag{4.85}$$

where the index * means relative to the controlling sequence $\{\tau_j^e, \tau_j^{*s}\}$. The inequality (4.87) follows from the monotonicity of $\mathbf{f}$ relative to τ_n^s in (4.66).

Thus we have

LEMMA 4. *Alongside W_n a majorant for $[\mathbf{w}_n]$ is also any stationary proper sequence W_n^* constructed relative to the sequence $\{\tau_j^e, \tau_j^{*s}\}$, $\tau_j^{*s} = \max(\tau_j^s, \varepsilon)$, $\varepsilon \leqslant -\mathbf{E}\xi_0/3$.*

The inequalities

$$w_{n,j} \leqslant \frac{1}{m-j+1} \sum_{i=1}^{m} w_{n,i} \leqslant \frac{1}{m-j+1} W_n \equiv v_{n,j}$$

complete the construction of the majorant for $\mathbf{w}_n$.

The event A_n defined by (cf. (4.76), (4.77))

$$\tau_{n-1}^e > W_{n-1}, \quad \tau_n^e + \cdots + \tau_{n+m-2}^e > \tau_n^s \tag{4.86}$$

obviously, will be a stationary renewing event for systems $\langle G, 1, G/m, 1\rangle$ with zero initial conditions.

For *nonzero initial conditions* there will be changes to the majorant W_n in (4.79) and the renewing event A_n quite analogous to those which emerged in (4.77) and (4.78). The verification of this is left to the reader.

As a result we may state the following ergodic theorem for systems with waiting $\langle G, 1, G/m, 1\rangle$. Let A_n denote the stationary event defined by (4.86).

THEOREM 2. *If* $\mathbf{P}(A_0) > 0$ *and* $\mathbf{E}\tau_0^s < m\mathbf{E}\tau_0^e$, *then* (4.58) *holds for the vectors of waiting times* $\mathbf{w}_n$ *with any initial condition* $\mathbf{w}_1$.

Of course in ergodic theorems it is possible to use other majorants or renewing events.

In Akhmarov and Leont'eva (1976) it was shown that for systems $\langle G_J, 1, G_J/m, 1\rangle$ with $\mathbf{w}_1 = \mathbf{0}$, that the condition $\mathbf{E}\tau_0^s < m\mathbf{E}\tau_0^e$ is necessary and sufficient for $\mathbf{P}(A_0) > 0$, where A_n is defined by (4.68), which is close to the result obtained in Leont'eva (1977, 1978). It has been reported to us by S. G. Foss that he has recently obtained criteria, different from theorem 2, for the strong convergence (4.58a) of $\mathbf{w}_n$ to stationary for systems with waiting.

We will say that *condition* (A) *is satisfied if the stationary sequence* $\mathbf{v}^n$ *satisfying* (cf. (4.66))

$$\mathbf{v}^{n+1} = \mathbf{R}(\mathbf{v}^n + \tau_n^s \mathbf{e} - \tau_n^e \mathbf{i})^+$$

is unique. One of these criteria is the following. *For any initial condition* $\mathbf{w}_1$ *the convergence* (4.58a), *for systems with waiting* $\langle G, 1, G/m, 1\rangle$, *holds if and only if condition* (A) *is satisfied.*

Another condition ensuring (4.58a) is the requirement of *monotone convergence* $U^{-n}\mathbf{w}_n \to \mathbf{w}^0$ a.s.

Remark. If it is already known that there is a proper stationary sequence $\{\mathbf{w}^n\}$ and that $\mathbf{P}_{\mathfrak{F}_{-\infty,n-1}}(\tau_n^e > x) > 0$, almost everywhere, for any $x > 0$, then the condition $\mathbf{P}(A_n) > 0$ for a *stationary* sequence $\{\mathbf{w}^n\}$ will be satisfied for

$$A_n = \{w_1^n = 0, w_m^n < \tau_n^e\}.$$

This follows at once from the fact that $\mathbf{P}(w_1^n = 0) > 0$ and, consequently, $\mathbf{P}(w_1^n = 0, w_m^n < x) > 0$ for sufficiently large x.

Theorems 1 and 2 imply the following result which gives simple sufficient conditions for ergodicity of systems.

THEOREM 3. *Let* $\mathbf{E}\tau_j^s < \infty$ *for systems* $\langle G, 1, G/m, 1\rangle_R$ *and* $\mathbf{E}\tau_j^s < m\mathbf{E}\tau_j^e$ *for systems* $\langle G, 1, G/m, 1\rangle$. *Suppose that there is a set* $\Omega \in \mathfrak{F}_{-\infty,0}$ *of positive probability such that for* $\omega \in \Omega$ *and all* $x > 0$

$$\mathbf{P}_{\mathfrak{F}_{-\infty,0}}(\tau_1^e > x, \tau_1^e + \cdots + \tau_m^e > \tau_1^s) > 0.$$

Then the results of theorems 1 *and* 2 *are valid.*

Proof. We consider, for example, a system with refusals and let A_n denote the event (4.76). Since $\mathbf{E}\tau_j^s < \infty$, then the variable V_n in (4.76) will be finite with probability 1. Let $\mathbf{P}(\Omega) = \varepsilon$. Choose x so large that the probability $\mathbf{P}(H_x)$ of the event $H_x = \{\zeta_0^+ + V_0 < x\} \in \mathfrak{F}_{-\infty,0}$ is larger than $1 - \varepsilon/2$ (ζ_n^+ and V_n are defined in lemma 1). Then we will have

$$\begin{aligned}\mathbf{P}(A_2) &= \mathbf{P}(\tau_1^e > \zeta_0^+ + V_0, \tau_1^e + \cdots + \tau_m^e > \tau_1^s)\\ &\geqslant \mathbf{P}(A_2 H_x) \geqslant \mathbf{P}(\tau_1^e > x, \tau_1^e + \cdots + \tau_m^e > \tau_1^s, H_x)\\ &= \mathbf{E}[\mathbf{P}_{\mathfrak{F}_{-\infty,0}}(\tau_1^e > x, \tau_1^e + \cdots + \tau_m^e > \tau_1^s); H_x]\\ &\geqslant \mathbf{E}[\mathbf{P}_{\mathfrak{F}_{-\infty,0}}(\tau_1^e > x, \tau_1^e + \cdots + \tau_m^e > \tau_1^s); H_x\Omega]\\ &> 0.\end{aligned}$$

The last inequality follows because, under the conditions of the theorem, the random variable under the expectation is positive on Ω, $P(\Omega) = \varepsilon$, and

$$\begin{aligned}\mathbf{P}(H_x\Omega) = 1 - \mathbf{P}(\bar{H}_x \cup \bar{\Omega}) &\geqslant 1 - \mathbf{P}(\bar{H}_x) - \mathbf{P}(\bar{\Omega})\\ &\geqslant \varepsilon - \frac{\varepsilon}{2} = \frac{\varepsilon}{2} > 0.\end{aligned}$$

Therefore the conditions of theorem 3 are satisfied. For systems with queues the proof is similar. □

From theorem 3 follows

THEOREM 4. *Let* (τ_j^e, τ_j^s) *be independent for different* j, $\mathbf{E}\tau_j^s < \infty$ *for systems with refusals and* $\mathbf{E}\tau_j^s < m\mathbf{E}\tau_j^e$ *for systems with queues. Then, if the random variable* τ_1^e *is not bounded, the results of theorems* 1 *and* 2 *hold.*

In this result only the unboundedness condition on τ_1^e may turn out to be 'superfluous'. This shows

THEOREM 5. *For systems with queues* $\langle G_I, 1, G_I/m, 1\rangle$ *the following conditions are equivalent:*

(1) $\mathbf{E}(m\tau_j^e - \tau_j^s) > 0$.

(2) *The stationary sequence* $\{\mathbf{w}^n\}$ *has stationary renewing events*

$$A_n = \{w_1^n = w_1^{n+1} = \cdots = w_1^{n+m-1} = 0, w_m^n < \tau_n^e + \cdots + \tau_{n+m-1}^e\}$$

for which $\mathbf{P}(A_n) > 0$.

In fact it is possible to add a third equivalent condition to (1) and (2):

(3) *For arbitrary initial conditions the ergodicity condition of theorem* 1 *of section* 4.6 *is satisfied, that is, for renewing events* A_n *of the form* (4.68)

$$\mathbf{P}\left(\bigcap_{l=l_0}^{\infty} \bigcup_{j=n/2}^{n} A_j T^{-l} A_{j+l}\right) \to 1$$

as $n \to \infty$ *and for arbitrary* $l_0 > 0$.

Thus under any one of these conditions the result of theorem 1 on convergence of the sequence $\{\mathbf{w}_n\}$ will hold. The fact of the convergence of the distribution of $\mathbf{w}_n$ to stationary under condition (1) and for any initial value $\mathbf{w}_1$ was established in Kiefer and Wolfowitz (1955).

Below we will give the proof of theorem 5. The proof of an extended version of this theorem with condition (3) requires additional arguments, similar to those given after lemma 1 in the discussion of *nonzero* initial conditions. It is obvious that for $\mathbf{w}_0 = \mathbf{0}$ the result of condition (3) follows from (2) and the monotonicity of $\mathbf{w}_{n+1}$ as a function of $\mathbf{w}_n$.

Proof of theorem 5. The fact that (2) implies (1) follows from known theorems that $\mathbf{w}_n \to \infty$ for $\mathbf{E}(m\tau_j^e - \tau_j^s) \leqslant 0$ (to avoid trivialities we omit the case of deterministic control $m\tau_j^e = \tau_j^s$).

We will now show that (1) implies (2). Since τ_j^s and τ_j^e are independent then (1) implies that $\mathbf{P}(m\tau_j^e - \tau_j^s > 0) > 0$, and, consequently, there exist $a > 0$ and $\varepsilon > 0$ such that

$$\mathbf{P}(\tau_j^e \geqslant a) > 0, \quad \mathbf{P}(\tau_j^s < ma - \varepsilon) > 0.$$

We have used the fact that $\{w_m^n; n \geqslant 1\}$ forms a proper stationary sequence. Using the independence of τ_j^e and τ_j^s again we obtain, for some $x > 0$ and all N,

$$P(w_m^1 < x; \tau_1^e \geqslant a; \tau_j^s < ma - \varepsilon; 1 \leqslant j \leqslant N) > 0.$$

We denote the event under the probability sign by $A(N)$. It is clear that on $A(N)$ the trajectory $\{\mathbf{w}^n; n \geqslant 1\}$ is majorized by $\{\mathbf{w}_n^*; n \geqslant 1\}$ for a system with deterministic control, $\tau_j^{e*} = a$, $\tau_j^{s*} = ma - \varepsilon$, and initial value $\mathbf{w}_1^* = \mathbf{x} = (x, \ldots, x)$. But it is easy to see that for $n \geqslant 1$

(1) $w_{n,m}^* - w_{n,1}^* \leqslant ma - \varepsilon$

(2) $[\mathbf{w}_{n+1}^*] = [\mathbf{w}_n^*] - \varepsilon$, if $w_{n,1} > 0 \left([\mathbf{w}_n] = \sum_{i=1}^{m} w_{n,i}\right)$.

Hence it follows that for sufficiently large n_0 and all $n \geqslant n_0$

$$w^*_{n,1} = 0, \quad w^*_{n,m} < ma - \varepsilon < \tau^{e*}_n + \cdots + \tau^{e*}_{n+m-1}$$

and, consequently, for $n_0 < n < N - m$, $A(N) \subset A_n$. Since $\mathbf{P}(A(N)) > 0$, then $\mathbf{P}(A_n) > 0$ and the theorem is proved. □

Repeating almost verbatim the arguments in this proof (in the part where it is shown that (1) implies (2)), we may also obtain an analogue of theorem 5 for systems with *refusals*.

THEOREM 5A. *For systems with refusals* $\langle G_I, 1, G_I/m, 1\rangle_R$ *the condition* $\mathbf{P}(m\tau^e_j - \tau^s_j > 0) > 0$ *implies* (2) *of theorem* 5.

Assertion (2) of theorem 5 contains an assumption about the existence of a stationary sequence $\{\mathbf{w}^n\}$ which is very inconvenient. However, we have seen that under the condition $\mathbf{E}(m\tau^e_j - \tau^s_j) > 0$ this assumption will be automatically satisfied for systems with queues (see condition (3)).

For systems with refusals a similar situation holds. The assumption will be automatically satisfied if

$$\mathbf{P}(m\tau^e_j - \tau^s_j > 0) > 0$$

(cf. Akhmarov, 1979c, 1981; Akhmarov and Leont'eva, 1976). However, in contrast to the requirement $\mathbf{E}(m\tau^e_j - \tau^s_j) > 0$ in theorem 5, the requirement $\mathbf{P}(m\tau^e_j - \tau^s_j > 0) > 0$ in theorem 5A (that is, for systems with refusals) appears to be superfluous, as in the condition of unboundedness of τ^e_j in theorem 4.

We now pass on to *stability theorems* for systems $\langle G, 1, G/m, 1\rangle$ and $\langle G, 1, G/m, 1\rangle_R$. We will use theorem 2 of section 4.6 and the above majorants $\mathbf{v}_n \geqslant \mathbf{w}_n$ both for systems with queues and for systems with refusals. The main peculiarity of these majorants is the fact that they simultaneously majorize the stationary sequences and the sequences with zero initial conditions $\mathbf{w}_1 = 0$. This is immediately clear from their construction (see (4.75)).

We restrict ourselves here to stability theorems for *stationary sequences* $\{\mathbf{w}^k\}$ of vectors of waiting times for systems with queues and with refusals (as we have seen, theorem 2 of section 4.6 allows the possibility of studying stability even for prestationary sequences $\{\mathbf{w}_n\}$).

Let the following conditions be satisfied (the upper index (r) has the same role here as in section 4.6).

(A*) *The finite-dimensional distributions of the stationary sequences* $\{\tau^{(r)e}_j, \tau^{(r)s}_j\}$ *weakly converge as* $r \to \infty$ *to the finite-dimensional distributions of* $\{\tau^e_j, \tau^s_j\}$.

(C*) $\mathbf{P}(\tau^s_1 - \tau^e_1 - \cdots - \tau^e_k = 0) = 0$ *for all* $k \geqslant 1$.

Conditions (A*) and (C*) will guarantee conditions (A) and (C) of theorem 2 of section 4.6.

We consider first *systems with queues*. Let A_n be the stationary renewing event defined by (4.86).

THEOREM 6. *Let* (A*) *hold*, $\mathbf{P}(A_0) > 0$, $\mathbf{E}\tau_1^s < m\mathbf{E}\tau_1^e$, *and* $\mathbf{E}\tau_1^{(r)s} \to \mathbf{E}\tau_1^s$ *as* $r \to \infty$. *Then the finite-dimensional distributions of* $\{\mathbf{w}^{(r)k}\}$ *weakly converge as* $r \to \infty$ *to the distributions of* $\{\mathbf{w}^k\}$.

For *systems with refusals* the following result holds. Let A_n be the renewing event defined by (4.76).

THEOREM 7. *Let* (A*) *and* (C*) *hold*, $\mathbf{P}(A_0) > 0$, $\mathbf{E}\tau_1^s < \infty$, *and* $\mathbf{E}\tau_1^{(r)s} \to \mathbf{E}\tau_1^s$ *as* $r \to \infty$. *Then the finite-dimensional distributions of* $\{\mathbf{w}^{(r)k}\}$ *weakly converge as* $r \to \infty$ *to the distributions of* $\{\mathbf{w}^k\}$.

In these results the existence of stationary sequences $\{\mathbf{w}^{(r)k}\}$ and $\{\mathbf{w}^k\}$ is guaranteed by the conditions of the theorems (see theorems 1, 2).

Conditions sufficient for $\mathbf{P}(A_0) > 0$ and easy to verify are contained in theorems 3, 4. Of course, it is possible to use other renewing events A_n.

Proof of theorem 6. We use theorem 2 of section 4.6. It is sufficient to show that conditions (B) and (C) of that theorem are satisfied. Consider condition (B). Obviously, the stationary renewing events (4.86) are in the form (4.63) where $s = 2$,

$$\gamma_{k,1} = \tau_{k-1}^e - W_{k-1}, \quad \gamma_{k,2} = \tau_k^e + \cdots + \tau_{k+m-2}^e - \tau_k^s \tag{4.87}$$

W_k is defined in lemma 2, so that $\gamma_{k,i}$ is measurable with respect to $\mathfrak{F}_{-\infty,k+L}$, $L = m - 2$.

If $\mathbf{P}(\tau_j^s > \varepsilon) > 0$ for any $\varepsilon > 0$, then as renewing event we choose

$$A_k = \{\gamma_{k,1}^* > 0, \gamma_{k,2}^* > 0\}$$

where $\gamma_{k,i}^*$ are defined in (4.87) but relative to $\{\tau_j^e, \tau_j^{*s}\}$,

$$\tau_j^{*s} = \max(\tau_j^s, \varepsilon), \quad 0 < \varepsilon < \frac{\mathbf{E}(\tau_0^s - m\tau_0^e)}{3}$$

(see lemma 4).

Thus we may, without loss of generality, suppose that

$$\mathbf{P}(\tau_j^s \geqslant \varepsilon) = 1$$

for sufficiently small $\varepsilon > 0$. Then the convergence of the distributions of $\gamma_{k,i}^{(r)}$ will follow from lemma 3. In fact, it is sufficient for us to prove convergence of the distributions of the $W_n^{(r)}$ (see (4.87) and lemma 2), given by

$$W_n^{(r)} = \sup_{k \leqslant N} [(m-1)^2 \tau_{k-1}^{(r)s} + (m-1)v_k^{(r)} + \xi_k^{(r)} + \cdots + \xi_{n-1}^{(r)}]. \tag{4.88}$$

Let $\nu^{(r)}$ be the least distance from n to those k for which the sup on the right in (4.88) is attained. Then, according to lemma 3,

$$\mathbf{P}(\nu^{(r)} < N) \leqslant \mathbf{P}\left(\sup_{k \geqslant N} \sum_{j=1}^{k} (\xi_{-j}^{(r)} + \varepsilon) > 0\right) + \sum_{k=1}^{\infty} \mathbf{P}\left(\tau_0^{(r)s} > \frac{\varepsilon(N+k)}{(m-1)^2}\right). \tag{4.89}$$

Here the right-hand side converges to 0 uniformly in r as $N \to \infty$. This follows from the discussions in section 4.1 and the conditions

$$\mathbf{E}\tau_0^{(r)s} \to \mathbf{E}\tau_0^s, \quad \mathbf{E}(\xi_0 + \varepsilon) < 0.$$

Since the distribution of the random variables in square brackets in (4.88) converges as $r \to \infty$, then the distribution of the maximum of these variables over k in $n - N \leqslant k \leqslant n$ also converges. This together with (4.89) gives the required convergence of the distributions of $\gamma_{k,j}$ (a more detailed argument was given in section 4.1).

For the proof of the theorem it remains to verify condition (C) of theorem 2 of section 4.6. The existence of a point $\mathbf{a}$ satisfying condition (C) is obvious from the construction of the renewing events—it suffices to take $\mathbf{a} = \mathbf{0}$. We need to establish the almost sure continuity (relative to the distributions of $(\tau_1, \ldots, \tau_k)$) of the functions

$$\mathbf{f}_k(\mathbf{0}, \mathbf{y}_1, \ldots, \mathbf{y}_k) = \mathbf{f}(\mathbf{f}_{k-1}(\mathbf{0}, \mathbf{y}_1, \ldots, \mathbf{y}_{k-1}), \mathbf{y}_k), \quad k \geqslant 2 \tag{4.90}$$

where $\mathbf{f}_1(\mathbf{0}, \mathbf{y}) = \mathbf{f}(\mathbf{0}, \mathbf{y})$ (see condition (C) of theorem 2 of section 4.6). It will be convenient here to express $\mathbf{y}$ and $\boldsymbol{\tau}$ as two-dimensional vectors $\mathbf{y} = (x, z)$, $\boldsymbol{\tau} = (\tau^e, \tau^s)$.

Then the function $\mathbf{f}(\mathbf{a}, \mathbf{y})$, for systems $\langle G, 1, G/m, 1 \rangle$ (see (4.66)), where $\mathbf{a}$ is an m-dimensional vector, can be written

$$\mathbf{f}(\mathbf{a}, \mathbf{y}) = \mathbf{R}(\mathbf{a} + z\mathbf{e} - x\mathbf{i})^+.$$

But it is obvious that for such a function $\mathbf{f}_k(\mathbf{0}, \mathbf{y}_1, \ldots, \mathbf{y}_k)$ will be continuous everywhere. The theorem is proved. □

Proof of theorem 7. Here we must also show that conditions (B) and (C) of theorem 2 of section 4.6 hold. Condition (B) follows from the description of the renewing event in the form (4.76), similar to (4.87), and the fact that the distributions of $V_n^{(r)}$ (see the statement of lemma 1) weakly converge as $r \to \infty$. We now verify conditions (C). For systems $\langle G, 1, G/m, 1 \rangle_R$ the function $\mathbf{f}(\mathbf{a}, \mathbf{y})$ (see (4.65), (4.90)), with the understandings adopted in the proof of the previous theorem, can be written

$$\mathbf{f}(\mathbf{a}, \mathbf{y}) = \mathbf{R}(\mathbf{a} + z\mathbf{e}I(a_1 = 0) - x\mathbf{i})^+$$

so that

$$\mathbf{w}_{n+1} = \mathbf{f}_n(\mathbf{0}, \tau_1, \ldots, \tau_n) \quad \text{for } \mathbf{w}_1 = 0.$$

We have

$$\mathbf{f}_1(\mathbf{0}, \mathbf{y}_1) = \mathbf{R}(z_1\mathbf{e} - x_1\mathbf{i})^+$$

$$\mathbf{f}_2(\mathbf{0}, \mathbf{y}_1, \mathbf{y}_2) = \mathbf{R}(\mathbf{R}(z_1\mathbf{e} - x_1\mathbf{e} - x_1\mathbf{i})^+ + z_2\mathbf{e} - x_1\mathbf{i})^+ \quad \text{for } \mathbf{m} \geqslant 2$$

etc. It is easy to see that the coordinates of $\mathbf{f}_k(\mathbf{0}, \mathbf{y}_1, \ldots, \mathbf{y}_k)$ and, in particular, the first coordinate, will either by 0 or take the form of a sum $z_j - x_j - x_{j-1} - \cdots - x_{j-l}$ for distinct j and l. Since a point of discontinuity of $\mathbf{f}_k$ is a zero value of the first coordinate, then from here and condition (C) of the theorem the required a.s. continuity of $\mathbf{f}_k$ follows. The theorem is proved. □

It is easy to see that the theorems proved in this section for systems with waiting are valid for systems with bounded waiting times or with bounded queues, since the majorants constructed for the case of unbounded queues still hold. However, the condition $\mathbf{E}(\tau_j^s - \tau_j^e m) < 0$ in this connection becomes irrelevant, since for the proof of the theorem on convergence of the distributions of truncated majorants (see condition (B) of theorem 2 of section 4.6) we can use the results of section 4.4.

In general the class of service systems whose characteristics are described by equations of the form $\mathbf{w}_{n+1} = \mathbf{f}(\mathbf{w}_n, \tau_n)$, for comparatively simple functions $\mathbf{f}$, is very broad. This allows the possibility of wide utilization of the methods of sections 4.6, 4.7 for the proofs of ergodic theorems and stability theorems. In addition, this makes the subsequent study of the sequence $\{\mathbf{w}_n\}$ described by (4.55) an urgent problem.

It appears that the renewal method will carry over completely even to the case of continuous time.

As already remarked, this method also has the merit that it allows an estimate of the speed of convergence both in ergodic theorems and in stability theorems. Here for independent τ_n the estimates obtained are almost best possible (see Akhmarov, 1979b, 1981).

To conclude this section we will quote without proof some results on estimates of speed of convergence, which were obtained in Akhmarov (1979b, c), by the renewal method, for systems $\langle G_I, 1, G_I/m, 1\rangle$ with queues and systems $\langle G_I, 1, G_I/m, 1\rangle_R$ with refusals.

We have already remarked that estimates of speed of convergence in ergodic theorems and stability theorems are of their very nature closely connected. We recall also that the absolute value of the difference

$$|\mathbf{P}(\mathbf{w}_n^{(r)} \in B) - \mathbf{P}(\mathbf{w}^0 \in B)|$$

can be estimated by the sum of the values

$$|\mathbf{P}(\mathbf{w}_n^{(r)} \in B) - \mathbf{P}(\mathbf{w}^{(r)0} \in B)|$$

$$|\mathbf{P}(\mathbf{w}^{(r)0} \in B) - \mathbf{P}(\mathbf{w}^0 \in B)|$$

the first of which is determined by the speed of convergence in ergodic theorems and the second by the speed of convergence in stability theorems. Therefore we give results relating to estimates of these two forms.

We begin with estimates in *ergodic theorems.*

THEOREM 8. *Let a system with waiting* $\langle G_I, 1, G_I/m, 1\rangle$ *satisfy*

(1) $\mathbf{w}_1 = 0$
(2) $\mathbf{E}(m\tau_j^e - \tau_j^s) > 0$
(3) $\mathbf{E}(\tau_j^s)^\alpha < \infty$ *for some* $\alpha > 1$.

Then for any measurable $B \subset R^m$

$$|\mathbf{P}(\mathbf{w}_n \in B) - \mathbf{P}(\mathbf{w}^0 \in B)| \leqslant \frac{c(\log n)^\alpha}{n^{\alpha-1}} \tag{4.91}$$

where c does not depend on B and n.

But if (3) *is replaced by*

(4) $\mathbf{E}\, e^{\mu\tau_j^s} < \infty$ *for some* $\mu > 0$

then

$$|\mathbf{P}(\mathbf{w}_n \in B) - \mathbf{P}(\mathbf{w}^0 \in B)| \leqslant c\, e^{-\gamma\sqrt{n}} \tag{4.92}$$

where c and $\gamma > 0$ *also do not depend on B and n.*

In a similar way we find a corresponding result for systems with refusals.

THEOREM 9. *Let a system with refusals* $\langle G_I, 1, G_I/m, 1\rangle_R$ *satisfy*

(1) $\mathbf{w}_1 = 0$
(2) $\mathbf{P}(m\tau_j^e - \tau_j^s > 0) > 0$
(3) $\mathbf{E}(\tau_j^s)^\alpha < \infty$ *for some* $\alpha > 1$.

Then (4.91) *holds for any measurable* $B \subset R^m$.

But if instead of (3) *we require*

(4) $\mathbf{E}\, e^{\mu\tau_j^s} < \infty$ *for some* $\mu > 0$

then (4.92) *will hold.*

The proofs of these results are contained in Akhmarov (1979b).

The estimates of speed of convergence in (4) of theorems 8, 9 can, it appears, be improved up to exponential. Comparison with the results for single-channel systems (see Borovkov, 1976a) shows that (4.91) can be improved only by a logarithmic multiplier.

Some estimates for the speed of convergence to the limit distributions have also been obtained for single-channel systems with *bounded waiting times* (see

Akhmarov, 1979a, b, c, 1981) (the restrictions on waiting times are similar to those described in section 4.4).

We pass on now to estimates in *stability theorems*. We first consider systems with waiting and denote by $\lambda(\xi, \eta)$ the Levy–Prokhorov distance (see section 4.3) between the distributions of the random variables ξ and η.

THEOREM 10. *Let the controlling sequences* $\{\tau_j^e, \tau_j^s\}$ *and* $\{\tau_j^{(r)e}, \tau_j^{(r)s}\}$ *of the systems with waiting* $\langle G_I, 1, G_I/m, 1\rangle$ *and* $\langle G_I, 1, G_I/m, 1\rangle^{(r)}$ *respectively, satisfy*

(1) $\mathbf{E}(m\tau_j^e - \tau_j^s) > 0$

(2) $\varepsilon = \varepsilon_r = \max(\lambda(\tau_j^{(r)e}, \tau_j^e), \lambda(\tau_j^{(r)s}, \tau_j^s)) \to 0$ *as* $r \to \infty$

(3) $\mathbf{E}(\tau_j^s)^\alpha < \infty, \quad \sup_r \mathbf{E}(\tau_j^{(r)s})^\alpha < \infty$ *for some* $\alpha > 1$.

Then

$$\lambda(\mathbf{w}^{(r)0}, \mathbf{w}^0) \leqslant c\varepsilon^{1-1/\alpha} \log \frac{1}{\varepsilon}.$$

If instead of (3) *it is required that*

$$(4)\ \mathbf{E}\, e^{\mu\tau_j^s} < \infty, \quad \sup_r \mathbf{E}\, e^{\mu\tau_j^{(r)s}} < \infty \tag{4.93}$$

for some $\mu > 0$, *then*

$$\lambda(\mathbf{w}^{(r)0}, \mathbf{w}^0) \leqslant c\varepsilon\left(\log \frac{1}{\varepsilon}\right)^2.$$

A similar result on estimates of speed of convergence in the stability theorem for systems with refusals contains stricter requirements on the controlling sequences: some of these conditions (see, for example, (4) and (5)) may be excessive.

THEOREM 11. *Let the controlling sequences* $\{\tau_j^e, \tau_j^s\}$ *and* $\{\tau_j^{(r)e}, \tau_j^{(r)s}\}$ *of systems with refusals* $\langle G_I, 1, G_I/m, 1\rangle_R$ *and* $\langle G_I, 1, G_I/m, 1\rangle_R^{(r)}$ *respectively, satisfy*

(1) $\mathbf{P}(m\tau_j^e - \tau_j^s > 0) > 0$.

Conditions (2) *and* (3) *of this theorem coincide with conditions* (2) *and* (3) *of theorem* 10.

(4) $\mathbf{P}(x \leqslant \tau_j^s < x + l) \leqslant \dfrac{c_1 l}{x^{\alpha+1}}$

for all $l > 0$ *and all sufficiently large* x: $c_1 = \text{const}$.

(5) *The random variables* $\tau_j^e - \tau_j^s$ *have bounded densities.*

Under these conditions

$$\lambda(\mathbf{w}^{(r)0}, \mathbf{w}^0) \leqslant c\varepsilon\left(\log\frac{1}{\varepsilon}\right)^3$$

If condition (3) (*for its statement see theorem* 10) *is replaced by* (4.93), *then*

$$\lambda(\mathbf{w}^{(r)0}, \mathbf{w}^0) \leqslant c\varepsilon\left(\log\frac{1}{\varepsilon}\right)^4.$$

The results still hold if the controlling sequences $\{\tau_j^e, \tau_j^s\}$ and $\{\tau_j^{(r)e}, \tau_j^{(r)s}\}$ are given on the same probability space and throughout theorems 10 and 11 the distance $\lambda(\xi, \eta)$ is replaced by the distance $\rho(\xi, \eta)$, defined in subsection 4.3.3:

$$\rho(\xi, \eta) = \inf\{\delta: \mathbf{P}(|\xi - \eta| > \delta) < \gamma\}$$

where $|\xi - \eta|$, in the many-dimensional case, denotes the Euclidean norm.

To be precise, theorem 11 is proved in Akhmarov (1981) just for the distance ρ. However, in view of the theorem of Strassen quoted in subsection 4.3.3, given the distances $\lambda(\tau_j^e, \tau_j^{(r)e})$, $\lambda(\tau_j^s, \tau_j^{(r)s})$ we can construct new controlling sequences $\{\tau_j^{e*}, \tau_j^{s*}\}$ and $\{\tau_j^{(r)e*}, \tau_j^{(r)s*}\}$, with the same distributions as the original sequences, but on the same probability space and having the properties

$$\rho(\tau_j^{e*}, \tau_j^{(r)e*}) \leqslant \lambda(\tau_j^e, \tau_j^{(r)e})$$

$$\rho(\tau_j^{s*}, \tau_j^{(r)s*}) \leqslant \lambda(\tau_j^s, \tau_j^{(r)s}).$$

We then obtain, under the conditions of theorem 11 (for λ), that the corresponding estimate for $\rho(\mathbf{w}^{(r)0*}, \mathbf{w}^{0*})$ holds (the index * denotes the correspondence with the controlling sequences $\{\tau_j^{(r)e*}, \tau_j^{(r)s*}\}$, $\{\tau_j^{e*}, \tau_j^{s*}\}$), from which, by the inequality $\lambda(\xi, \eta) \leqslant \rho(\xi, \eta)$, we obtain the required estimate for

$$\lambda(\mathbf{w}^{(r)0}, \mathbf{w}^0) = \lambda(\mathbf{w}^{(r)0*}, \mathbf{w}^{0*}).$$

Some estimates of the speed of convergence in stability theorems have also been obtained for systems with an *infinite number of service channels* (see Akhmarov, 1980; Zolotarev, 1977).

References

Afanas'eva, L. G. (1965). On the existence of a limit distribution in a queuing system with bounded sojourn time. *Teor. Veroyatnost i ee Primenen*, **10**, 570–578 [in Russian]. MR 32 #8410.

Afanas'eva, L. G., and Martynov, A. V. (1969). The ergodic properties of a queuing system with bounded sojourn time. *Teor. Veroyatnost i ee Primenen*, **14**, 102–112 [in Russian]. MR 40 #2173.

Akhmarov, I. (1979a). Stability of a random walk in a strip with retarding boundaries. *Sibirsk. Mat. Zh.*, **20**, No. 3, 645–650 [in Russian]. MR 80k: 60086.

Akhmarov, I. (1979b). The rate of convergence in the ergodicity and continuity theorems for multichannel queuing systems. *Teor. Veroyatnost i ee Primenen*, **24**, 418–424 [in Russian]. MR 80m: 60093.

Akhmarov, I. (1979c). Ergodicity and stability of multichannel queuing systems with a limited waiting time. *Sibirsk. Mat. Zh.*, **20**, No. 4, 911–916 [in Russian]. MR 80m: 60092.

Akhmarov, I. (1980). The convergence rate in continuity theorems for systems with an infinite number of service channels. *Sibirsk. Mat. Zh.*, **21**, 16–21 [in Russian]. MR 81j: 60099.

Akhmarov, I. (1981). The rate of convergence in ergodicity and continuity theorems for systems with refusals. *Teor. Veroyatnost i ee Primenen*, **26**, 182–189 [in Russian].

Akhmarov, I., and Leont'eva, N. P. (1976). Conditions for convergence to limit processes and the strong law of large numbers for queuing systems. *Teor. Veroyatnost i ee Primenen*, **21**, 559–570 [in Russian]. MR 58 #24597.

Arndt, U. (1978). A continuity theorem for the stationary distributions in general *m*-server queues. *Elektron. Informationsverarb. Kybernet.*, **14**, 395–402. MR 80a: 60116.

Barrer, D. Y. (1957). Queuing with impatient customers and indifferent clerks. *Operations Res.*, **5**, 644–649. MR 19 #779.

Bernstein, S. N. (1944). Distribution limit theorems of probability theory on sums of independent variables. *Uspekhi Mat. Nauk*, **10**, 65–115 [in Russian].

Billingsley, P. (1968). *Convergence of probability measures*, Wiley, New York. MR 38 #1718.

Blomqvist, N. (1973). A heavy traffic result for the finite dam. *J. Appl. Probability*, **10**, 223–228. MR 50 #11463a.

Borisov, I. A., and Borovkov, A. A. (1980). Asymptotic behaviour of the number of free channels for systems with refusals. *Teor. Veroyatnost i ee Primenen*, **25**, 449–463 [in Russian].

Borovkov, A. A. (1964a). Asymptotic methods in queuing theory. (Winter School in Probability Theory and Math. Statistics held in Uzhgorod 1964, pp. 3–40.) *Izdat. Akad. Nauk Ukrain. SSR, Kiev, 1964* [in Russian]. MR 31 #4094.

Borovkov, A. A. (1964b, 1965). Some limit theorems of queuing theory: I, II. *Teor. Veroyatnost i ee Primenen*, **9**, 608–625; **10**, 409–437 [in Russian]. MR 30 #2564; MR 32 #8411.

Borovkov, A. A. (1966a). Asymptotic analysis of certain queuing systems. *Teor. Veroyatnost i ee Primenen*, **11**, 675–682 [in Russian]. MR 35 #7433.

Borovkov, A. A. (1966b). Conditions for convergence to diffusion processes and asymptotic methods in queuing theory. *Proc. Internat. Congr. Math.* (*Moscow, 1966*), 533–538 [in Russian]. MR 39 #6415.

Borovkov, A. A. (1967a). Convergence of weakly dependent processes to the Wiener process. *Teor. Veroyatnost i ee Primenen*, **12**, 193–221 [in Russian]. MR 35 #6192.

Borovkov, A. A. (1967b). Convergence to diffusion processes. *Teor. Veroyatnost i ee Primenen*, **12**, 459–482 [in Russian]. MR 36 #964.

Borovkov, A. A. (1967c). Limit laws for queuing processes in multichannel systems. *Sibirsk. Mat. Zh.*, **8**, No. 5, 983–1004 [in Russian]. MR 36 #6022.

Borovkov, A. A. (1970). Theorems on the convergence to Markov diffusion processes. *Z. Wahrscheinlichkeitstheorie und verw. Gebiete*, **16**, 47–76. MR 44 #2277.

Borovkov, A. A. (1972a). *Stochastic processes in queuing theory*, Izdat 'Nauka', Moscow [in Russian]. MR 47 #4349. (Trans.: see Borovkov, 1976a.)

Borovkov, A. A. (1972b). Certain properties of the supremum of a sum of stationarily connected variables. *Teor. Veroyatnost i ee Primenen*, **17**, 147–150 [in Russian]. MR 47 #4325.

Borovkov, A. A. (1972c). Convergence of the distributions of functionals of stochastic processes. *Uspekhi Mat. Nauk*, **27**, No. 1 (163), 3–41 [in Russian]. MR 53 #4160.

Borovkov, A. A. (1972d). Convergence of the distributions of functionals of sequences and processes given on the entire axis. *Trudy Mat. Inst. Steklov*, **128**, 41–65 [in Russian]. MR 47 #7782.

Borovkov, A. A. (1972e). Continuity theorems for multichannel systems with breakdowns. *Teor. Veroyatnost i ee Primenen*, **17**, 458–468 [in Russian]. MR 47 #1155.

Borovkov, A. A. (1973). Conditions for convergence to degenerate processes. *Teor. Veroyatnost i ee Primenen*, **18**, 449–456 [in Russian]. MR 48 #5149.

Borovkov, A. A. (1975a). Random walk in a strip with impeding boundaries. *Matem. Zametki*, **17**, 649–657 [in Russian]. MR 52 #15683.

Borovkov, A. A. (1975b). General asymptotic methods of queuing theory. *Management Sciences*, **22**. (XXII TIMS Internat. Meeting in Kioto, July 1975.)

Borovkov, A. A. (1976a). *Stochastic processes in queueing theory*, Springer-Verlag, New York–Berlin, 1976. MR 52 #12118.

Borovkov, A. A. (1976b). Convergence of measures and random processes. *Uspekhi Mat. Nauk*, **31**, 3–68 [in Russian]. MR 53 #11688.

Borovkov, A. A. (1976c). *Probability theory*, Izdat 'Nauka', Moscow [in Russian]. MR 56 #6767.

Borovkov, A. A. (1977a). Multichannel servicing processes with an intensive input flow: I, II. *Sibirsk. Mat. Zh.*, **18**, 966–986; 1220–1245 [in Russian]. MR 58 #18769a, 18769b.

Borovkov, A. A. (1977b). Some estimates for the rate of convergence in stability theorems. *Teor. Veroyatnost i ee Primenen*, **22**, 689–699 [in Russian]. MR 57 #1690.

Borovkov, A. A. (1978a). Ergodic theorems and stability theorems for a class of stochastic equations and their applications. *Teor. Veroyatnost i ee Primenen*, **23**, 241–262 [in Russian]. MR 58 #7810.

Borovkov, A. A. (1978b). Ergodicity and stability theorems for random walks in a strip and their applications. *Teor. Veroyatnost i ee Primenen*, **23**, 705–714 [in Russian]. MR 80d: 60090a.

Borovkov, A. A. (1980). Continuity theorems and estimates of the rate of convergence of the components of factorization for random walks defined on a Markov chain. *Teor. Veroyatnost i ee Primenen*, **25**, 329–338 [in Russian]. MR 81f: 60094.

Cohen, J. W. (1973). Asymptotic relations in queuing theory. *Stochastic Processes Appl.*, **1**, 107–124. MR 51 #2005.

Cramer, H., and Leadbetter, M. (1967). *Stationary and related stochastic processes*, Wiley, New York. MR 36 #949.

Dao Kha Fuk, and Nagaev, S. V. (1971). Stochastic inequalities for sums of independent random variables. *Teor. Veroyatnost i ee Primenen*, **16**, 660–674 [in Russian].

Delbrouch, L. E. N. (1973). A set of Wiener–Hopf integral equations with common solution in fluctuation theory. *J. Math. Anal. Appl.*, **44**, 100–112. MR 49 #4088.

Dellacherie, C. (1972). *Capacités et processus stochastiques*, Springer-Verlag, New York–Berlin. MR 56 #6810.

Doob, J. L. (1953). *Stochastic processes*, Wiley, New York. MR 15 #445.

Dudley, R. M. (1968). Distances of probability measures and random variables. *Ann. Math. Statist.*, **39**, 1563–1572. MR 37 #5900.

Dynkin, E. B. (1963). *Markov processes*, Izdat 'Nauka', Moscow [in Russian]. MR 33 #1886. (Trans.: Academic Press and Springer-Verlag, New York, 1965. MR 33 #1887.)

Eidel'man, S. D. (1964). *Parabolic systems*, Izdat 'Nauka', Moscow [in Russian]. MR 29 #4998.

Essen, M. (1973). Banach algebra methods in renewal theory. *J. Analyse Math.*, **26**, 303–336. MR 53 #14705.

Feller, W. (1950, 1966). *Introduction to probability theory and its applications*, Vols. 1, 2, Wiley, New York. MR 12 #424, MR 35 #1048.

Foss, S. G. (1980). On the approximation of multichannel service systems. *Sibirsk. Mat. Zh.*, **21**, 132–140 [in Russian].

Foss, S. G. (1981). Comparison of queuing disciplines in multichannel systems with waiting. *Sibirsk. Mat. Zh.*, **22**, 190–197 [in Russian].

Franken, P. (1970). Ein Stetigkeitssatz für Verlustsysteme. *Operationsforschung und Mathematische Statistik*, **11**, 9–23 (Akademie-Verlag, Berlin, 1970). MR 41 #4690.

Franken, P., and Kalähne, U. (1978). Existence, uniqueness and continuity of stationary distributions for queuing systems without delay. *Math. Nachr.*, **86**, 97–115. MR 80h: 60113.

Franken, P., Konig, D., Arndt, U., and Schmidt, V. (1981). *Queues and point processes*, Akademie-Verlag, Berlin.

Franken, P., and Stoyan, D. (1974). Stabilitätssatze für eine Klasse homogener Markowschen Prozesse. *Math. Nachr.*, **61**, 311–316. MR 50 #8705.

Gaver, D. P. (1968). Diffusion approximations and models for certain congestion problems. *J. Appl. Probability*, **5**, 607–623. MR 38 #6686.

Geza Schay (1974). Nearest random variables with given distributions. *Ann. Probability*, **2**, 163–166.

Gikhman, I. I., and Skorokhod, A. V. (1965). *Introduction to the theory of random processes*, Izdat 'Nauka', Moscow [in Russian]. MR 33 #6689. (Trans.: W. B. Saunders, Philadelphia, 1969. MR 40 #923.)

Gikhman, I. I., and Skorokhod, A. V. (1971, 1973, 1974). *The theory of random processes*, Vols. 1, 2, 3, Izdat 'Nauka', Moscow [in Russian]. MR 49 #6287; 6288; MR 58 #31323a. (Trans.: Springer-Verlag, Berlin, 1974, 1975, 1979. MR 49 #11603; MR 51 #11656; MR 58 #31323b.)

Gikhman, I. I. (1973). Limit theorems for sums of infinitely small random vectors. *Internat. conf. on probability theory and statistics, Vil'nyus, 1973*, pp. 165–168 [in Russian].

Gittins, J. C. (1978). A comparison of service disciplines for $GL/G/m$ queues. *Math. Operationsforsch. Statist. Ser. Optim.*, **9**, No. 2, 255–260. MR 80c: 60109.

Gnedenko, B. V. (1970). On some unsolved problem in queuing theory. *Sixth Internat. Telegraphic Congr., Munich, 1970* [in Russian].

Gnedenko, B. V., and Kovalenko, I. N. (1966). *Introduction to queuing theory*, Izdat 'Nauka', Moscow [in Russian]. MR 37 #5957. (Trans.: Israel Program for Scientific Translations, Jerusalem, 1968. MR 39 #2229.)

Ibragimov, I. A., and Linnik, Yu. V. (1965). *Independent and stationarily connected random variables*, Izdat 'Nauka', Moscow [in Russian]. MR 34 #2049.

Iglehart, D. L. (1965). Limiting diffusion approximations for the many server queue and the repairman problem. *J. Appl. Probability*, **2**, 429–441. MR 32 #1775.

Iglehart, D. L. (1973). Weak convergence in queuing theory. *Advances Appl. Probability*, **5**, 570–594. MR 49 #6404.

Iglehart, D. L., and Whitt, W. (1970). Multiple channel queues in heavy traffic: I. *Advances Appl. Probability*, **2**, 150–177. MR 42 #1237.

Il'in, A. M., Kalshnikov, A. S., and Oleinik, O. A. (1962). Second-order linear equations of parabolic type. *Uspekhi Mat. Nauk*, **17**, 3–146 [in Russian]. MR 25 #2328.

Kalashnikov, V. V. (1977). Analysis of stability in queuing problems by a method of trial functions. *Teor. Veroyatnost i ee Primenen*, **22**, 89–105 [in Russian]. MR 57 #1701.

Kalashnikov, V. V. (1978). *Qualitative analysis of complex systems by the method of trial functions*, Izdat 'Nauka', Moscow [in Russian].

Kalashnikov, V. V., and Tsitsiashvili, G. Sh. (1972). On stability of queuing systems with respect to disturbances of their distribution functions. *Izv. Akad. Nauk SSSR, Tekhn. Kibernet.*, **2**, 41–49 [in Russian]. MR 49 #8133.

Kac, M. (1947). Random walk and the theory of Brownian motion. *Amer. Math. Monthly*, **54**, 369–391. MR 9–46.

Kennedy, D. (1972). The continuity of the single-server queue. *J. Appl. Probability*, **9**, 370–381. MR 49 #6406.

Khinchin, A. Ya. (1936). *Asymptotic laws of probability theory*, Ob. Nauch. Tekh. Izdat., Moscow–Leningrad [in Russian].

Kiefer, J., and Wolfowitz, J. (1955). On the theory of queues with many servers. *Trans. Amer. Math. Soc.*, **78**, 1–18. MR 16–601.

Kingman, J. F. C. (1961). The single server queue in heavy traffic. *Proc. Cambridge Phil. Soc.*, **57**, 902–904. MR 24 #A1150.

Kingman, J. F. C. (1962). On queues in heavy traffic. *J. Roy. Statist. Soc., Ser. B*, **24**, 383–392. MR 26 #5654.

Köllerström, J. (1974). Heavy traffic theory for queues with several servers. *J. Appl. Probability*, **11**, 544–552. MR 51 #2014.

Kolmogorov, A. N. (1931). *Eine Verallgemeinerung der Laplace–Liapounoffschen Satzes*, Ukrain Acad. Sci, Dept. Math. & Nat. Sci, pp. 959–961.

Kolmogorov, A. N. (1958. Analytic methods in probability theory. *Uspekhi Mat. Nauk*, **5**, 5–41 [in Russian].

Kotzurch, M., and Stoyan, D. (1976). A quantitative continuity theorem for the mean stationary waiting time in *GL/GL/*1. *Math. Operationsforsch. Statist.*, **7**, 595–599. MR 54 #14154.

Kovalenko, I. I. (1961). Some queuing problems with restrictions. *Teor. Veroyatnost i ee Primenen*, **6**, 222–228 [in Russian]. MR 26 #5651.

Krupin, V. G. (1976). A continuity theorem for queuing systems with bounded waiting time. *Trudy III All-Union School on Queuing Theory, Pushchino, Moscow State Univ., 1976*, pp. 36–40 [in Russian].

Kyprianou, E. (1971). The virtual waiting time of the *GL/G/*1 queue in heavy traffic. *Advances Appl. Probability*, **3**, 249–269. MR 45 #1287.

Leont'eva, N. P. (1977). Queuing systems with arbitrary initial conditions. *Teor. Veroyatnost i ee Primenen*, **22**, 831–837 [in Russian]. MR 58 #3112.

Leont'eva, N. P. (1978). Convergence on limit processes in many server queuing systems. *Sibirsk. Mat. Zh.*, **19**, 793–814 [in Russian]. MR 58 #24610.

Lindley, D. V. (1959). Contribution to the discussion of the paper 'Geometric distributions in the theory of queues' by C. B. Winsten. *J. Roy. Stat. Soc. Ser. B*, **21**, 1–22.

Lindvall, T. (1973). Weak convergence of probability measures and random functions in the function space $D(0, \infty)$. *J. Appl. Probability*, **10**, 109–121. MR 50 #14870.

Lipster, R. Sh., and Shiryaev, A. N. (1974). *Statistics of random processes*, Izdat 'Nauka', Moscow [in Russian]. MR 55 #4365.

Lisek, B. (1982). A method for solving a class of recursive stochastic equations. *Z. Wahrscheinlichkeitstheorie verw. Gebiete*, **60**, 151–161.

Loeve, M. (1962). *Probability theory*, Springer-Verlag, Berlin. MR 16–598.

Loulou, R. (1973). Multi-channel queues in heavy traffic. *J. Appl. Probability*, **10**, 769. MR 50 #14985.

Loynes, R. (1962). The stability of a queue with non-independent inter-arrival and service times. *Proc. Cambridge Phil. Soc.*, **58**, 497–520. MR 25 #4581.

Miller, D. R., and Sentilles, F. D. (1975). Translated renewal processes and the existence of a limiting distribution for the queue length of the *GI/G/s* queue. *Ann. Probability*, **3**, 424–439. MR 53 #6789.

Mogul'skii, A. A. (1973). Absolute estimates for the moments of certain boundary functionals. *Teor. Veroyatnost i ee Primenen*, **18**, 350–357 [in Russian]. MR 47 #9724.

Nagaev, A. V. (1969). Limit theorems tha take into account large deviations when Cramer's condition is violated. *Izv. Akad. Nauk UzSSR, Ser. Fiz-Mat. Nauk*, **13**, 17–22 [in Russian]. MR 43 #8108.

Nagaev, S. V. (1983). Asymptotic behaviour of probabilities of one-sided large deviations. *Teor. Veroyatnost i ee Primenen*, **26**, 369–372 [in Russian].

Newell, G. F. (1968). Queues with time dependent arrival rates: I, II, III. *J. Appl. Probability*, **5**, 436–451; 579–590; 591–606. MR 40 #2176; 6659; 6660.

Pakes, A. G. (1978). On the maximum and absorption time of left continuous random walk. *J. Appl. Probab.*, **15**, No. 2, 292–299.

Phatarfod, R. M., Speed, T. P., and Walker, A. M. (1971). A note on random walks. *J. Appl. Probab.*, **8**, 198–201.

Prokhorov, Yu. V. (1956). Convergence of random processes and limit theorems of probability theory. *Teor. Veroyatnost i ee Primenen*, **1**, 177–238 [in Russian]. MR 18–943.

Prokhorov, Yu. V. (1963). Transition phenomena in queuing processes. *Litovsk. Mat. Sb.*, **3**, 199–205 [in Russian]. MR 29 #5308.

Rogozin, B. A. (1973). Asymptotic behaviour of the coefficients in the Levi–Wiener theorems on absolutely converging trigonometric series. *Sibirsk. Mat. Zh.*, **14**, 1304–1312 [in Russian]. MR 49 #7684.

Rossberg, H. J. (1965). Über die Verteilung von Wartezeiten. *Math. Nachr.*, **30**, 1–16. MR 32 #6573.

Rozanov, Ju. A. (1975). Some system approaches to water resources problems: II. Statistical equilibrium of processes in dam storage. *Internat. Inst. Appl. Systems Anal. Research Report, Feb. 1975.*

Sakhanenko, A. I. (1974). The convergence of the distributions of functionals of processes given on the whole axis. *Sibirsk. Mat. Zh.*, **15**, 102–119 [in Russian]. MR 49 #9940.

Samandarov, E. G. (1963). Service systems under heavy traffic conditions. *Teor. Veroyatnost i ee Primenen*, **8**, 327–330 [in Russian]. MR 27 #4292.

Sevast'yanov, B. A. (1971). *Branching processes*, Izdat 'Nauka', Moscow [in Russian]. MR 49 #9968.

Skorokhod, A. V. (1956). Limit theorems for stochastic processes. *Teor. Veroyatnost i ee*

Primenen, **1**, 289–319 [in Russian]. MR 18–943.

Skorokhod, A. V. (1961). *Research into the theory of random processes*, Izdat Kiev Univ. [in Russian].

Skorokhod, A. V. (1964). *Random processes with independent increments*, Izdat 'Nauka', Moscow [in Russian].

Stone, C. (1963). Weak convergence of stochastic processes defined on semi-infinite time intervals. *Proc. Amer. Math. Soc.*, **14**, 694–696. MR 27 #3015.

Stoyan, D. (1972). Ein Stetigkeitssatz für einlinige Wartemodelle der Bedienungstheorie. *Math. Operationsforsch. Statist.*, **3**, 103–111. MR 49 #1619.

Stoyan, D. (1976). A critical remark on a system approximation in queuing theory. *Math. Operationsforsch. Statist.*, **7**, 953–956. MR 54 #14160.

Stoyan, H. (1973). Monotonie- und Stetigkeits eigenschaften mehrliniger Wartesysteme der Bedienungstheorie. *Math. Operationsforsch. Statist.*, **4**, 155–163. MR 53 #9429.

Strassen, V. (1965). The existence of probability with given marginals. *Ann. Math. Statist.*, **36**, 423–439. MR 31 #1693.

Szcotha, W. (1976). An invariance principle for queues in heavy traffic. *Preprint No. 91, Inst. of Math. Polish Akad. Sci.*

Vasicer, O. A. (1977). Inequality for the variance of waiting time under general queuing discipline. *Operat. Research*, **1977**, 879–884.

Viskov, O. U., and Prokhorov, Yu. V. (1964). The probability of loss of a call in heavy traffic. *Teor. Veroyatnost i ee Primenen*, **9**, 99–104 [in Russian]. MR 29 #1678.

Whitt, W. (1970). Weak convergence of probability measures on the function space $C(0, \infty)$. *Ann. Math. Statist.*, **41**, 939–944. MR 41 #6259.

Whitt, W. (1974a). Heavy traffic limit theorems for queues: A survey. *Lecture Notes in Economics and Math. Systems*, 98, Berlin–Heidelberg–New York.

Whitt, W. (1974b). The continuity of queues. *Advances Appl. Probability*, **6**, 175–183. MR 49 #6417.

Whitt, W. (1982). Existence of limiting distributions in the GI/G/s queue. *Math. of Oper. Res.*, **7**, No. 1, 88–94.

Wolff, R. W. An upper bounds for multichannel queues. *J. Appl. Probab.* **14**, No. 4, 884–888.

Zolotarev, V. M. (1975a). Quantitative estimates in continuity problems for queuing systems. *Teor. Veroyatnost i ee Primenen*, **20**, 215–218 [in Russian]. MR 51 #4471.

Zolotarev, V. M. (1975b). The continuity of stochastic sequences that are generated by recurrent procedures. *Teor. Veroyatnost i ee Primenen*, **20**, 834–847 [in Russian]. MR 53 #4199.

Zolotarev, V. M. (1976). On the stochastic continuity of queuing systems of type $G/G/1$. *Teor. Veroyatnost i ee Primenen*, **21**, 260–279 [in Russian]. MR 54 #893.

Zolotarev, V. M. (1977). Quantitative estimates for the continuity property for queuing systems of type $G/G/\infty$. *Teor. Veroyatnost i ee Primenen*, **22**, 700–711 [in Russian]. MR 58 #31463.

Subject Index

Applied Probability and Statistics (Continued)

FLEISS • Statistical Methods for Rates and P…
FRANKEN, KÖNIG, ARNDT, and SCHM… Processes
GALAMBOS • The Asymptotic Theory…
GIBBONS, OLKIN, and SOBEL • Se… New Statistical Methodology
GNANADESIKAN • Methods f… Observations
GOLDBERGER • Economet…
GOLDSTEIN and DILLON…
GREENBERG and WEBST… Literature
GROSS and CLARK • Survival … the Biomedical Sciences
GROSS and HARRIS • Fundamentals …
GUPTA and PANCHAPAKESAN • Multip… and Methodology of Selecting and Ranking P…
GUTTMAN, WILKS, and HUNTER • Introductor… *Third Edition*
HAHN and SHAPIRO • Statistical Models in Engineerin…
HALD • Statistical Tables and Formulas
HALD • Statistical Theory with Engineering Applications
HAND • Discrimination and Classification
HILDEBRAND, LAING, and ROSENTHAL • Prediction Analysis… Cross Classifications
HOAGLIN, MOSTELLER, and TUKEY • Understanding Robust and Exploratory Data Analysis
HOEL • Elementary Statistics, *Fourth Edition*
HOEL and JESSEN • Basic Statistics for Business and Economics, *Third Edition*
HOLLANDER and WOLFE • Nonparametric Statistical Methods
IMAN and CONOVER • Modern Business Statistics
JAGERS • Branching Processes with Biological Applications
JESSEN • Statistical Survey Techniques
JOHNSON and KOTZ • Distributions in Statistics
Discrete Distributions
Continuous Univariate Distributions—1
Continuous Univariate Distributions—2
Continuous Multivariate Distributions
JOHNSON and KOTZ • Urn Models and Their Application: An Approach to Modern Discrete Probability Theory
JOHNSON and LEONE • Statistics and Experimental Design in Engineering and the Physical Sciences, Volumes I and II, *Second Edition*
JUDGE, HILL, GRIFFITHS, LÜTKEPOHL and LEE • Introduction to the Theory and Practice of Econometrics
JUDGE, GRIFFITHS, HILL and LEE • The Theory and Practice of Econometrics
KALBFLEISCH and PRENTICE • The Statistical Analysis of Failure Time Data
KEENEY and RAIFFA • Decisions with Multiple Objectives
LAWLESS • Statistical Models and Methods for Lifetime Data
LEAMER • Specification Searches: Ad Hoc Inference with Nonexperimental Data
McNEIL • Interactive Data Analysis
MAINDONALD • Statistical Computation
MANN, SCHAFER and SINGPURWALLA • Methods for Statistical Analysis of Reliability and Life Data
MARTZ and WALLER • Bayesian Reliability Analysis
MIKÉ and STANLEY • Statistics in Medical Research: Methods and Issues with Applications in Cancer Research
MILLER • Survival Analysis
MILLER, EFRON, BROWN, and MOSES • Biostatistics Casebook
MONTGOMERY and PECK • Introduction to Linear Regression Analysis